Scientific and Engineering Computations for the 21st Century —Methodologies and Applications

Proceedings of the 15th Toyota Conference

Scientific and Engineering Computations for the 21st Century —Methodologies and Applications

Edited by

M. Mori
Department of Mathematical Sciences
Tokyo Denki University
Ishizaka, Hatoyama-cho
Hiki, Saitama 350-0349
Japan

and

T. Mitsui
Graduate School of Human Informatics
Nagoya University
Furo-cho, Chikusa-ku
Nagoya 464-8601
Japan

2002

ELSEVIER
Amsterdam – Boston – London – New York – Oxford – Paris – San Diego – San Francisco – Singapore – Sydney – Tokyo

ELSEVIER SCIENCE B.V.
Sara Burgerhartstraat 25
P.O. Box 211, 1000 AE Amsterdam, The Netherlands

Reprinted from *Journal of Computational and Applied Mathematics*, volume 149/1

First edition 2002

British Library Cataloguing in Publication Data
Toyota Conference (15th : 2001 : Mikkabi, Shizuoka, Japan)
Scientific and engineering computations for the 21st century : methodologies and applications : proceedings of the 15th Toyota Conference
1.Numerical analysis - Congresses 2.Engineering mathematics - Congresses
I.Title II.Mori, M. III.Mitsui, T.
519.4

ISBN 0444509941

Library of Congress Cataloging in Publication Data
Toyota Conference (15th : 2001 : Shizuoka-shi, Japan)
Scientific and engineering computations for the 21st century : methodologies and applications : proceedings of the 15th Toyota Conference / edited by M. Mori and T. Mitsui.--1st ed.
p.cm.
ISBN 0-444-50994-1 (hbk. : alk. paper)
1.Numerical analysis--Congresses 2.Mathematics--Data processing--Congresses. 3. Engineering mathematics--Data processing--Congresses. I. Mori, Masatake. II.Mitsui, T. (Taketomo) 111. Title.

QA297 .T66 2001
519.4--dc21

2002029963

ISBN: 0-444-50994-1

♾ The paper used in this publication meets the requirements of ANSI/NISO Z39.48-1992 (Permanence of Paper).
Printed in The Netherlands.

JOURNAL OF COMPUTATIONAL AND APPLIED MATHEMATICS

Special Issue

Proceedings of the 15th Toyota Conference:
Scientific and Engineering Computations for the 21st Century —Methodologies and Applications
Mikkabi, Shizuoka, Japan, 28–31 October 2001

Guest editors:

M. Mori
Department of Mathematical Sciences, Tokyo Denki University
Ishizaka, Hatoyama-cho, Hiki, 350-0349 Saitama, Japan

T. Mitsui
Graduate School of Human Informatics, Nagoya University
Furo-cho, Chikusa-ku, 464-8601 Nagoya, Japan

ELSEVIER Amsterdam – London – New York – Oxford – Paris – Shannon – Tokyo

JOURNAL OF COMPUTATIONAL AND APPLIED MATHEMATICS

Volume 149, Number 1, 1 December 2002

Contents

ELSEVIER Journal of Computational and Applied Mathematics 149 (2002) ix–x

JOURNAL OF COMPUTATIONAL AND APPLIED MATHEMATICS

www.elsevier.com/locate/cam

Preface

It is a great privilege for us to serve as the Guest Editors for this special Proceedings issue of the Journal of Computational and Applied Mathematics which features articles on

"Scientific and Engineering Computations for the 21st Century
—Methodologies and Applications"
Presented at The Fifteenth Toyota Conference

held on October 28–31, 2001 in Mikkabi, Shizuoka, Japan. The conference participants included experts on advanced scientific and engineering computations from numerous countries. As members of the organizing committee, we invited them to present their thoughts and views on the past, present and future of their respective fields of specialization as we step forth into a new century.

The 20th century saw tremendous achievements and progress in science and technology. Undoubtedly, computers and computer-related technologies acted as one of vital catalysts for accelerating this progress in the latter half of the century. We believe that the contributions of mathematical sciences have been equally profound. The synergy between mathematics and computer science has played a key role in accelerating the progress of both fields as well as science and engineering.

Mathematical sciences will undoubtedly continue to play this vital role in this new century. In particular, mathematical modeling and numerical simulation will continue to be among the essential methodologies for solving massive and complex problems that arise in science, engineering and manufacturing. Underpinning this all from a sound, theoretical perspective will be *numerical algorithms*. In recognition of this observation, we decided to focus the conference on the following specific topics:

- Fundamental numerical algorithms,
- Applications of numerical algorithms, and
- Emerging technologies.

Fundamental numerical algorithms (e.g., the multipole expansion and iterative linear solvers) have a long history that can be traced back to the beginnings of mathematics itself. Tremendous leaps in the development of new approaches in this field have taken place in the past few years due, in no small part, to improvements in inexpensive and powerful computational hardware. Applications of numerical algorithms (e.g., computational fluid dynamics (CFD) and material design through computation) have revolutionized the physical sciences as well as in industry in the latter half of the former century with the use of computers and computer simulations. Emerging technologies (e.g., computer graphics, visualization and problem solving environments) include exciting and innovative

PII: S0377-0427(02)00516-2

approaches for mathematical modeling in scientific and engineering computations. These off-beat approaches may one day be accepted as the mainstream way of thinking.

The articles included in this special issue elucidate state-of-the-art achievements in these three topics from various angles and suggest the future directions. Although we cannot hope to cover all the aspects in scientific and engineering computations, we hope that the articles will interest, inform and inspire members of the science and engineering community.

In closing, we would like to acknowledge those who contributed to the success of the conference and this special Toyota Conference Proceedings issue. They include: the members of the Organizing Committee of the Toyota Conference:

Hideo Kawarada, Chiba University
Mei Kobayashi, IBM Research, Tokyo Research Laboratory
Yoshio Oyanagi, University of Tokyo
Masaaki Sugihara, Nagoya University
Tetsuro Yamamoto, Ehime University
Riichi Takahashi, Toyota Central R & D Labs., Inc.
Kazufumi Morimoto, Toyota Central R & D Labs., Inc.

and ourselves; the conference speakers and contributing authors; participants in the poster and discussion sessions; and the anonymous referees who reviewed preliminary drafts of the articles and suggested many improvements. Finally, we would like to express our gratitude to Toyota Motor Corporation and Toyota Central Research and Development Laboratories, Inc. for their generous financial support of this conference as well as this conference series over a 15 year period. Their kind and thoughtful hospitality made for a very memorable and warm atmosphere in which many new friendships were nurtured and many long-time friendships were renewed.

Guest editors:
Masatake Mori
Department of Mathematical Sciences,
Tokyo Denki University, Ishizaka, Hatoyama-cho, Hiki,
350-0349 Saitama, Japan

Taketomo Mitsui
Graduate School of Human Informatics,
Nagoya University, Furo-Cho, Chikusa-ku,
464-8601 Nagoya, Japan
E-mail address: tom.mitsui@cc.nagoya-u.ac.jp

PII: S0377-0427(02)00610-6

ELSEVIER Journal of Computational and Applied Mathematics 149 (2002) xiii–xiv

JOURNAL OF COMPUTATIONAL AND APPLIED MATHEMATICS

www.elsevier.com/locate/cam

Organizing Committee of the 15th Toyota Conference

Chairperson:

Professor Masatake Mori
Department of Mathematical Sciences
Tokyo Denki University
Ishizaka, Hatoyama-cho, Hiki, Saitama 350-0349, Japan

Vice-chairperson:

Professor Taketomo Mitsui
Graduate School of Human Informatics
Nagoya University
Furo-cho, Chikusa-ku, Nagoya 464-8601, Japan

Members:

Professor Hideo Kawarada
Department of Urban Environment Systems, Faculty of Engineering
Chiba University
1-33 Yayoi-cho, Inage-ku, Chiba 263-8522, Japan

Dr. Mei Kobayashi
Tokyo Research Laboratory
IBM Japan, Ltd.
1623-14 Shimotsuruma, Yamato 242-8502, Japan

Professor Yoshio Oyanagi
Department of Computer Science
The University of Tokyo
7-3-1 Hongo, Bunkyo-ku, Tokyo 113-8656, Japan

Professor Masaaki Sugihara
Department of Computational Science and Engineering
Graduate School of Engineering
Nagoya University
Furo-cho, Chikusa-ku, Nagoya 464-8603, Japan

PII: S0377-0427(02)00612-X

ELSEVIER Journal of Computational and Applied Mathematics 149 (2002) xv–xvi

JOURNAL OF COMPUTATIONAL AND APPLIED MATHEMATICS

www.elsevier.com/locate/cam

Toyota Conference Sponsored by Toyota Motor Corporation

For contribution to the growth and advancement of emerging fundamental science and technology, the first Toyota Conference was organized in 1987 in celebration of the 50th anniversary of the TOYOTA MOTOR CORPORATION. Since then, the conference has been held every year, dealing with a broad range of subjects in different fields. This has always been an informal conference with a relaxed atmosphere which has provided the latest information on the state of the art in each field. Intensive discussions are created for an exchange of ideas among more than 60 leading researchers from all over the world.
Sponsored by Toyota Motor Corporation

History of the Toyota Conference

The 1st Toyota Conference
Molecular Conformation and Dynamics of Macromolecules in Condensed Systems, September 28–October 1, 1987 (Aichi, Japan), organized by M. Nagasawa

The 2nd Toyota Conference
Organization of Engineering Knowledge for Product Modelling in Computer Integrated Manufacturing, October 2–5, 1988 (Aichi, Japan), organized by T. Sata

The 3rd Toyota Conference
Integrated Micro Motion Systems—Micromachining, Control & Application, October 22–25, 1989 (Aichi, Japan), organized by F. Harashima

The 4th Toyota Conference
Automation in Biotechnology, October 21–24, 1990 (Aichi, Japan), organized by I. Karube

The 5th Toyota Conference
Nonlinear Optical Materials, October 6–9, 1991 (Aichi, Japan), organized by S. Miyata

The 6th Toyota Conference
Turbulence and Molecular Processes in Combustion, October 11–14, 1992 (Shizuoka, Japan), organized by T. Takeno

PII: S0377-0427(02)00611-8

The 7th Toyota Conference
Toward the Harnessing of Chaos, October 31–November 3, 1993 (Shizuoka, Japan), organized by M. Yamaguti

The 8th Toyota Conference
Toward Global Planning of Sustainable Use of the Earth—Development of Global Eco-Engineering, November 8–11, 1994 (Shizuoka, Japan), organized by S. Murai

The 9th Toyota Conference
Brain & Mind—for Better Understanding of the Dynamic Function of Mind and its Supporting Brain Mechanism, December 5–8, 1995 (Shizuoka, Japan), organized by M. Ito

The 10th Toyota Conference
Atomic, Molecular and Electronic Dynamic Processes on Solid Surfaces, November 5–8, 1996 (Shizuoka, Japan), organized by M. Aono

The 11th Toyota Conference
Nanostructured Materials in Biological and Artificial Systems, November 5–8, 1997 (Shizuoka, Japan), organized by A.Yamagishi

The 12th Toyota Conference
Challenge of Plant and Agricultural Sciences to the Crisis of Biosphere on the Earth in the 21st Century, November 25–28, 1998 (Shizuoka, Japan), organized by A. Komamine

The 13th Toyota Conference
Affective Minds, November 29–December 2, 1999 (Shizuoka, Japan), organized by G. Hatano

The 14th Toyota Conference
Present and Future of Modeling Global Environmental Change—Toward Integrated Modeling—, October 3–6, 2000 (Shizuoka, Japan), organized by T. Matsuno

The 15th Toyota Conference
Scientific and Engineering Computations for the 21st Century—Methodologies and Applications—, October 28–31, 2001 (Shizuoka, Japan), organized by M. Mori

ELSEVIER Journal of Computational and Applied Mathematics 149 (2002) 1–12

JOURNAL OF COMPUTATIONAL AND APPLIED MATHEMATICS

www.elsevier.com/locate/cam

Advances in multidimensional integration

Ronald Cools

Department of Computer Science, Katholieke Universiteit Leuven, Celestijnenlaan 200A, B-3001 Heverlee, Belgium

Received 27 October 2001; received in revised form 6 May 2002

Abstract

This paper gives a personal bird's eye view of some aspects of multivariate numerical integration. It will sketch what happened during the past 50 years and point the reader's attention to what did not happen.

Keywords: Cubature; Quadrature; Numerical integration

1. Introduction

In this paper, I will give a very personal view on some aspects of the approximations of multivariate integrals

$$I[f] := \int_{\Omega} w(\mathbf{x}) f(\mathbf{x})\,\mathrm{d}\mathbf{x}, \quad \Omega \subset \mathbb{R}^d, \quad w(\mathbf{x}) \geqslant 0 \text{ for all } \mathbf{x} \in \Omega. \tag{1}$$

Such integrals appear in many mathematical models and can seldom be calculated analytically. This is especially true for multivariate integrals. The numerical approximation on integrals is one of the corner stones of numerical analysis. We encounter it in software for finite and boundary element problems. Statistical software also contains techniques to approximate integrals. The approximation of integrals is also an important step in methods to solve integral equations, e.g., in computer graphics. An application area that recently became very popular is financial mathematics: determining the value of sophisticated financial derivatives, such as exotic options, and determining the value at risk.

We are used to approximating functions by weighted sums of "easier" functions, such as monomials, splines or wavelets. Similarly, we are used to approximating functionals by weighted sums of "easier" functionals. An integral is typically approximated by a weighted sum of integrand

E-mail address: ronald.cools@cs.kuleuven.ac.be (R. Cools).

PII: S0377-0427(02)00517-4

evaluations. One must keep in mind that one may also use other functionals, such as derivatives of the integrand or even integrals of lower-dimensional projections. Here, I will only consider approximations of the form

$$Q[f]:=\sum_{j=1}^{N} w_j f(\mathbf{y}^{(j)}) \quad \text{with } \mathbf{y}^{(j)} \in \Omega. \tag{2}$$

If the dimension of the integration region $d=1$ the approximation Q is called a *quadrature formula*. If $d \geqslant 2$ the approximation Q is called a *cubature formula*. The term *integration rule* is also used.

Basic integration rules (cubature formulas) for the approximation of multiple integrals are essential building blocks in many applications. In this text, I will almost entirely restrict myself to this topic. I will ignore almost completely the important aspect of error estimators, convergence acceleration by extrapolation and adaptive subdivisions strategies.

How should the points and weights of a basic integration rule be chosen? The choice of the points $\mathbf{y}^{(j)}$ and weights w_j is independent of the function f. There is no unique best criterion for this choice. There are several criteria to specify and classify integration rules based on their behaviour for specific classes of functions. People seem to agree that the approximation should be exact for constant functions, that is,

$$I[1]=\int_{\Omega} w(\mathbf{x})\,\mathrm{d}\mathbf{x}=\sum_{j=1}^{N} w_j = Q[1].$$

Roughly speaking, one can distinguish two classes of cubature formulas: *polynomial-based methods* and *number theoretic methods*. The first class contains rules designed to be exact for some set of polynomials, while the second contains rules based on uniformly distributed points.

The polynomial-based methods received most attention during the past century and that is the topic I have to say most about. I devote some space to quasi-Monte Carlo methods because the future of these methods looks very bright.

2. Will faster computers make everything trivial eventually?

In a “Present state of the art” section in their survey book [8], Davis and Rabinowitz wrote that

> Given the present state of theory and hardware (1983), one can distinguish three dimensional ranges in addition to the two-dimensional case which deserves separate treatment:
>
> *Range* I: integrals of dimension 3 to about 6 or 7,
> *Range* II: dimensions 7 or 8 to about 15,
> *Range* III: dimensions greater than 15.

In two dimensions, they considered the situation satisfactory. In Range I, they considered adaptive methods based on rules exact for polynomials as the most important tools. Range II was considered the borderline range. Here, much depends on the smoothness of the integrand. In this range, adaptive Monte Carlo and quasi-Monte Carlo methods definitely become competitive, but only low accuracy is achievable. Range III was considered "really high" dimensional. Monte Carlo and quasi-Monte Carlo are indicated. One should not hope for more than 2 digits accuracy. Is this much different today?

To put everything in perspective, let us turn back the clock about 20 years, when the routine TRIEX [9] was developed to approximate integrals with singularities over a triangle. The authors put an upper limit of 23 000 function evaluations on their experiments. The experiments for TRIEX were done on an IBM 370/158 which was a 1 MIPS machine. Today a Pentium III is about 1000 MIPS. One cpu-hour of 1980 is devaluated to about 4 cpu seconds in the year 2000. (The real speedup will be less since the measure "mips" cannot really be used to compare machines developed 20 years apart.) So, current machines can afford to spend more function evaluations on a problem to have it completed in the same amount of wall clock time.

Will faster computers be sufficient to solve our future problems even if we do not make any other progress? I believe not. As computers become more powerful every day, researchers start thinking about problems nobody dreamed about some years ago: integrals with hundreds of variables. Solving these problems remains very difficult or even impossible. Application areas will demand more and more. The growing speed of computers alone will not help. Indeed, we are working with a lot of techniques that were developed with low dimensions in mind and some of these cannot directly be applied in high dimensions. That is the main reason why the ranges indicated by Davis and Rabinowitz apply almost unchanged.

For the approximation of low-dimensional integrals a reasonable collection of techniques and software is nowadays available. The approximation of multivariate integrals is however a fundamentally different problem. In this research area, one suffers from the so-called curse of dimensionality: the computational complexity grows exponentially with the dimension.

3. Polynomial-based methods

Polynomials are very popular for approximations. The oldest criterion for the approximation of integrals is the *algebraic degree* of a quadrature or cubature formula. A quadrature or cubature formula has algebraic degree m if it is exact for all polynomials or (total) degree at most m.

One can think about several variants of this criterion, e.g., one can consider the trigonometric degree or one can consider other spaces of polynomials.

By $\mathbb{P}^d$ we denote the vector space of all polynomials in d variables. This space is spanned by the monomials

$$\mathbf{x}^{\mathbf{k}} := x_1^{k_1} \cdots x_d^{k_d}, \quad \mathbf{k} := (k_1, \ldots, k_d) \in \mathbb{N}^d.$$

The (total) degree of a monomial is

$$\deg(\mathbf{x}^{\mathbf{k}}) := |\mathbf{k}| := \sum_{j=1}^{d} k_j.$$

By $\mathbb{P}_m^d$ we denote the subspace of all polynomials in d variables of degree at most m. The definition of degree of a cubature formula is equivalent with demanding that

$$Q[f_j] = I[f_j], \quad j = 1,2,\ldots,\dim \mathbb{P}_m^d, \tag{3}$$

where the f_j form a basis for $\mathbb{P}_m^d$.

If f_j and N are fixed, then (3) form a system of nonlinear equations

$$\sum_{i=1}^{N} w_i f_j(\mathbf{y}^{(i)}) = I[f_j], \quad j = 1,2,\ldots,\dim \mathbb{P}_m^d. \tag{4}$$

Each point introduces $d+1$ unknowns: the d coordinates of the point $\mathbf{y}^{(i)}$ and the weight w_i. Actually, each equation in (4) is a polynomial equation.

One can distinguish between two approaches to construct cubature formulas:

(1) search for polynomials that vanish at the points of the formula;
(2) solve the system of nonlinear equations (4) directly.

In Sections 3.1 and 3.2, I will briefly introduce these approaches. For a more through introduction I refer to [3] and its references.

3.1. Vanishing polynomials

The approximation of one-dimensional integrals using a quadrature formula based on the zeros of an orthogonal polynomial is a very classical mathematical subject. Only very few mathematicians tried to generalise this to higher dimensions prior to the middle of the previous century. A very important milestone was Radon's paper [17] which triggered a lot of research [6]. Only in the 1960s the effort increased and significant progress was made by, amongst others, [1] Stroud and Mysovskikh. (See their survey books [21] and [14].) They introduced elements from algebraic geometry. Möller [13] was the first to recognise that this connection could be presented more transparently by using ideal theory. Significant theoretical progress was made in the 1970s and 1980s with the introduction of ideal theory. The practical results were however disappointing.

An apparently simple question such as "What is the minimal number of points required by a cubature formula of degree m for the integral I" does not have a simple answer.

A very general upper and lower bound is given in the following theorem, which combines the easy-to-prove lower bound for positive functionals and Tchakaloff's upper bound, see e.g., [21].

Theorem 1. *A cubature formula Q of the form* (2) *of degree m for an integral* (1) *exists with its number of points N satisfying*

$$\dim \mathbb{P}_{\lfloor m/2 \rfloor}^d \leqslant N \leqslant \dim \mathbb{P}_m^d.$$

[1] I restrict myself to the two most mentioned names in this area.

Actually, no cubature formula can exist with less points than $\dim \mathbb{P}^d_{\lfloor m/2 \rfloor}$. The common proof of this lower bound is not constructive. If one has a cubature formula with more than $\dim \mathbb{P}^d_m$ points, then one can always delete points and recompute the weights by solving a system of linear equations. Furthermore, it is known that a cubature formula attaining this upper bound exists with only real points inside Ω and all weights positive.

Observe that the lower bound of Theorem 1 is equal for $m = 2k$ and $2k + 1$. So, it will not come as a surprise that in general this lower bound cannot be attained for odd degrees. There is a connection with orthogonal polynomials which in general shows this.

In the multivariate case there is more than one orthogonal polynomial of a given degree. It is common to normalise the orthogonal polynomials of degree k to the form

$$p_k = \mathbf{x}^{\mathbf{k}} + q_k, \quad \mathbf{k} \in \mathbb{N}^d, \ |\mathbf{k}| = k, \ \deg(q_k) \leqslant k - 1. \tag{5}$$

Orthogonal polynomials in the form (5) are called *basic orthogonal polynomials*. Each basic orthogonal polynomial of degree k has only one term of degree k, with constant 1.

As an illustration we formulate a generalisation by Mysovskikh of a property of Gauss quadrature formulas.

Theorem 2. *A necessary condition for the existence of a cubature formula of degree* $2k + 1$ *with* $N = \dim \mathbb{P}^d_k$ *points is that the basic orthogonal polynomials of degree* $k + 1$ *have* N *common zeros.*

It is nowadays known that the condition of this theorem does not hold for standard regions such as a square or triangle with a constant weight function.

Consider the vector space of all polynomials that vanish in a given set of points. This is more than simply a vector space. Indeed, if one multiplies a polynomial that vanishes at all points of the cubature formula by an arbitrary polynomial, the product also vanishes at all points. The set of all polynomials that vanish in a given set of points is a polynomial ideal.

The following theorem, first proven by Möller [13], shows the relation between ideals and cubature formulas.

Theorem 3. *Let* I *be an integral of the form* (1) *over a d-dimensional region. Let* $\{\mathbf{y}^{(1)}, \ldots, \mathbf{y}^{(N)}\} \subset \mathbb{C}^d$ *and* $\mathfrak{A} := \{f \in \mathbb{P}^d \colon f(\mathbf{y}^{(i)}) = 0, \ i = 1, \ldots, N\}$. *Then the following statements are equivalent.*

- $f \in \mathfrak{A} \cap \mathbb{P}^d_m$ *implies* $I[f] = 0$.
- *There exists a cubature formula* Q *of the form* (2) *such that* $I[f] = Q[f]$, *for all* $f \in \mathbb{P}^d_m$, *with at most* $\dim \mathbb{P}^d_m - \dim(\mathfrak{A} \cap \mathbb{P}^d_m)$ (*complex*) *weights different form zero.*

Using ideal theory, a theorem such as Theorem 2 becomes very easy to prove, see [3].

Theorem 3 allows that the cubature formulas have complex or multiple points. Schmid proposed to avoid this by considering *real ideals*.

If the common zeros of an ideal $\mathfrak{A}$ are denoted by $V(\mathfrak{A})$ and the ideal of all polynomials which vanish at a finite set $X \subset \mathbb{R}^d$ by $\mathfrak{A}_X$, then obviously $X \subset V(\mathfrak{A}_X)$. An ideal $\mathfrak{A}$ is called real if $X = V(\mathfrak{A}_X)$. Using real ideals Schmid obtained a complete characterisation of cubature formulas with real points. His characterisation was successfully applied to construct cubature formulas for several two-dimensional regions, see, e.g., [18,19].

Meanwhile higher lower bounds are known. In [3] all these are collected together with all known formulas that attain these bounds. As an illustration, I mention the improved lower bound for regions such as a square and a triangle with a constant weight function.

Theorem 4. *The number of points N in a cubature formula* (2) *of degree $2k-1$ for an integral* (1) *where Ω is a square or a triangle and $w(x)=1$, satisfies*

$$N \geqslant \frac{k(k+1)}{2} + \left\lfloor \frac{k}{2} \right\rfloor .$$

We know nowadays that the minimal number of points depends on the region of integration *and* on the weight function. We still do not know the lowest possible number of points for simple regions such as a square or triangle with a constant weight function. For these simple regions there is a gap between the highest known lower bound for the number of points, given in Theorems 1 and 4, and the lowest number of points in a known formula already for degree resp. 13 and 9. In Table 1 we summarise what is known about lower bounds and lowest number of points in a known cubature formula for these two standard two-dimensional regions. Observe that for a square for even degrees $\geqslant 10$ no formula is known (except of course those of higher odd degree).

Although some mathematicians continue this line of research, I believe it is exhausted for standard regions and I expect that only very few (if any) new cubature formulas will be constructed during the following years based on this approach. Actually, for standard regions most known cubature formulas exact for polynomials were not constructed with the aid of this theory.

Looking for cubature formulas using orthogonal polynomials will become of practical interest again only if we better understand the relation between the minimal number of points in a cubature formula and the given integral. The construction methods based on this theory strongly depend on the quality of the known lower bounds and that restricts their applicability.

3.2. Direct approach

One can construct cubature formulas exact for a space of polynomials by solving the large system of polynomial equations associated with it. As computers continue to become more powerful, one can give them larger systems and occasionally they will chew out a useful (real points inside the integration region) cubature formula. Although significant progress was made in recent years on methods to solve systems of polynomial equations, these methods still cannot deal with systems associated to cubature formulas. No good news on this is expected in the near future.

We should however keep in mind that most known formulas were first obtained by solving such a nonlinear system. This was not done by specialised techniques for polynomial equations but by iterative zero finders. To make this approach work, it is essential to restrict the search to cubature formulas with a certain structure. The more symmetry imposed on the cubature

Table 1
Overview of known results

Degree	$N_{\min}$	$\tilde{N}_{\min}$	□	◺
1	1	1	1	1
2	3		3	3
3	3	4	4	4
4	6		6	6
5	6	7	7	7
6	10		10	10
7	10	12	12	12
8	15		15	15
9	15	17	17	19
10	21			22
11	24	24	24	27
12	28			33
13	28	31	33	36
14	36			42
15	36	40	44	48
16	45			52
17	45	49	57	61
18	55			70
19	55	60	68	73
20	66			79

$N_{\min}$ is the lower bound of Theorem 1; $N_{\min}$ is the lower bound of Theorem 4; □ denotes a square with constant weight function; ◺ denotes a triangle with constant weight function.

formula, the smaller the system becomes. There are systematic ways to deal with symmetries in cubature formulas but this approach still requires a good part of luck. For an introduction, see [3].

3.3. *Consolidation phase*

The time has come to consolidate what is known about cubature formulas exact for polynomials and to make this knowledge more easily available to potential users. The Handbook of Mathematical Functions [1] contains Radon's cubature formula of degree 5 with 7 points and this probably explains the widespread use of that formula. The National Institute of Standards and Technology (NIST) is working on a new (digital) version of this widespread Handbook. (See [12] and `http://dlmf.nist.gov/` for more about this project.) Whatever it will contain, will become widespread! It is not expected to include many tables of cubature formulas however and so there is room and need for another initiative in this specialised area. We collected the references to known cubature formulas [7,4] and started to make these available electronically. Meanwhile the webpage at, `www.cs.kuleuven.ac.be/~nines/research/ech/ecf.html` also contains the points and weights of many cubature formulas. The formulas that are included are recomputed and presented with 16 and 32 significant digits. The files are computer generated, avoiding human typing errors.

3.4. Is there a need for higher degree cubature formulas?

In many two- and three-dimensional applications, practitioners use very low-degree cubature formulas (even degree 1) and they deal with their problem by using a finer triangularisation. A significant number of users of the above webpage say they want to apply higher-order methods, which we all hope will require less subregions.

An example illustrating the positive effect of using higher-degree cubature formula is the following. Consider

$$\int_0^1 \int_0^1 \int_0^1 r^{-0.5} \log r \exp(xy+z)\, \mathrm{d}x\, \mathrm{d}y\, \mathrm{d}z \quad \text{with } r := \sqrt{x^2+y^2+z^2}. \tag{6}$$

We approximate this integral using CUBPACK [5]. In Figs. 1 and 2 the results are presented when a formula of degree 7 (diamond), degree 9 (square) and degree 13 (circle) is used. These are connected by a different type of line for clarification. In Fig. 1 the estimated error is presented. In

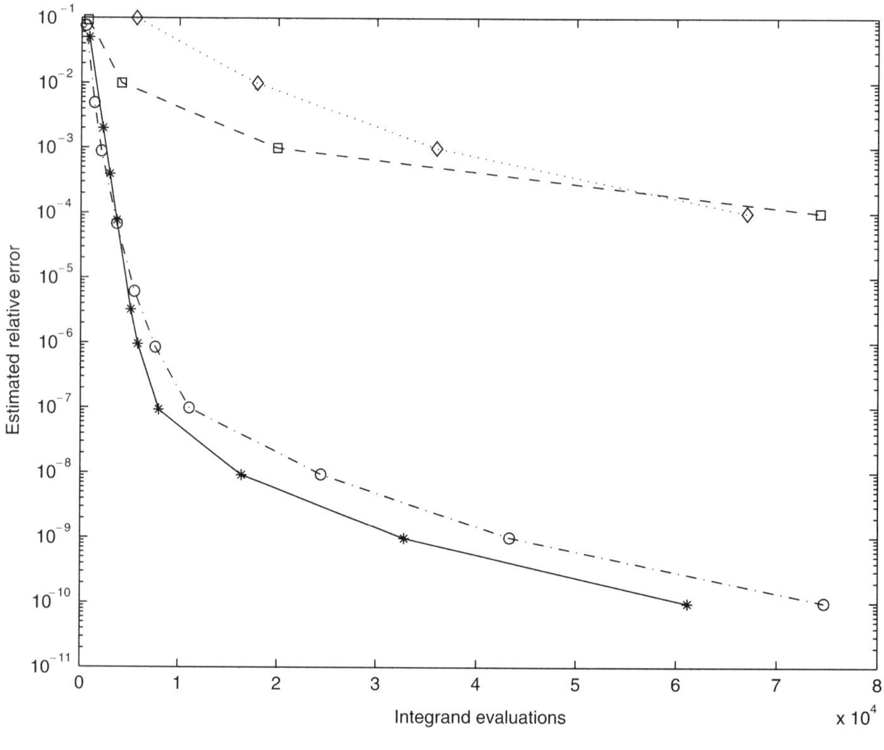

Fig. 1. Comparison of estimated errors for the integral (6).

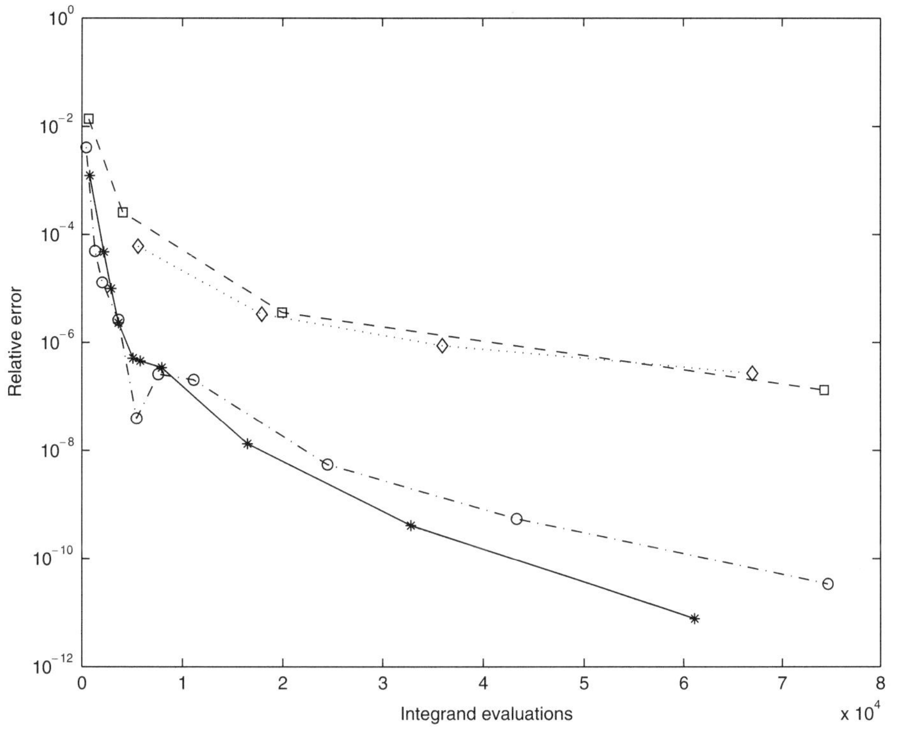

Fig. 2. Comparison of real errors for the integral (6).

Fig. 2 the real error is presented. In all cases the same global adaptive strategy with subdivision in 2 at each step was used. The stars represent results obtained using the degree 13 rule in combination with an improved subdivision strategy. This clearly shows that high accuracy can only be obtained using the high-degree rule.

4. Quasi-Monte Carlo methods

For higher-dimensional problems, practitioners use Monte Carlo methods since many years. In many situations it is the method of last resort. The weight function $w(\mathbf{x})$ in I (1) is now interpreted as a probability density function (pdf) and the value of the integral is interpreted as the mean or expected value of the integrand f. If the points $\mathbf{y}^{(j)}$ are random variables selected according to the pdf $w(\mathbf{x})$ then, given an $\varepsilon > 0$, we have

$$\lim_{N\to\infty} \text{probability}\left(I[f] - \varepsilon \leqslant Q[f] := \frac{1}{N}\sum_{j=1}^{N} f(\mathbf{y}^{(j)}) \leqslant I[f] + \varepsilon \right) = 1.$$

The probability that the sample average is close to the mean value I can theoretically be made arbitrarily close to 1 if we take sufficiently large samples.

An advantage of a Monte Carlo method is that one obtains an error estimate almost for free. The central limit theorem gives a probabilistic error bound:

$$\lim_{N\to\infty} \operatorname{prob}\left(\left|\frac{1}{N}\sum_{j=1}^{N} f(\mathbf{y}^{(j)}) - I[f]\right| \leqslant \frac{\lambda\sigma}{\sqrt{N}}\right) = \frac{1}{\sqrt{2\pi}}\int_{-\lambda}^{\lambda} \mathrm{e}^{-t^2/2}\,\mathrm{d}t.$$

The variance of the integral

$$\sigma^2 = \int_\Omega (f(\mathbf{x}) - I[f])^2 w(\mathbf{x})\,\mathrm{d}\mathbf{x} = \int_\Omega f^2(\mathbf{x}) w(\mathbf{x})\,\mathrm{d}\mathbf{x} - I[f]^2$$

can with almost no extra cost be estimated from the function values. This very general and easy applicable approach has two drawbacks. First, its convergence is very slow (proportional to $N^{1/2}$) and the method is only useful to meet low accuracy requests. Second, the error estimate is only probabilistic.

It is important to note that the error bound consists of two parts: the variance (which is a property of the integrand) and the number of samples (which is a property of the set of points used). There is half-a-century experience in variance reduction techniques which is the main reason these methods are so successful!

A variant of the Monte Carlo method called *quasi-Monte Carlo*, received increasing attention from mathematicians after the work by (to mention just tow names) Sobol and Korobov. In a quasi-Monte Carlo method the random points of a Monte Carlo method are replaced by deterministic points that are distributed “better than randomly”. Moreover, quasi-Monte Carlo methods provide the user with deterministic error bounds.

These theoretical bounds are in general very pessimistic and that is one of the reasons for which these methods did not receive much attention in the past. The by now classical result is the Koksma–Hlawka error bound, see, e.g., [15].

Theorem 5. *If f has bounded variation $V(f)$ on $[0,1]^d$ in the sense of Hardy and Krause, then, for any $\mathbf{y}^{(1)},\dots,\mathbf{y}^{(N)} \in [0,1]^d$, we have*

$$\left|\frac{1}{N}\sum_{j=1}^{N} f(\mathbf{y}^{(j)}) - \int_{[0,1]^d} f(\mathbf{x})\,\mathrm{d}\mathbf{x}\right| \leqslant V(f) D_N^*(\mathbf{y}^{(1)},\dots,\mathbf{y}^{(N)}).$$

Also this bound is the product of a measure of the quality of the points used, the *discrepancy* D_N^*, and a factor that depends on the function, the variation $V(f)$. Computational experience confirms that discrepancy is a good indicator of quasi-Monte Carlo performance, while variation is not a typical indicator. Furthermore, one should keep in mind that this bound is not easily computable. The Koksma–Hlawka inequality is a worst-case bound.

The discrepancy typically contains a term $(\log N)^d/N$. For very large N, the Koksma–Hlawka bound nicely goes down. However, it only starts to go down if $N > \exp(d)$, and so one did not expect these methods to be applicable for, say $d = 30$, because then the number of points should be larger than 10^{13}, which will continue to be unpractical for many years to come.

Thanks to some successes in financial applications in the last years of the previous century, quasi-Monte Carlo methods nowadays receive the attention they deserve. The standard example used to show that quasi-Monte Carlo methods can work very well in high-dimensional practical problems is a 360-dimensional integral coming from finance.

One distinguishes to classes: (1) lattice rules and (2) methods based on low-discrepancy sequences. Lattice rules are designed especially for periodic functions. So far quasi-Monte Carlo methods are derived for integrals over a unit cube with constant weight function.

The many activities in this area in recent years become obvious if one looks at the many recent surveys, e.g., [15,20,22,10,2]. To have an impression of what goes on nowadays in the research community, I invite the reader to have a look at some recent conference proceedings, e.g., [16,11].

The direct application of quasi-Monte Carlo methods is nowadays limited to integrals on hypercubes. These methods consider each dimension of a problem to be of equal importance, i.e., they are designed for isotropic problems. High-dimensional practical problems tend to be anisotropic however.

Because of the unexpected successes in some applications, research on quasi-Monte Carlo methods now concentrates on the fast generation of very high-dimensional low-discrepancy sequences and on transferring techniques originally designed to improve Monte Carlo methods. Searching for explanations why these methods work unexpectedly well for some classes of practical problems, is also a research topic.

Practical error estimation is nowadays based on ideas from Monte Carlo methods: one includes randomisations. A lot of effort will be devoted in the near future to obtain other (faster and deterministic) error estimators. Even more effort will be devoted to developing variants of these methods for anisotropic problems and for other regions than hypercubes. What is needed are adaptive algorithms that detect the anisotropy in a given problem and exploit this. This feature is available in current automatic integration routines for low-dimensional integration but remains a challenge for, e.g., 30-dimensional problems. It is a challenge researchers should accept.

5. Final remarks

We observe a stagnation of progress in the area of constructing cubature formulas exact for polynomials. During the past five years almost no theoretical progress was made and only a few new cubature formulas were constructed. There is still work to do here! Surprisingly, almost no attention has been paid to the construction of cubature formulas for four- and higher-dimensional regions. Consequently, the available adaptive software uses cubature formulas which are probably much more expensive than really necessary. New cubature formulas will probably extend Range I.

For lower dimensions a reasonable collection of techniques is available and combined with the growing speed of computers, adaptive software will evolve to more complex algorithms. Several strategies will be embedded in the same piece of software. The software will try them all and will proceed with the most promising ones until the requested accuracy is attained.

For higher dimensions, I expect a lot will happen in the near future, but unfortunately my crystal ball does not give a clear view on the outcome of this.

References

[1] M. Abramowitz, I.A. Stegun (Eds.), Handbook of Mathematical Functions with Formulas, Graphs and Mathematical Tables, National Bureau of Standards Applied Mathematics Series, Vol. 55, U.S. Government Printing Office, Washington, DC, 1964.

[2] R.E. Caflisch, Monte Carlo and Quasi-Monte Carlo Methods, Acta Numerica, Vol. 7, Cambridge University Press, Cambridge, 1998, pp. 1–49.

[3] R. Cools, Constructing Cubature Formulae: The Science Behind the Art, Acta Numerica, Vol. 6, Cambridge University Press, Cambridge, 1997, pp. 1–54.

[4] R. Cools, Monomial cubature rules since "Stroud": a compilation—part 2, J. Comput. Appl. Math. 112 (1–2) (1999) 21–27.

[5] R. Cools, A. Haegemans, CUBPACK: a package for automatic cubature; framework description, submitted for publication.

[6] R. Cools, I.P. Mysovskikh, H.J. Schmid, Cubature formulae and orthogonal polynomials, J. Comput. Appl. Math. 127 (2001) 121–152.

[7] R. Cools, P. Rabinowitz, Monomial cubature rules since 'Stroud': a compilation, J. Comput. Appl. Math. 48 (1993) 309–326.

[8] P.J. Davis, P. Rabinowitz, Methods of Numerical Integration, Academic Press, London, 1984.

[9] E. de Doncker, I. Robinson, TRIEX: integration over a TRIangle using nonlinear EXtrapolation, ACM Trans. Math. Software 10 (1984) 17–22.

[10] M. Drmota, R.F. Tichy, Sequences, Discrepancies and Applications, Lecture Notes in Mathematics, Vol. 1651, Springer, Berlin, 1997.

[11] K.-T. Fang, F. Hickernell, H. Niederreiter (Eds.), Monte Carlo and quasi-Monte Carlo methods 2000, Proceedings of the Fourth International Conference on Monte Carlo and Quasi-Monte Carlo Methods in Scientific Computing, Springer, Berlin, 2002.

[12] D.W. Lozier, B.R. Miller and B.V. Saunders, Design of a digital mathematical library for science, technology and education, Proceedings of the IEEE Forum on Research and Technology Advances in Digital Libraries; IEEE ADL 1999, Baltimore, MD, May 19, 1999.

[13] H.M. Möller, Polynomideale und Kubaturformeln, Ph.D. Thesis, Universität Dortmund, 1973.

[14] I.P. Mysovskikh, Interpolatory Cubature Formulas, Izdat. Nauka, Moscow, 1981 (Russian).

[15] H. Niederreiter, Random Number Generation and Quasi-Monte Carlo Methods, CBMS-NSF Regional Conference Series in Applied Mathematics, Vol. 63, SIAM, Philadelphia, 1992.

[16] H. Niederreiter, J. Spanier (Eds.), Monte Carlo and quasi-Monte Carlo methods 1998, Proceedings of the Third International Conference on Monte Carlo and Quasi-Monte Carlo Methods in Scientific Computing, Springer, Berlin, 2000.

[17] J. Radon, Zur mechanischen Kubatur, Monatsh. Math. 52 (1948) 286–300.

[18] H.J. Schmid, Interpolatorische Kubaturformeln, Dissertationes Math., Vol. CCXX, Polish Scientific Publishers, Warszawa, 1983.

[19] H.J. Schmid, Two-dimensional minimal cubature formulas and matrix equations, SIAM J. Matrix Anal. 16 (3) (1995) 898–921.

[20] I.H. Sloan, S. Joe, Lattice Methods for Multiple Integration, Oxford University Press, Oxford, 1994.

[21] A.H. Stroud, Approximate Calculation of Multiple Integrals, Prentice-Hall, Englewood Cliffs, NJ, 1971.

[22] S. Tezuka, Uniform Random Numbers: Theory and Practice, Kluwer Academic Publishers, Boston, 1995.

ELSEVIER

Journal of Computational and Applied Mathematics 149 (2002) 13–25

JOURNAL OF COMPUTATIONAL AND APPLIED MATHEMATICS

www.elsevier.com/locate/cam

Material modeling platform

Masao Doi

Department of Computational Science and Engineering, Nagoya University, Furo-cho Chikusa, 464-8603 Nagoya, Japan

Received 31 December 2001; received in revised form 8 February 2002

Abstract

We have conducted a polymer modeling project which aims bridging the molecular characteristics of polymers with macroscopic properties. This requires an integration of various physical models (atomistic, continuum and other mesoscopic models) in a single system, and has a difficulty of multi-scale, multi-physics problem. Here I discuss the strategy we took to achieve the aim. We developed a system called "simulation platform" which provides a common user interface for all simulation programs. The simulation platform provides a common service of data viewing, editing, plotting and graphics, and can be useful for both simulation programmers and simulation users. For constructing such system, we designed a data format called UDF (User Definable Format).

1. Introduction

Understanding how material properties depend on its micro-structures has been a central problem in physics, chemistry and material science. It is also a crucial question in industries since this understanding gives the technological basis for controlling material properties and creating new materials. As computer is becoming a useful tool in many areas of engineering (such as architectural, mechanical and electrical engineering), there is a growing expectation for a similar simulation tool for material engineering. The key word "computer aided material engineering" has been used for more than 10 years. However, although it has seen some successes in some areas, the technology is still at a primitive stage. The major reason for this is that material properties depend on many physical phenomena which cannot be handled by a single simulator.

We faced this problem in our project, the polymer modeling project, which was jointly conducted by industries and academia supported by the government from 1998 to 2002. The objective of this project is to construct a polymer modeling system which bridges the molecular characteristics of polymers with macroscopic properties. This objective is not easy to achieve.

E-mail address: doi@cse.nagoya-u.ac.jp (M. Doi).

0377-0427/02/$ - see front matter
PII: S0377-0427(02)00518-6

Polymer molecules are made of millions of atoms, and polymeric materials are the mixtures of them. The material properties of polymers depend on many factors in the manufacturing process: polymerization process (monomers, catalysts, molecular weight distribution and degree of branching), the blending condition (controlling the interface and dispersion states), and process condition (controlling chain stretching and crystallization, etc). The problem involves many length scales and many physics, and is one of the multi-scale, multi-physics problem. How to handle such problem is one of the major challenge in current computational science and engineering. Here, I would like to talk how we tackled the problem, and what kind of software we made towards our goal.

2. Seamless zooming

2.1. The ultimate form of multi-scale modeling

Multi-scale modeling is a method which models a large, complex system by combining various simulation models each dedicated to certain physical aspects characterized by a certain length scale. Such method is definitely needed in our simulator for polymeric materials which have many meso structures at the intermediate levels between the atomistic level and the macroscopic level. Ideally, the system we would like to have is a system by which we can "zoom-in" at any length scale of the material, and can study the phenomena occurring there. We called such system "seamless zooming" system. It represents our image of the ultimate system for multi-scale modeling. It is easy to envision such a system, but the question is how to realize it.

The seamless zooming system, or any multi-scale modeling system, needs two components. One is a set of the simulation programs each of which is dedicated to the simulation at a certain length scale. The other is the mechanism which integrates these simulation programs.

Designing the simulation programs for our system was not difficult. Scientists have proposed theories to model the dynamics of polymers at various length scales, and have written many programs based on them. These programs were made for research purpose only, and were not intended to be used by the other people. It is therefore useful to make simulation programs to be used for others based on them. We called such programs engines since their role is to repeat calculations many times. We chose to develop four engines listed in Table 1. However, our objective is to make an integrated system which allows users to go from one engine to the other. For this purpose, we need some mechanism of integrating engines. Fig. 1 shows a sketch of our system: the engines run on a common base (called modeling platform) which provides a framework of integration. The difficult problem was how to design the platform.

2.2. Design problem of integration

The role of the platform is to integrate engines. For this to be done, engines have to exchange the information they have. Multi-scale modeling must be based on a consistent expression for physical objects. Therefore, there must be some common framework to express states and properties of physical objects. At first we tried to set up some standard format for the data of physical objects commonly used by the engines, such as "monomers", "polymers" or "molecular weight". However, this

Table 1
Simulation engines

Molecular dynamics (COGNAC)[a]	A molecular dynamics simulation engine for "coarse grained models" of polymers. It can deal with a general molecular model (including fully atomistic model), and calculates the time evolution of the system in equilibrium and nonequilibrium conditions
Repitation dynamics (PASTA)[b]	An engines calculating the rheological response of entangled polymers for given molecular weight distribution. It is based on the repitation dynamics and the dual slip-link model of entanglement
Interfacial dynamics (SUSHI)[c]	An engine calculating the equilibrium and nonequilibrium structures of block polymers and polymer blends near interfaces by solving the self-consistent field equation of Edwards
Micro-fluidics and Micro-elastics (MUFFIN)[d]	A generic engine solving the partial differential equations appearing in the problem of soft materials. It actually consists of five simulation programs, each dealing with (1) phase separation, (2) electrolyte dynamics, (3) micro-fluidics, (4) elastic deformation of solids and (5) deformation and swelling of gels

[a]COarse Grained molecular dynamics program by NAgoya Cooperation.
[b]Polymer rheology Analyzer with Slip-link model of enTAnglement.
[c]Simulation Utilities for Soft and Hard Interfaces.
[d]MUlti-Farious FIeld simulator for Nonequilibrium systems.

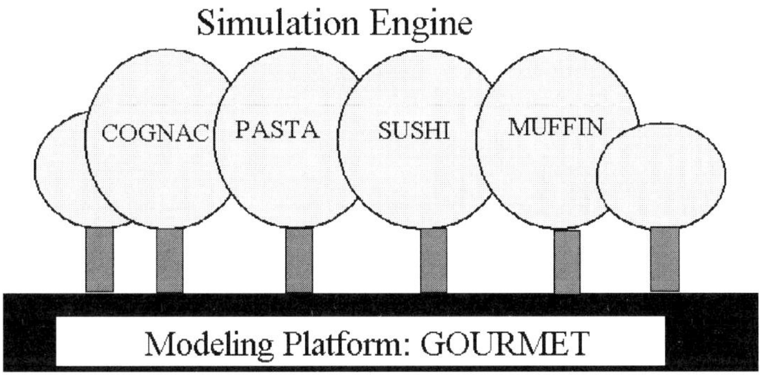

Fig. 1. Outline of the system.

approach turned out to be impossible. The engine programmers have different background, and have different views for the definitions for "monomers" and "polymers". Settling the terminology was difficult, and settling the data structure was even more difficult. After long discussions, we realized that even if we were able to invent some data format, nobody would be happy, and nobody would use it.

We abandoned the idea of defining a common data format. We realized that the integration must be done not by force but by voluntary collaboration of people: it is more important to set up a framework with which people can understand the meaning of the data provided by other engines, than to invent some data format (understood by computers, but in a very restrictive context). If people can understand the meaning of the data made by engines, they can transform the data from one engine to the other, and if such work is accumulated, it will eventually become the "seamless zooming". Therefore the key to "seamless zooming" is the collaboration of many engines, which, eventually, means the collaboration of people. This is the conclusion we arrived at finally. This consideration lead us to the following principle for the design of the platform:

(1) We leave the task of defining the data structures to those who write simulation programs.
(2) We ask them to state the definition of the data structure in a certain format which we define.
(3) We encourage them to use this format by providing various services.

Engine programmers can design the data structure they use. To connect the engine to the platform, they have to state the data structure in a certain format called user definable format (UDF). The platform then provides various services to the engines which use this format. This mechanism will be explained in more detail in the following.

3. UDF: a self-descriptive data format

UDF is a data file format which is needed to connect the engines to the platform. In fact, the only requirement of the platform for the engine programmers is to use this format in their input and output files. Engine programmers can define their own data structure, but they are asked to state the structure in the data file. UDF file is a self-descriptive text file: it consists of two parts, the data definition part and the data part. The data definition part states the name and the unit of the data written in the data part. This part is essential for the file to be understood by the platform, and by other users. An example of the UDF file is the following:

```
---------------------------------------------
//datafile spring.udf
\begin{def}
    input: {
           spring_const : double(kg/s^2)
           mass         : double (kg)
           }
    output: {
           periodicity: double (s)
           result[]:{ time            : double (s),
                      displacement    : double (m)
                      velocity        : double (m/s)
                      }
           }
\end{def}
\begin{data}
    input: {0.05, 0.1}
\end{data}
---------------------------------------------
```

The part between `\begin{def}` and `\end{def}` is the definition part. In this example, the definition part states that the data following the key word "input" has two members, `spring_const` and `mass`, both being double type (in C/C++ notation), and having unit ($\mathtt{kg/s^2}$) and (`kg`), respectively. The actual data is written later, between `\begin{data}` and `\end{data}`. From this definition, one can understand that the number 0.05 written after `input:` have the name `input.spring_const`, and it represents 0.05 ($\mathtt{kg/s^2}$). In this example, the file does not have the data for `output`; the data may be filled later by some engine.

The syntax `result[]` indicates that it is an array of data. Each element of the array can be referred to as `result[0]`, `result[1]`,.... Thus an element, for example, `result[0]` has three data members, each referred to as `result[0].time`, `result[0].displacement` and `result[0].velocity`.

All data in the UDF file have names, and can be quoted by their names. Therefore, once the UDF file is opened in the C/C++ program. Individual data can be accessed by their names. The following is the example of the C++ program which handles the above file.

```
---------------------------------------------
# include ''platform.lib''
uf = UDFManager(''spring.udf'');
double m; uf.get(''input.mass'',m);
double k;uf.get(''input.spring_const'', k);
uf.put(''output.periodicity'', 6.28*sqrt(m/k));
---------------------------------------------
```

This program changes the file "spring.udf" to the following:

```
---------------------------------------------
//datafile spring.udf
...the data definition part is unchanged and omitted.
\begin{data}
    input:{0.05,0.1}
    output:{4.44,[]}
\end{data}
---------------------------------------------
```

UDFManager provides various services for programmers concerning the file IO. UDFManager supports the IO of simple data type (int, float, double, string, etc.), and their arrays (`int[]` or `vector<int>`, etc.). C++ programmers can write a code to read and write complex data structure such as `result[]` at once. There is a utility program which generates such C++ code automatically from the definition of the UDF file.

The unit in the definition of UDF is an option. If the data does not have unit, UDFManager regard it as a nondimensional data. Using nondimensional data seems to be a common practice of scientific programmers, but we encourage them to put unit to all data. Unit is an essential quantity needed in data transformation. If programmers use nondimensional quantities in their programs, they should state the basic physical quantities they use to make the nondimensional expression. To encourage them to do it, we implemented the following syntax:

```
---------------------------------------------
// Unit example1
\begin{unit}
    [M] = 5 (kg)
    [L] = 10 (m)
    [T] = 60 (s)
\end{unit}
\begin{def}
    input: { mass        : double (M),
            spring_const: double (M/T^2)
         initial_condition:{
             displacement: double [L],
             velocity : double [L/T]
          }
         }
\end{def}
\begin{data}
    input:{0.05, 0.1, {0.1, 0.3} }
\end{data}
---------------------------------------------
```

As it is demonstrated by this example, programmers can define their own unit system between `\begin{unit}` and `\end{unit}`. The data are passed to the engines in the unit defined by the engine programmer, but they can be translated to the other unit system like SI. For example, `input.mass` is shown as `0.05[M])` in the default, but if the standard SI unit is chosen, `input.mass` is shown as `0.25(kg)`.

If the engine programmer likes to change the value of the basic physical quantity, he can use the following syntax.

```
---------------------------------------------
// Unit example2
\begin{unit}
    [K0]  = {$basic_quantity.k0} (kg/s^2)
    [M]   = basic_quantity.m0 (kg)
    [T]   = [(M/K0)^0.5]
\end{unit}
\begin{def}
    basic_quantity: {k0: double (kg/s^2)
                     m0: double (kg)...
                    }
\end{def}
---------------------------------------------
```

In this case, users can see the data in SI unit by putting certain numbers to `basic_quantity.k0` and `basic_quantity.m0`. If the data for `basic_quantity.k0` is empty or has value 0, UDFManager does not do the unit transformation, so the result is equivalent to use the nondimensional data. This syntax allows programmers to write their codes in dimensionless form, simultaneously allowing other users to see the data with dimensions.

The following example shows a more complex syntax for the data definition in the UDF file:

```
-------------------------------------------
// data file atom.udf
\begin{unit}
    [L] = 1(nm), [T] = 1(fs), [M] = 1.66E- 27(kg)
\end{unit}
\begin{def}
    class Vector:{x:double(unit), y:double(unit),z:double(unit)} (unit)
    class AtomType:{name:KEY, mass:double [M]}
    atomType[]: AtomType
    atom[]:    {position : Vector [L],
                velocity : Vector [L/T],
                type : <AtomType, KEY>}
\end{def}
\begin{data}
atomType[]: [ {''Ar'', 40}, {''Ne'', 20}]
atom[]:[{{1.1 1.3 0.2} {1.0, - 2.0, 3.0} ''Ar''}
     { {0.1 0.3 1.2} {- 1.0, 0.0, 4.0} ''Ne''} ]
\end{data}
-------------------------------------------
```

Here `class Vector` defines a new data type `Vector` which has three members `x,y,z` all having the same unit. The defined data type can be used in the definition of other data object as it is demonstrated by the example of `atom[]`, where `atom[i].position` and `atom[i].velocity` are the vector type each having unit `[L]` and `[L/T]`, respectively. Likewise `AtomType` has two members, `name` and `mass`. The data type KEY in the definition of `AtomType` is a way of mapping the pointer in C/C++ to the text file. It states that the data of this class can be uniquely identified by the value of the data placed here. The definition of `atom[]` demonstrates how KEY is used: the syntax `type: <atomType, KEY>` indicates that `type` is used to specify a data of `AtomType`. In this example, `atom[0].type` is “Ar”, so it refers to a data of `AtomType` with the KEY value equal to “Ar”, which is `atomType[0]`.

Note that KEY is introduced to express the logical structure of the data explicitly. The actual data of KEY is a text string, so the definition part of the above UDF file can be replaced by the following:

```
-------------------------------------------
\begin{def}
    class Vector:{ x:double (unit), y:double(unit), z:double(unit)} (unit)
    atomType[]:{name:string, mass:double [M]}
    atom[] :{position:Vector [L], velocity:Vector [L/T], type:string}
\end{def}
-------------------------------------------
```

This definition is perfectly valid for the data part shown above. However, it blurs the fact that the `atom[].type` indicates a certain data in `atomType[]`. The definition using KEY is a better way to express the idea of the programmer. UDF has two ways of mapping pointers: one is KEY which uses text string, and the other is ID which uses integer. Like unit, KEY and ID are introduced to make the file readable for others, and to reduce misunderstanding between researchers.

4. Modeling platform

UDF file is a text file, and can be edited by any text editor. The operation, however, will become cumbersome as the data size increases (typically tens of mega bytes). In such a case it is more convenient to handle the UDF file by a special program.

We have developed a program GOURMET (Graphical open user interface for material design environment) for handling the UDF file. GOURMET offers various services for programmers and users such as:

(1) Browsing and editing the UDF file.
(2) Data handling by script language.
(3) Data plotting and making animations.
(4) Running and monitoring simulation engines.

These services are explained below.

4.1. Browsing and editing the UDF file

The UDF file can be regarded as a mini database file: the structure of the database is defined in the definition part. GOURMET may be regarded as a mini-database engine for the UDF files. User can view the contents of the UDF file in a tree view or a table view (see Fig. 2). The displayed data can be exported to and imported from text editors or other applications by the operation of cut and paste.

The data are initially shown in the unit defined in the UDF file, but it can be shown by other unit like SI or cgs by the unit menu. Whichever units are used in GOURMET, the data are translated to the unit defined by the UDF file and passed to the engine.

4.2. Data handling by script language

The data in the UDF can be handled by script language. At present, GOURMET supports the script language python [[2]]. GOURMET has a window where user can write python program and execute it. GOURMET introduces a small extension for python: all data in the UDF file can be handled like variables in python. For example, the following script is valid in GOURMET:

```
print $atom[0].position.x
print $atom[0].position
print $atom[].type
```

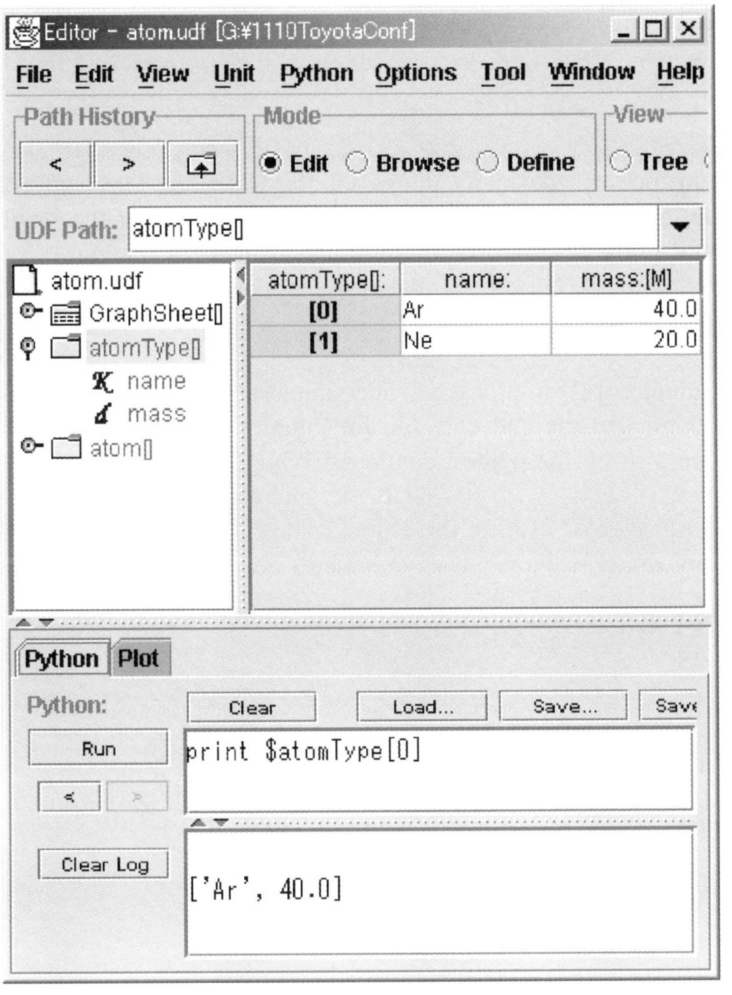

Fig. 2. Data viewing in GOURMET.

The variable starting with the symbol $ is a UDF variable. The syntax `$atom[0]` represents an array element, while `$atom[]` represents the whole array. The result of the above program is

```
1.1
[1.1, 1.3, 0.2]
['Ar', 'Ne']
```

Note that the structured data in UDF file can be mapped to the array data in python. Conversely, the data in the UDF file can be changed by the python program:

```
$atom[0].positing[0.0, 0.0, 0.0]
$atom[1] = $atom[0]
print $atom[].type
```

The second command replaces `atom[1]` with the data of `atom[0]`, so that the output of the above program is

```
['Ar', 'Ar']
```

Using python program, one can do many things for the UDF file. One can use python program to create large input file, or do statistical analysis for output file. In fact the flexibility of GOURMET owes greatly to python. This will become more apparent in the following.

4.3. Data plotting and making animations

GOURMET uses gnuplot [[4]] to plot data. The simplest usage is to plot the data written in table. GOURMET creates command file and data file for gnuplot and invokes the gnuplot program. This procedure can be done also by the python command. Thus, one can make a python program which conducts certain statistical analysis for the UDF file, and show the result by gnuplot.

GOURMET has a function of making three-dimensional graphics and animation using Open GL. The following program draws a dumbbell (two spheres connected by a cylinder):

```
center1 = [0,0,0]
center2 = [1,1,1]
sphere(center1, 1)
sphere(center2, 1)
cylinder(center1, center2, 1)
```

The following python program display all atoms in atom.udf:

```
i = size ($atom[])
while i > 0:
    i = i-1
    sphere($atom[i].position, 0)
```

If one executes such a program for each time step, one can create an animation to see the time evolution of the system. Various functions are available to show the configurations of molecules or fields in three-dimensional graphics (see Fig. 3).

4.4. Running and monitoring simulation engines

GOURMET offers services for users to operate engines. The basic service is to send the input UDF file to the engine server (the computer in which the engine is implemented), to start the engine, and to get the output file when the calculation ends. This service can be enhanced by the engine programmer if he utilizes the other UDF files which GOURMET are exchanging with the engine server. These files include information on (1) the output of engines during calculation, (2) the instruction of users (pause and restart), and (3) the parameters which may be changed by users while the engine is paused. GOURMET has windows to display these files. Utilizing these files, a programmer can make his engine work interactively with users. For example, he can make the

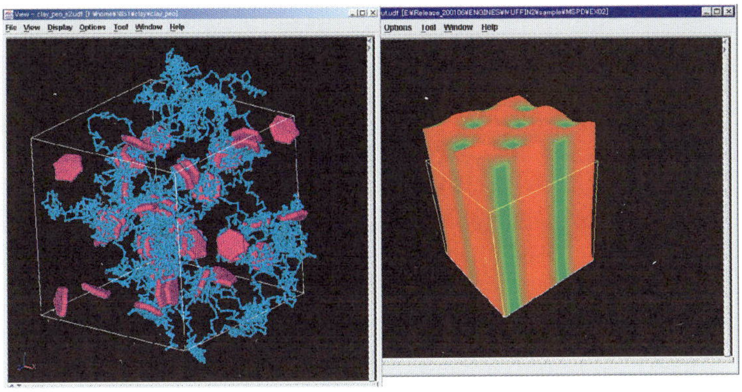

Fig. 3. Examples of the three-dimensional graphics.

minimum search engine in such a way that user can change the convergence parameters during the calculation by monitoring the results of engines.

4.5. Customization

GOURMET has several options to allow users to customize the interface for their own use. The following are the typical examples.

(1) *Wrapping*: The simulation engines shown in Table 1 (and many other engines designed for general use) have many input parameters, but usually only a small subset of them are needed in a particular application. In such a case, the user can create his own input file, and write a python program to generate the input file for the simulation engine.
(2) *Action*: This is a function of assigning python programs to data objects. For example, a user can write a program to calculate the mean position of `atom[]`, and assign it to the data item `atom[]`. He can then start this program by selecting the field `atom[]` in the view window of GOURMET. The "action" is similar to the "methods" in C++ class, but unlike C++ class, the assignment is done by writing the python program in a text file. Thus a user can define his own action independently of the original UDF file.

5. Collaborative operation of engines

The platform can be used as a base of operating various engines collaboratively. Such activity may be regarded as a prototype of zooming, namely going from one level of description to the other. In our project, we made several prototypes of zooming

An example is to predict the mechanical properties of block polymers (see Fig. 4, [[1]]). Block polymer is a polymer made of two or more blocks each consisting of certain monomeric species. Block polymers spontaneously form micro-domain structures consisting of spherical, cylindrical and plate-like domains. The structure depends on many parameters such as the characteristics of

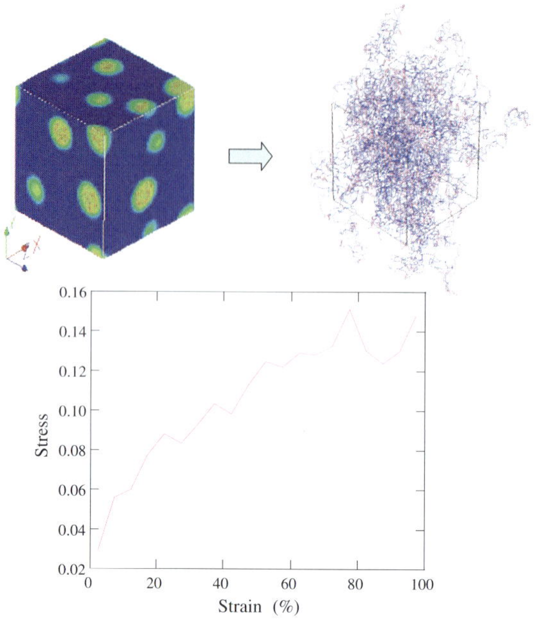

Fig. 4. The procedure of calculating the stress–strain curve of block copolymers.

monomeric species, the length of a block of monomers, the topology of the connection of the blocks, the mixing ratio, etc. Predicting the equilibrium structure by molecular dynamics is practically impossible since the ordering takes place with the time of the order of 10^3 s, while the characteristic time scale of molecular dynamics is 10^{-12} s. We thus used the interfacial engines (SUSHI) to calculate the spatial distribution of monomer density at equilibrium. We developed a program, which is implemented in our molecular dynamics engine (COGNAC), to generate the equilibrium configuration of polymer chains for given spatial distribution of monomer density. We then calculate the stress–strain curve of the block-copolymers by the molecular dynamics. We applied similar techniques to predict the stress–strain behavior of crystalline polymers.

Another example of the collaborative operation of engines is to calculate the permeability of gas molecules through the film of block copolymers. Using the interfacial dynamics engine (SUSHI), one can calculate the spatial structure of micro-domains. He can then calculate the diffusion constant of a gaseous molecule in each domains using the molecular dynamics engine (COGNAC). Finally, using the micro-fluidics and micro-elastics engine(MUFFIN), he can calculate the effective permeability of the gaseous molecules.

So far such operations must be done manually. User has to create the input UDF file for certain engine using the results of the other engines. The work may become easier with GOURMET, but one has to write certain programs (by python or C/C++). In future, the operations may be automated, but, at this stage we left the task to users. We did this deliberately as it is an operation which needs deep understanding of the problem, and is too dangerous to do automatically.

6. Conclusion

The platform described here was created in our project aiming at seamless zooming. We purposely tried to remove any mechanism which forces engine programmers to participate in zooming. The platform only provides an environment which facilities engine programmers to make his program understandable for others, and for them to use the programs made by others.

From the view point of computational science and engineering, there are many challenging problems regarding the procedure of zooming such as: (1) Is there any general framework representing physical concepts for simulation engines? (2) Is it possible to check the consistency of the information dealt with different engines? (3) Is it possible to design a system which secures the consistency of information in the procedure of zooming? All are very challenging. Such research will be done in future, perhaps, by the collaboration of physicists, chemists and software engineers. At this stage, however, our view is that what is currently needed is a mechanism which facilitates the collaboration of many simulation programs, which actually means the collaboration of people. The software package we developed is called OCTA [[3]], and the whole system including source codes will become downloadable in April 2002. We hope that it will be a step towards this direction.

Acknowledgements

This work is supported by the national project, which has been entrusted to the Japan Chemical Innovation Institute (JCII) by the New Energy and Industrial Technology Development Organization (NEDO) under METI's Program for the Scientific Technology Development for Industries that Creates New Industries.

References

[1] T. Aoyagi, T. Honda, M. Doi, Micro structural study of mechanical properties of ABA triblock copolymer using self-consistent field and molecular dynamics, J. Chem. Phys., in press.
[2] "Python" is a software originally developed by Guido van Rossum. See http://www.python.org.
[3] The information on OCTA is obtained at http://octa.jp.
[4] The information on the software "gnuplot" is obtained at http://www.cs.dartmouth.edu/gnuplot_info.html.

ELSEVIER

Journal of Computational and Applied Mathematics 149 (2002) 27–56

JOURNAL OF COMPUTATIONAL AND APPLIED MATHEMATICS

www.elsevier.com/locate/cam

Materials by design and the exciting role of quantum computation/simulation

A.J. Freeman

Department of Physics and Astronomy, Materials Research Center, Northwestern University, 2739 Ridge Avenue, Evanston, IL 60208-3112, USA

Received 1 December 2001; received in revised form 6 February 2002

Abstract

It is now well-recognized that we are witnessing a golden age of innovation with novel materials, with discoveries important for both basic science and device applications—some of which will be treated at this Workshop. In this talk, we discuss the role of computation and simulation in the dramatic advances of the past and those we are witnessing today. We will also describe the growing acceptance and impact of computational materials science as a major component of materials research and its import for the future. In the process, we will demonstrate how the well-recognized goal driving computational physics/computational materials science—simulations of ever-increasing complexity on more and more realistic models—has been brought into greater focus with the introduction of greater computing power that is readily available to run sophisticated and powerful software codes like our highly precise full-potential linearized augmented plane wave (FLAPW) method, now also running on massively parallel computer platforms.

We will then describe some specific advances we are witnessing today, and computation and simulation as a major component of quantum materials design and its import for the future, with the goal—to synthesize materials with desired properties in a controlled way via materials engineering on the atomic scale. The theory continues to develop along with computing power. With the universality and applicability of these methods to essentially all materials and properties, these simulations are starting to fill the increasingly urgent demands of material scientists and engineers.

E-mail address: art@freeman.phys.nwu.edu (A.J. Freeman).

0377-0427/02/$ - see front matter
PII: S0377-0427(02)00519-8

1. Introduction and background perspective

1.1. Computational physics—as the third branch of physics

As fields of science advance there is a natural evolution in theory from analytic to numerical solutions. Once satisfactory theoretical formulations have been established and their validity demonstrated for simple test cases, the next natural step is the application of the theory to realistic, i.e., more complex systems. At this point, large-scare computation on computers is necessary; one moves from investigating systems with a few degrees of freedom to those with many degrees of freedom. Thus, complexity is the key to understanding computational science [27].

The step from the simple analytical to the complex numerical domain is particularly interesting in physics where theories such as relativity and quantum mechanics have revolutionized our understanding of the macrocosmos and microcosmos. Hence, the application of such powerful theories has greatly stimulated the development of computational physics.

By its nature, methods and needs, computational physics is so different and separate from the two well-known and well recognized branches of physics—analytic theoretical physics and experimental physics—as to constitute the third branch of physics. Basically, computational physics is the numerical investigation of complex systems; it permits theory to be extended well beyond analytic limitations. As with other areas of computational science, the development of this new field continues to be intimately connected to the existence and evolution of supercomputing power.

Condensed matter theory is an area where computational physics has had a wide impact. This field includes such diverse subdisciplines as the electronic and geometric structure of perfect and imperfect solids, disordered and amorphous materials, liquid metals and alloys, surfaces, interfaces, and artificial (laboratory made) exotic materials such as superlattices. Condensed matter theory ranges from direct applications of principles for device fabrication to complete formal theories to explain and predict new properties of matter. It is one of the rare areas where the science is as basic as science gets, yet the ideas developed are often easily applicable to practical problems. Since all these developments are based on the underlying quantum theory of matter we describe this first before discussing numerical approaches and examples of applications.

1.2. Quantum theory of matter—a new dimension to materials science

The development of quantum mechanics in the early 1920s provided the whole of chemistry and most of physics with a rigorous, quantitative theoretical foundation. Indeed, as P.A.M. Dirac put it in 1926, chemistry had "become a branch of applied mathematics". Indeed, it is possible, in principle at least, to determine theoretically all the properties of matter—electromagnetic, mechanical, thermal, and chemical. It was quickly realized however, that a major problem still existed: the quantum mechanical equations governing the properties of molecules and materials were much too complex to be solved exactly.

Nevertheless, during the past half century, considerable progress has been made in the development of theoretical and computational methods to solve approximately the quantum mechanics of atomic, molecular, and solid state systems. The real breakthrough occurred when electronic computers made it possible to carry out the monumental numerical work required to solve the quantum mechanical problem.

Today, we have reached a threshold: advanced theoretical-computational techniques combined with the growing power of computers provide an understanding of matter at the atomic-scale with an unprecedented level of detail and accuracy. This capability has given birth to a new branch of scientific endeavor: computational physics and materials science. Not only can we simulate experiment, but using the computational approach, we have begun to design new molecules and materials and to predict their properties without actually synthesizing them. In contrast to an analytical–theoretical approach, which isolates and idealizes real systems to unravel fundamental relations and laws, the computational approach is synthetic: its goal is to simulate more and more detailed systems including as much of the environment as possible.

1.3. The computational task in electronic structure theory

For describing the electronic structure of real materials, and hence, all of their properties, one needs to first reduce the set of coupled differential equations for the orbitals to a manageable size and then to devise algorithms and mathematical procedures for obtaining accurate numerical solutions.

Two different approaches exist for reducing the size of the set of differential equations: either one considers the solid to be represented by a finite but reasonably large cluster of atoms or one makes use of the translational symmetry which exists for a perfect infinite crystal. In the latter case, the so-called energy band approach makes use of the fact that the size of a macroscopic solid is very large compared to atomic dimensions to justify the mathematical description of a solid as an infinite repetition of small, identical unit cells. Since the time average of the electronic orbitals in all these unit cells is the same, one only has to solve the differential equations for the orbitals in one of the unit cells. In the case of perfect crystals, the unit cell contains relatively few atoms; however, when the effects of impurities, vacancies, defects, or dislocations are included to investigate more realistic situations, the size of the cell rapidly becomes very large.

A major development in electronic structure theory was the discovery and formulation of density functional theory by Hohenberg, Kohn, and Sham in the mid-1960s. This is based on the important theorem that all ground state properties of matter can be derived from the charge density distribution. All material systems are considered as formed from atomic nuclei and swarms of strongly interacting electrons. The electrons are defined at any point in space by a collective probability density given by the charge distribution arising from all electrons in the system. The conceptually crucial aspect of density functional theory is this collective approach. From this and the so-called local density approximation, one can derive a new set of simplified decoupled differential equations which is formally equivalent to the original set of equations of motion for the orbitals. The great advantage is that these new (local density) equations are completely determined by the charge density distribution and can be solved independently for each orbital.

1.4. Solution of the local density equations

Local density theory permits one to solve a relatively small number of decoupled partial differential equations to obtain the charge densities. The catch in all this is that the charge densities are required to construct the local density equations themselves. The way out of this dilemma is to set up an iterative procedure until “self-consistency” is achieved.

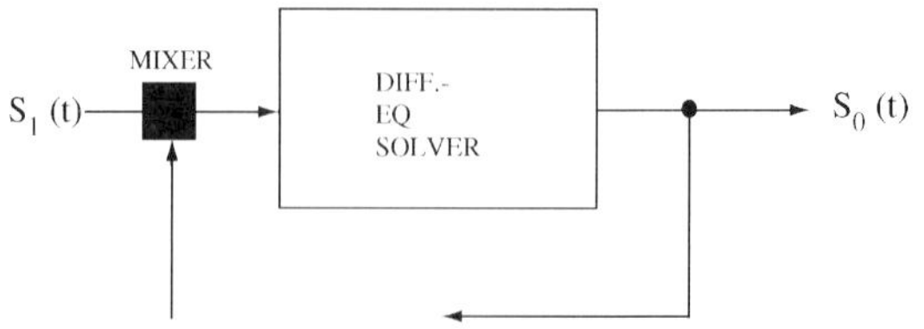

Fig. 1. Schematic of the feed-back systems for producing self-consistency.

The basic idea in such an iterative cycle is to start with assumed approximate values for the orbitals and to use this input to decouple the differential equations. The solution of this simplified system of equations then provides us with new approximate values for the orbitals (i.e., charge density) and the iteration procedure is continued to self-consistency, (i.e., until the input and output charge densities are identical within a prescribed precision). This iterative technique has a nice analogue in electronic circuits. To see this, one replaces the set of differential equations by some electronic device (a "black box") which takes an input signal $S_I(t)$ (input orbitals) and produces an output signal $S_0(t)$ (output orbitals). (See Fig. 1.) One is looking for a special value of the input signal, S_c, for which the output signal is also equal to S_c. Unfortunately, this device is an amplifier; any deviation from S_c in the input will give rise to a larger deviation of the output. However, the sign of the deviation is reversed and hence one can use standard feedback techniques to drive the device to the constant self-consistent solution S_c. (See Fig. 1.) In our case, this feedback is equivalent to some mixing of input and output charge densities.

At this point, the set of coupled differential equations has been replaced by simple equations plus an iterative feedback loop. The main numerical effort is in solving the differential equation; the feedback implies that this step has to be repeated many times. The number of iterations depends on the complexity of the system under consideration. The ten or so iterations needed for a simple mono-atomic solid like aluminum increases to 500–100 for more complex systems. The final results obtained are considered to be accurate "self-consistent" solutions which are then independent of any starting assumptions. Obviously the complexity of the physical system to be modeled dictates a massive computational requirement regardless of the numerical approach.

1.5. Electronic structure of real materials

The last two decades have witnessed dramatic advances in condensed matter theory, driven, in large part, by new and sophisticated experiments on high purity materials which have been well and carefully characterized. In most areas of condensed matter theory, and particularly in electronic structure, these advances may be attributable directly to the close collaboration of theoretical and experimental research. Indeed, the new found ability to apply fundamental theoretical concepts to real materials (rather than to simple model systems) made possible by utilizing the continued rapid development of computer power has served to fill the increasingly urgent demand of experimentalists for theoretical interpretation of their data. Also, in some cases these computational efforts can be used to provide data which would be impossible or impractical to obtain experimentally. This development has been an essential element in the phenomenal growth seen in this area of materials science.

The range and variety of materials under investigation is enormous. One of the recent important developing trends lies in the preparation of synthetic structures on the submicron level, which permits new scientific phenomena to be investigated and novel device applications to be made on artificial materials not found in nature. In addition to the technological reason for focusing attention on submicron problems such as MOS devices, there are interesting phenomena to study which are of fundamental and physical interest in themselves. Indeed our attention has spread from consideration of bulk properties to surfaces and interfaces for obtaining a better fundamental understanding of interfacial and reduced dimensionality phenomena.

2. Density functional theory and the FLAPW method

2.1. Introduction

The full-potential spin-polarized linear augmented plane wave (FLAPW) method is presently considered to be the most accurate electronic structure calculation scheme. It has its origin in the augmented plane wave (APW) method introduced by Slater [94] (details about this method can also be found in Ref. [60]). In this approach, real space is partitioned into spherical regions around atoms ("muffin-tins" or "atomic spheres") and interstitial regions between the spheres. Computationally, the APW method is demanding since the basis functions are energy dependent and the eigenvalue problem nonlinear. The subsequent linearization of the APW method (LAPW) [4,51], where the energy dependence is removed by selecting a fixed set of suitable muffin-tin radial functions and their energy derivatives, represented an important development. In the FLAPW method there is no shape approximation for either the charge density or the potential, and all electrons are treated in the self-consistent process; the core electrons are treated fully relativistically and the valence electrons are treated semi-relativistically. Further details can be found in original papers [48,108] and in recent reviews [93,29].

Aside from the early important introduction of a total energy capability [107], recent improvements and extensions to the present FLAPW calculation scheme include: (i) evaluation of forces on the atoms, which affords automatic optimization of the atomic geometry [115,116,95], (ii) methodological developments resulting in significant speed-ups [111], (iii) spin–orbit coupling affording calculation of magnetic properties, namely, magnetocrystalline anisotropy (MCA), surface magneto-optic Kerr effect (SMOKE), and magnetic circular X-ray dichroism (MCD), (iv) optical properties, (v) the option of several different exchange-correlation functionals, i.e., the local density approximation (LDA) and the generalized gradient approximation (GGA) and (vi) the treatment of excited states via screened exchange (sX), as well as model *GW* approaches , and (vii) the calculation of vibrational frequencies. A very recent and important break-through is the successful implementation of a parallelized version of the code [16].

2.2. Density-functional theory

The FLAPW method employs density-functional theory (DFT), introduced by Hohenberg and Kohn [44] and Kohn and Sham [52]. The underlying theorem (Hohenberg–Kohn–Sham theorem) on which this theory rests is that the total energy, E, of an atomistic system can be expressed as a functional of its electron density, ρ, namely, $E = E[\rho]$, that E is at its minimum for the ground-state density, and is stationary with respect to first-order variations in the density.

Typically, the Born–Oppenheimer approximation [13] is employed which assumes that the motion of the nuclei are negligible with respect to that of electrons. This implies that the electronic structure is calculated for a given atomic geometry; the nuclei are then moved according to classical mechanics.

2.2.1. The Kohn–Sham equations

To obtain the ground-state density, the variational principle is applied with respect to the one-particle wave functions:

$$[(-\hbar^2/2m)\nabla^2 + V_{\text{eff}}(\mathbf{r})]\psi_i(\mathbf{r}) = \varepsilon_i\psi_i(\mathbf{r}), \tag{1}$$

where

$$V_{\text{eff}}(\mathbf{r}) = V_{\text{C}}(\mathbf{r}) + \mu_{xc}[\rho(\mathbf{r})] \tag{2}$$

is the effective potential and ε_i the effective one-electron eigenvalues. Eqs. (1) are the "Kohn–Sham equations" and the solutions, $\psi_i(\mathbf{r})$, form an orthonormal set, i.e., $\int \psi_i^*(\mathbf{r})\psi_j(\mathbf{r})\,\mathrm{d}\mathbf{r} = \delta_{ij}$. The Coulomb or electrostatic potential is given as

$$V_{\text{C}}(\mathbf{r}) = -e^2 \sum_{\alpha} \frac{Z_\alpha}{|\mathbf{r} - \mathbf{R}_\alpha|} + e^2 \int \frac{\rho(\mathbf{r}')}{|\mathbf{r} - \mathbf{r}'|}\,\mathrm{d}\mathbf{r}' \tag{3}$$

which can also be calculated using Poisson's equation, i.e.,

$$\nabla^2 V_{\text{C}}(\mathbf{r}) = -4\pi e^2 q(\mathbf{r}), \tag{4}$$

where $q(\mathbf{r})$ represents the electronic charge distribution *and* the positive point charges at position $\mathbf{R}_\alpha$. The exchange-correlation potential is given by

$$\mu_{xc} = \partial E_{xc}[\rho]/\partial\rho. \tag{5}$$

Because the exchange-correlation potential (and energy) are not known, approximations have to be made.

2.2.2. Spin-polarized density functional theory

The generalization of density-functional theory to spin-polarized systems has been made within the local spin density approximation (LSD) [104,37]. The important quantity, in addition to the electron density $\rho(\mathbf{r})$, is the spin density $\sigma(\mathbf{r})$ which is the density difference between the spin-up and spin-down configurations, i.e., $\sigma(\mathbf{r}) = \rho_\uparrow(\mathbf{r}) - \rho_\downarrow(\mathbf{r})$; the total density being given by $\rho(\mathbf{r}) = \rho_\uparrow(\mathbf{r}) + \rho_\downarrow(\mathbf{r})$. Because the exchange-correlation potential for spin-up and spin-down electrons is in general different, the spin-polarized form of the Kohn–Sham equations are

$$[(-\hbar^2/2m)\nabla^2 + V_{\text{eff}}^{\sigma}(\mathbf{r})]\psi_i^{\sigma}(\mathbf{r}) = \varepsilon_i^{\sigma}\psi_i^{\sigma}(\mathbf{r}), \quad \text{where } \sigma = \uparrow \text{ or } \downarrow, \tag{6}$$

and

$$V_{\text{eff}}^{\sigma}(\mathbf{r}) = V_{\text{C}}(\mathbf{r}) + \mu_{xc}^{\sigma}[\rho(\mathbf{r}), \sigma(\mathbf{r})]. \tag{7}$$

Thus there are two sets of single-particle wave functions, one for spin-up (or "majority") electrons and one for spin-down ("minority") electrons, each with corresponding one-electron eigenvalues.

2.2.3. Exchange-correlation functional

Local-density approximation (LDA). A very successful and widely used approximation for the exchange-correlation energy is the local-density approximation (LDA). Here the exchange-correlation energy is assumed to depend only on the local electron density of each volume element $d\mathbf{r}$:

$$E_{xc}[\rho] \approx \int \rho(\mathbf{r})\varepsilon_{xc}[\rho(\mathbf{r})]\,d\mathbf{r}. \tag{8}$$

$\varepsilon_{xc}[\rho]$ is the exchange-correlation energy per electron of a *homogeneous* electron gas and is expressed as an analytic function of the electron density, as is the exchange-correlation potential, μ_{xc}. There are various forms of the LDA in the literature; we refer to those of Hedin–Lunqvist [43] and Wigner (nonspin-polarized forms), and Barth–Hedin [104] (spin-polarized form) since they are the ones implemented in the present FLAPW program version.

Generalized gradient approximation (GGA). In recent years, the generalized gradient approximation (GGA) is being considered as a possible improvement over the LDA. The GGA has been found to generally improve the description of total energies, ionization energies, and electron affinities of atoms, atomization energies of molecules [81,49,11] and some solid state properties [31,50,19,54]. Adsorption energies of adparticles on surfaces are also reported to be improved [100,45] as are reaction energies [84,8]. Furthermore, the GGA leads to significantly better activation energy barriers for H_2 dissociation [40,41] and also the relative stability of structural phases appears to be better described for magnetic [57] and nonmagnetic systems [69,39].

There are a number of GGA approaches in the literature. The generic form of the GGA exchange-correlation energy may be written as

$$E_{xc}^{\mathrm{GGA}}[\rho] = \int \rho(\mathbf{r})\varepsilon_{xc}^{\mathrm{GGA}}(\rho(\mathbf{r}), \nabla\rho(\mathbf{r}))\,d\mathbf{r} \tag{9}$$

so that it depends locally on the electronic density $\rho(\mathbf{r})$ and its gradient.

The GGA developed by Perdew and Wang (PW) [81] is derived essentially from first principles, combining the gradient expansions of the exchange and correlation holes of a non-uniform electron gas with real-space truncations to enforce constraints imposed by properties of the physical exchange-correlation hole. The GGA developed by Becke and Perdew [10] on the other hand relies on fitted parameters. In the present program version of the FLAPW program the GGA implemented is that proposed by Perdew et al. (PBE) [80]. This functional is regarded to be conceptually more concise than the PW GGA but appears to yield very similar results [30].

2.3. The FLAPW basis-set

In the FLAPW method, in bulk material, real space is partitioned into spherical regions around atoms ("muffin-tins" or "atomic spheres") and interstitial regions between the spheres. In the spherical region, the basis functions are products of radial functions and spherical harmonics, and in the interstitial region plane waves are used. For the film geometry, there is a number of atomic layers surrounded by vacuum, thus in addition to spherical and interstitial regions, one defines a vacuum region, which starts at $\pm D/2$ and ends at $\pm\tilde{D}/2$ (see Fig. 2), where the wave functions are products of two-dimensional (2D) plane waves and z-dependent functions which are solutions of the one-dimensional Schrödinger equation of the (x, y)-averaged potential in the vacuum region.

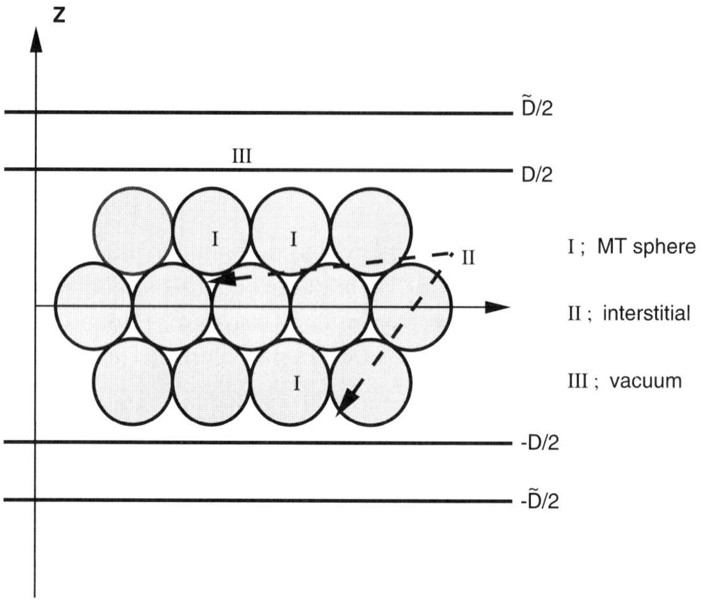

Fig. 2. Geometry for a film calculation showing division of space in a film geometry.

Specifically, the FLAPW one-particle wave functions in the film geometry are:

$$\psi_i(\mathbf{r},\mathbf{k}_{\|}) = \sum_j c_{ij}\phi(\mathbf{r},\mathbf{K}_j); \quad \mathbf{K}_j = \mathbf{k}^{\|} + \mathbf{G}_j, \tag{10}$$

where $\mathbf{k}^{\|}$ is an arbitrary vector of the 2D BZ and $\mathbf{G}_j$ is a three-dimensional (3D) reciprocal lattice vector (in the z-direction an artificial periodicity between the boundaries at $\pm\tilde{D}/2$ is imposed). The basis functions are:

$$\phi(\mathbf{r},\mathbf{K}_j) = \begin{cases} \Omega^{-1/2}\mathrm{e}^{\mathrm{i}\mathbf{K}_j.\mathbf{r}} & \text{interstitial,} \\ \sum_{lm} [A^{\alpha}_{lm}(\mathbf{K}_j)u_l(E^{\alpha}_l, r_\alpha) + B^{\alpha}_{lm}(\mathbf{K}_j)\dot{u}_l(E^{\alpha}_l, r_\alpha)]Y_{lm}(\hat{r}_\alpha) & \text{sphere,} \\ \sum_q [A_q(\mathbf{K}_j)u_{\mathbf{k}q}(E_v, z) + B_q(\mathbf{K}_j)\dot{u}_{\mathbf{k}q}(E_v, z)]\mathrm{e}^{\mathrm{i}(\mathbf{k}^{\|}+\mathbf{K}^{\|}_q).\mathbf{r}} & \text{vacuum.} \end{cases} \tag{11}$$

The Coulomb potential $V(\mathbf{r})$ is obtained by solving Poisson's equation in each of the three regions. At infinity the potential is gauged equal to zero in the case of the film geometry. The effective single-particle potential is constructed by adding the exchange-correlation potential, which is determined by the charge density in real space and transformation of it into each representation. The core electrons are treated fully relativistically and are updated at each iteration using a scheme for *free* atoms (only the spherical part of the potential is used). The valence electrons are expanded in a variational basis set and are treated scalar-relativistically.

The FLAPW technology has been implemented via extensive code developments (about 45 man years of effort since its original formulation and development at Northwestern University in 1981). It is the only electronic structure code that also treats 2D systems with single slab geometry, whereas other codes make an artificial 3D super slab treatment with a vacuum lesions. The code is unique in that it has combined, or integrated, the bulk and thin film capabilities into a single code, where both are treated to the same high precision. It has been written in Fortran 90 with full dynamic memory allocation and runs on many machines (PC, HP, SGI, Dec-Alpha, NEC, SX4, all Crays, IBM SP, etc). It has a short simple use of friendly input and has been fully parallelized with high efficiency and good scalability and has run on up to 512 processors [16].

3. Some recent applications

It needs to be emphasized that the DFT-FLAPW method is universal in that it can be applied to essentially all materials and their properties. Clearly, this follows from the fact that all properties of materials originate in the electronic structure and by that we mean structural, electronic, magnetic, optical, mechanical, and many others. As selected examples, we cite the following—all of which have been worked on in my own research group. Of course this is only a few of the examples that could be given based on the work of others: (i) Transparent conducting oxides, (ii) semiconductor hetero-structures, (iii) metal–ceramic interfaces, (iv) metal–semiconducting interfaces, (v) ferroelectrics and phase transitions, (vi) magnetism of transition and rare-earth metal surfaces and interfaces and multilayers, (vii) high T_c superconductors, (viii) embrittlement of grain boundaries in Fe (of importance for our steel research), and (ix) mechanical properties of Al intermetallics (which includes recent studies of the fundamentals of dislocation and fracture properties of intermetallics). Since we do not have the space to discuss all of these in any detail, the following gives some recent examples to show the power of the methods and their application to a number of important areas.

3.1. *Optical properties and electronic structures of semiconductors with screened-exchange LDA*

The optical properties of semiconductors have been the focus of both theoretical and experimental studies for many years. Recent major advances in quantitative computations related to the development of density-functional theory [44,52] in the local density approximation (LDA) have provided detailed calculations of the optical and dielectric properties based mostly on the independent-particle approximation [26,36] However, due to the major LDA shortcoming associated with excitation energies, [102,82,92], typically represented by band gap problems in semiconductors, the ab initio calculation of the optical properties still includes large discrepancies from experiment.

Several attempts have been undertaken to improve the theoretical/computational determination of optical properties. The simplest method is to apply a scissors operator, [9,21] which displaces the LDA eigenvalues for the unoccupied states by a rigid energy shift. The problem with this approximation is to find a proper way to choose the correction energy even when provided with the experimental band structure: One choice is to adjust the eigenvalues so that the main peak of the imaginary part of the dielectric function corresponds to the transition energy $X_{5v} \rightarrow X_{1c}$ [21]. The resulting peak position, however, is placed slightly higher in energy than in experiments because the region which contributes to the peak is not local but is rather widely spread out in the Brillouin zone [3].

On the other hand, direct calculations of quasiparticle energies are available within the *GW* approximation [42]. The lowest expansion of the self-energy is called the *GW* approximation, representing the Green's function G times the screened interaction W. Recently, optical properties have been calculated within the *GW* approximation, [21,2] and excellent agreement with experiment for Si has been demonstrated including excitonic and local-field effects [2]. It was also shown that the existence of nonlocality in the self-energy makes the relation $\mathbf{v}=\mathbf{p}/m$ incorrect to ensure gauge invariance and charge conservation [21,97].

Although the *GW* approximation removes most of the problems of LDA in treating excitation properties, its heavy computational demands have hampered determining properties self-consistently as well as applications to more complicated systems. One way to reduce the numerical effort of the *GW* approximation has been developed by its simplification, called the model *GW* method [38,63]. The model *GW* (sometimes called LDA$+\delta\Sigma_{\mathrm{COHSEX}}$) is considered to be one of the simplifications of the *GW* approximation, which reduces considerably the numerical effort. Here, Gygi and Baldereschi proposed [38] to focus on the discrepancy in the self-energy between metals and semiconductors. The self-energy is then rewritten as two terms: (i) the self-energy of a metallic inhomogeneous electron gas, which can be approximated by the LDA expression, $v_{xc}(r)\delta(r-r')$; and (ii) the main contribution to the difference potential, δW, which comes from the different screening behavior in the long wave length region from the metallic one. The static dielectric functions for the metals, semiconductors, and zero-gap semiconductor are asymptotically expressed in terms of k_{TF}, the Thomas–Fermi wave vector, and $\varepsilon(0)$, the dielectric constant of the semiconductor. In metals, the screened interaction decreases rapidly for $|r-r'| > k_{\mathrm{TF}}^{-1}$, but in insulators, it decreases as $(\varepsilon(0)|r-r'|)^{-1}$ for large $|r-r'|$ due to the incompleteness of the screening in insulators.

The results obtained with the model *GW* method for various nonmagnetic semiconductors [38] using the pseudopotential method, and transition metal oxides [63,64] using the full-potential linearized augmented-plane-wave (FLAPW) method [48,108,107], give generally good agreement with the results of full *GW* calculations. Recently, the model *GW* method using the FLAPW method was also applied to Er doped in Si which includes 256 atoms in unit cell [6], and to strained layer InAsSb/InAs superlattices [83].

The recently proposed screened-exchange LDA method [15,91] (sX-LDA) is one of the theories designed to find a better energy functional beyond LDA by modeling the exchange-correlation hole within *nonlocal* density schemes. Since the choice of the energy functional in density functional theory is rather arbitrary, one can consider the model system or the energy functional in which some part of the electron-electron interaction is taken into account.

The sX-LDA method was first proposed by Bylander and Kleinman [15] (who called the method the modified LDA) in order to describe a better band gap. Seidl et al. [91] showed that the method can actually be described in the generalized Kohn–Sham framework, using an effective potential formulated by LDA, and nonlocal and local screened exchange potentials. The virtue of this method is that the sX-LDA eigenvalues can describe the discontinuity of the exchange-correlation potential for the band gaps we obtain and so the sX-LDA is a promising method which can realize accurate ground-state total energies, ground-state charge densities, and band gaps. Note that even though the sX-LDA is expected to have good band gaps, it does not mean that all the sX-LDA eigenvalues correspond to the quasiparticle energies. The meaning of the sX-LDA eigenvalues is still nothing but Lagrange parameters as in the LDA. However, providing that the LDA can yield the excited states with a certain accuracy except the band gap, one expects equivalent or better results for the sX-LDA eigenvalues in the exicted states.

Encouraging results for the band gaps and structural properties by sX-LDA using [91] the plane-wave pseudopotential method and [32,109] the FLAPW method were indeed demonstrated for several semiconductor materials. The advantages of the sX-LDA over the *GW* approximation are that it is much less computationally demanding—almost equivalent with the model *GW* method—and that it permits the self-consistent determination of the ground-state properties. In particular, the lattice constants obtained for Si, Ge and GaAs show [91] better agreement with experiment than LDA; therefore, the ground states of sX-LDA are expected to be better than those of LDA.

We have performed self-consistent sX-LDA calculations [5] as implemented in the FLAPW method for Si, Ge, InSb and GaAs. After checking the calculated electronic structures, we determined the linear optical properties for these materials using the independent-particle approximation and no adjustable parameters. We also demonstrated the effect of the nonlocality of the Hamiltonian and present dielectric constants and effective plasma frequencies obtained from the sum rules.

The calculated sX-LDA band structures for cubic Si, Ge, InSb, and GaAs are shown in Fig. 3 in comparison with the LDA (FLAPW) results [5]. We used experimental lattice constants for all materials: 10.26 (Si), 10.69 (Ge), 12.24 (InSb), and 10.68 (GaAs) in atomic units. In Table 1, we summarize and compare the calculated band gaps and valence band widths (VBW) with the self-consistent sX-LDA results obtained using the plane-wave pseudopotential method [91] and with experiment. The improvement of the band gaps by the sX-LDA method is clearly shown for all materials, and agreement with experiment is good. Although our FLAPW results give a larger discrepancy of the band gaps from experiment than the more approximate plane-wave pseudopotential method, [91] the corrected sX-LDA results show overall agreement with each other; we have results that are closer to experiment at the Γ point in Ge and at the X point in GaAs and a larger discrepancy for InSb.

More recently, accurate band structures of binary III–V semiconductors AB ($A = Al, Ga, In$ and $B=P, As, Sb$) and selected ternary III–V semiconductors have been calculated using the all-electron ab initio screened-exchange approach [32]. The method was validated on 24 semiconducting materials. Typical deviations between experimental band gaps and sX-FLAPW predictions are a few tenths of one eV as shown in Fig. 4 for 24 elemental and binary III–V and II–VI semiconductors of the zincblende and wurtzite structures. Critically for narrow band gap materials like InAs and InSb, the relative error of the method is the same for semiconductors with wide and narrow band gaps. Thus, the method is particularly well suited for modeling thermophotovoltaic and infrared detector materials.

The linear electronic response to a longitudinal external perturbation is described in the independent particle approximation, or the random-phase approximation (RPA), by the dielectric matrix,

$$\varepsilon(\mathbf{q}+\mathbf{G},\mathbf{q}+\mathbf{G}',\omega) = \delta_{\mathbf{G},\mathbf{G}'} - v(\mathbf{q}+\mathbf{G})\chi_0(\mathbf{q}+\mathbf{G},\mathbf{q}+\mathbf{G}',\omega), \tag{12}$$

with $v(\mathbf{q}) = 4\pi e^2/q^2$, and χ_0 is the polarization function of the independent particle. Neglecting local-field effects, the imaginary part of the dielectric function in the long-wavelength limit, $q \to 0$, is written as

$$\varepsilon_2(\omega) = \frac{8\pi^2 e^2}{V} \lim_{q\to 0} \sum_{c,v} \sum_{\mathbf{k}} \frac{1}{q^2} |\langle c,\mathbf{k}+\mathbf{q}|e^{i\mathbf{q}\cdot\mathbf{r}}|v,\mathbf{k}\rangle|^2 \delta[E_c(\mathbf{k}+\mathbf{q}) - E_v(\mathbf{k}) - \hbar\omega], \tag{13}$$

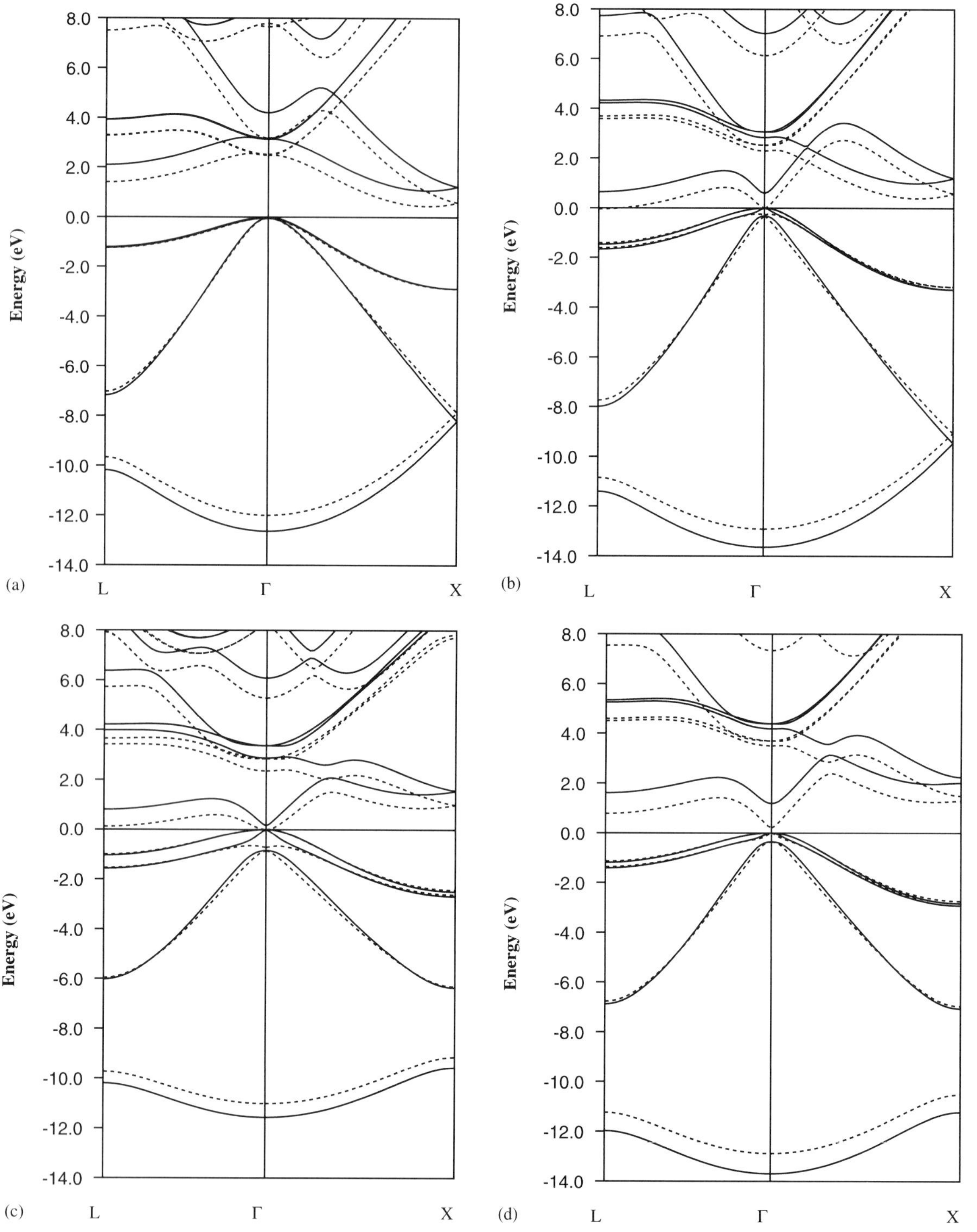

Fig. 3. Self-consistent band structures of (a) Si, (b) Ge, (c) InSb, and (d) GaAs by the sX-LDA method (solid line) and LDA (dashed line).

Table 1
Energy gaps at L, Γ, and X, measured from the top of the valence band, and valence band width (VBW) of Si, Ge, InSb, GaAs in eV calculated by the sX-LDA, which are compared with the LDA results (FLAPW)

		L	Γ	X	VBW
Si	FLAPW	1.40	2.48	0.58	12.08
	sX-LDA	2.17	3.22	1.30	12.61
	sX(PP)	2.18	3.37	1.55	12.47
	Expt	1.65 ± 0.01 $2.1, 2.4 \pm 0.15$	3.05, 3.4	1.25	12.5 ± 0.6
Ge	FLAPW	−0.04	−0.27	0.58	12.92
	sX-LDA	0.68	0.63	1.22	13.65
	sX(PP)	0.66	0.28	1.45	13.41
	Expt	0.74	0.89	1.3 ± 0.2	$12.6, 12.9 \pm 0.3$
InSb	FLAPW	0.12	−0.72	0.99	11.01
	sX-LDA	0.82	0.18	1.55	11.57
	sX(PP)	1.02	0.21	1.82	11.02
	Expt		0.24	1.79	11.7, 11.2
GaAs	FLAPW	0.78	0.21	1.29	12.89
	sX-LDA	1.64	1.19	2.01	13.70
	sX(PP)	1.70	1.11	2.35	13.40
	Expt	1.81	1.52	1.98	13.1

The sX-LDA results using the pseudopotential method from Ref. [91] (sX(PP)), and experiment (Expt) taken from Ref. [91] and references therein.

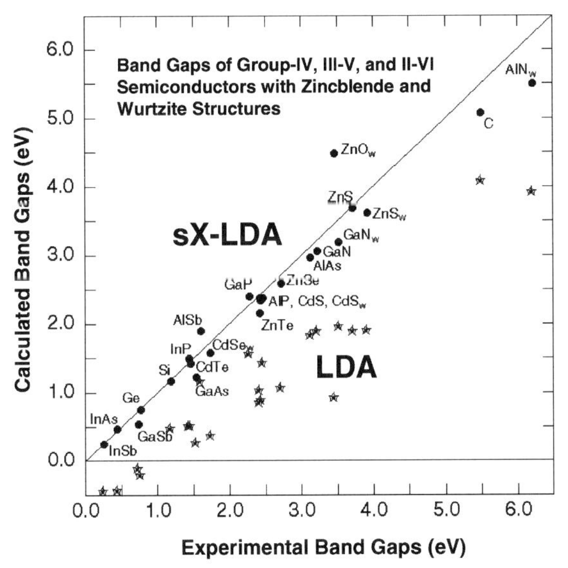

Fig. 4. Calculated vs. experimental energy band gaps for 24 elemental and binary semiconductors. Results from sX-FLAPW calculations are shown as solid circles. Values obtained with the conventional local density approximation LDA are represented as stars. Note the larger scatter of the LDA results and the qualitative failure of LDA for narrow band gap materials.

where c and v denote conduction- and valence-band. The direction of $\mathbf{q}$ defines one of the diagonal elements of the dielectric tensor $\varepsilon_{\alpha\alpha}$ which are all equal for cubic crystals. Taking the trace, therefore, makes it possible to restrict the calculations to the irreducible Brillouin zone (IBZ) defined by the crystal symmetry group.

If we use the relation $\mathbf{p}/m = \mathrm{i}[H, \mathbf{r}]/\hbar$, Eq. (13) is expressed as

$$\varepsilon_2(\omega) = \frac{8\pi^2 e^2}{\omega^2 m^2 V} \sum_{c,v} \sum_{\mathbf{k}} |\langle c, \mathbf{k} | \hat{\mathbf{e}} \cdot \mathbf{p} | v, \mathbf{k} \rangle|^2 \delta[E_c(\mathbf{k}) - E_v(\mathbf{k}) - \hbar\omega]. \tag{14}$$

The calculation using Eq. (14) is much faster than using Eq. (13) since only wavefunctions at $\mathbf{k}$ are needed instead of pairs at $\mathbf{k}$ and $\mathbf{k}+\mathbf{q}$. As pointed out in Refs. [21,97,47], however, expression (14) is not correct when the Hamiltonian includes nonlocal potentials. Rather than the momentum operator, one needs to use the velocity operator defined by the Heisenberg equation of motion,

$$\mathbf{v} = \frac{\mathrm{d}\mathbf{r}}{\mathrm{d}t} = \frac{\mathrm{i}}{\hbar}[H, \mathbf{r}] = \frac{\mathbf{p}}{m} + \frac{\mathrm{i}}{\hbar}[v_{sx}^{\mathrm{NL}}(\mathbf{r}, \mathbf{r}'), \mathbf{r}]. \tag{15}$$

Due to the nonlocality of the Hamiltonian, the gauge invariance requires the light-matter interaction in terms of $\mathbf{v} \cdot \mathbf{A}$ instead of $\mathbf{p} \cdot \mathbf{A}$ [97]. Then it is easy to show that the dipole transition probability is equivalent to the longitudinal expression, Eq. (13), in the long wavelength limit through the relation,

$$i\langle i|[H, \mathbf{r}]|j\rangle = \lim_{q \to 0} \left(\frac{1}{q}\right)(E_i - E_j)\langle i|\mathrm{e}^{\mathrm{i}\mathbf{q}\cdot\mathbf{r}}|j\rangle. \tag{16}$$

In other words, Eq. (13) is appropriate whether or not the Hamiltonian is nonlocal; however, the transverse expression Eq. (14), does not satisfy gauge invariance or equivalently charge conservation.

The calculated $\varepsilon_2(\omega)$ for Si is shown in Fig. 5. In Fig. 5(a), we show the results using the longitudinal expression, Eq. (13), and the transverse one, Eq. (14), calculated by the FLAPW method without any corrections. The two expressions yield an almost identical dielectric function, as expected because the Hamiltonian is local. The shape of the spectrum and the intensity of the peaks agree with other first-principles calculations [46,1].

The results calculated by the sX-LDA method are shown in Fig. 5(b). A substantial reduction in $\varepsilon_2(\omega)$ is obtained with Eq. (14) due to the improper treatment of the nonlocal Hamiltonian. In fact, some of the off-diagonal coefficients for the sX-LDA show a large value; for example, $X_{1v} \to X_{1c}$ and $L_{1v} \to L_{1c}$ transitions in Si. This fact is contrary to the usual scissors-operator approximation which assumes that the wavefunctions do not change for the quasi-particle states [58,1]. Thus, we conclude that Eq. (14) is not appropriate for evaluating optical properties within the sX-LDA method.

We show in Fig. 6 the imaginary part of the dielectric functions calculated by the sX-LDA method using the longitudinal expression, Eq. (13), for Si, Ge, InSb, and GaAs in comparison with experimental data measured at room temperature [7]. Since, as is well-known, the dielectric function depends strongly on temperature, we should compare with experiments undertaken at low temperature for the detailed peak positions of $\varepsilon_2(\omega)$. Overall the calculated peak positions are very close to experiments at low temperature. In particular, we obtained excellent agreement for the main peak, E_2, except for InSb which can be understood from an underestimation of the band gap at the X point even by the sX-LDA method (Table 1). The overestimation of the intensity of E_2, which usually appears in LDA calculations, is not so large here but is reasonable considering lifetime effects [105]. Slightly higher peak positions for E_1 were obtained, but the peak structure is actually not clear due to the underestimation of the intensity around E_1. A recent calculation [2] of $\varepsilon_2(\omega)$

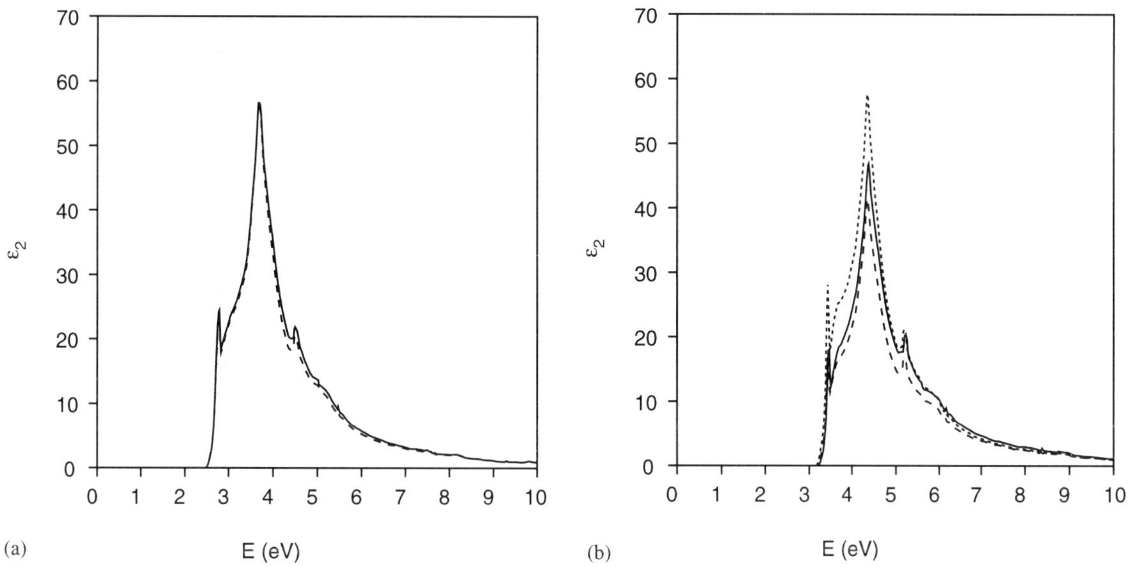

Fig. 5. The imaginary part of the dielectric functions of Si calculated by (a) LDA and (b) sX-LDA. Solid line: the longitudinal expression [Eq. (13)]; dashed line: the transverse expression [Eq. (14)]; dotted line: the transverse expression with the renormalization factor introduced by Levine and Allan [58] and Del Sole and Girlanda [21].

for Si within the *GW* approximation with corrections for local-field and excitonic effects shows the latter effect to increase the intensity of E_1 by roughly twice as much as without corrections. Thus, we believe the discrepancy from experiment should be mostly corrected by excitonic effects which are neglected in the present calculations.

In summary, we have presented first-principles calculations of the electronic and optical properties of Si, Ge, InSb, and GaAs with the sX-LDA method implemented in the FLAPW method. We obtained self-consistent band structures in overall good agreement with experiment and with previously reported sX-LDA results using the pseudopotential method [91]. The dielectric functions evaluated with the longitudinal expression show good agreement with experiment, considering the neglect of lifetime, local-field, and excitonic effects. The method employed to calculate the optical properties has the following features: (i) it is based *purely* on first-principles in that we do not use any parameters to adjust the experimental results; (ii) the nonlocality of the Hamiltonian makes the transverse or momentum-matrix formula invalid; (iii) the difference of the wavefunctions between the LDA and self-consistent sX-LDA yields inappropriate results when calculated with the transverse expression. In particular, feature (i) gives us an opportunity to study and predict detailed optical properties related to structures, e.g., the pressure dependence. Also, it is necessary to go beyond the scissors-operator approximations in the case that the LDA calculations give very small or even negative band gaps, as in Ge and InSb, where we sometimes can hardly distinguish which states should be occupied. Finally, we stress that the computational demands for the self-consistent sX-LDA are much less than for the full *GW* approximation, so that more complex problems such as optical properties with impurities in bulk and at surfaces or interfaces are possible.

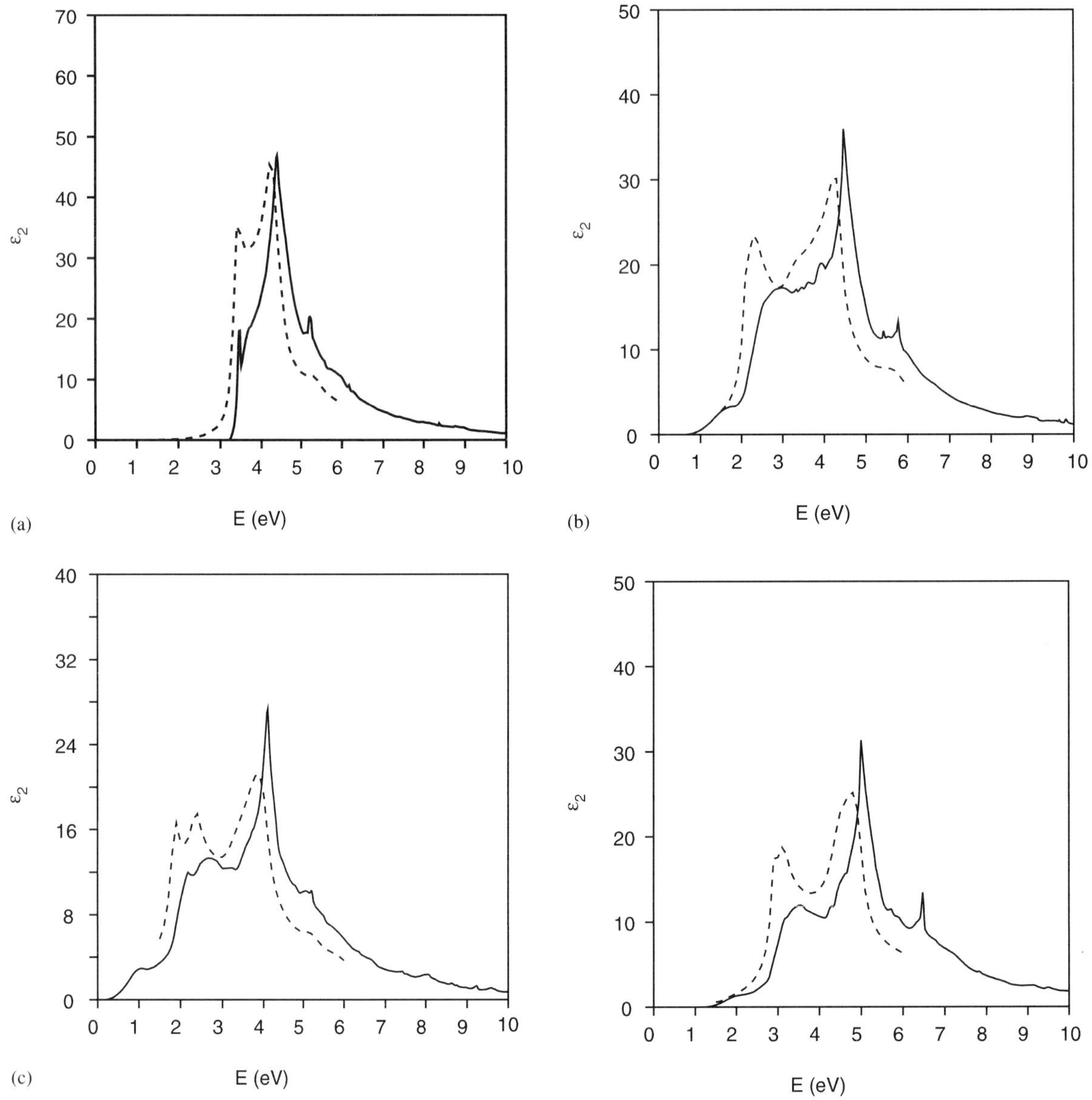

Fig. 6. The imaginary part of the dielectric functions of (a) Si, (b) Ge, (c) InSb, and (d) GaAs calculated by the sX-LDA method using the longitudinal expression [Eq. (13)] (solid line) in comparison with experiments measured at room temperature (dashed line).

3.2. Spintronics prediction of a new class of new temperature ferromagnetic semiconductors

The exploration of new magnetic semiconductors for possible future magneto-electronic ("spintronic") applications, which might realize in one material both information processing (from the charge degree of freedom) and storage (from the spin degree of freedom), has become a subject of great interest [74]. The well developed technology of growing these semiconductors allows for tuning their magnetic properties not only by an external magnetic field but by varying the band structure and/or carrier, impurity and magnetic ion concentrations [76]. One important strategy is to employ elements and crystal structures that avoid defect formation and yield ferromagnetism and spin-injection at much higher temperatures. Thus, while the ferromagnetic III–V and II–VI based semiconductors have been investigated intensively, the Curie temperature is still limited to 110K in (Ga,Mn)As [74,75] although room temperature ferromagnetism in Mn doped GaN has been predicted [23]. The low critical temperature remains a serious bottleneck for their practical application.

Now, investigations have been extended to chalcopyrite semiconductors [67,117]. The chalcopyrite differs from the zinc-blende crystal structure by a doubling of the unit cell along a four-fold axis, which renders the system body-centered tetragonal, and thus provides more options for host semiconductors in the search for room temperature ferromagnetic semiconductors. It was reported that room-temperature ferromagnetism (FM) is realized in $CdGeP_2$ when Mn is diffused into its surface following vacuum deposition [67]. However, strong theoretical evidence indicates that no carriers are introduced by Mn in $Cd_{1-x}Mn_xGeP_2$ at $T=0$ and that it prefers antiferromagnetism (AFM) rather than FM [117]. Very recently, it was found [18] that another II–IV–V_2 chalcopyrite, $Zn_{1-x}Mn_xGeP_2$, is a room temperature FM semiconductor and that it undergoes a FM to AFM transition for $T<47$ K. This finding confirms the accuracy of our predictions for $Cd_{1-x}Mn_xGeP_2$, and may resolve the disagreement with experiment.

In the search for ferromagnetic semiconductors, we turned to a new class of I–III–VI_2 chalcopyrite semiconductors, as exemplified by $CuGa_{1-x}Mn_xSe_2$. Our first-principles calculations, performed with both the full-potential linearized augmented plane wave (FLAPW) and $DMol^3$ methods [22] within the generalized gradient approximation (GGA) [80] to density functional theory, predict it to be a ferromagnetic semiconductor with a Curie temperature of ~ 150 K. While still too low a T_C for applications, this finding opens the possibility of discovering other I–III–VI_2 materials with higher Curie temperatures.

The reason we chose the I–III–VI_2 chalcopyrite as a host semiconductor is based on the contrast between $Ga_{1-x}Mn_xAs$ and $Cd_{1-x}Mn_xGeP_2$. It is well known that hole-like carriers are necessary to make the Mn atoms couple ferromagnetically in semiconductors [77]. Clearly, the Mn in $Ga_{1-x}Mn_xAs$, which goes in at the +3 cation site, forms defect bands and provides holes to make the nonmagnetic semiconductor ferromagnetic [118]. In the case of $Cd_{1-x}Mn_xGeP_2$, the valence of the Mn ion matches that of the host, and results in AFM coupling. In the case of $CuGaSe_2$, we expected that Mn would substitute on the Ga site and so provide holes and make the host semiconductor ferromagnetic—as in GaAs. This conjecture was confirmed using first-principles total energy calculations of the energy of formation, E^f.

To determine the favored magnetic ordering and Curie temperature, one has to calculate Heisenberg exchange constants J_n (exchange interaction energy between nth nearest-neighbors). One way to calculate J_n is to employ a linear response technique to obtain directly an analytic expression for the exchange [73]. An alternative approach, adopted here, is to calculate numerically the change in total energy to high accuracy for different assumed spin configurations. These total energy differences are then used to estimate J_n.

For each concentration x, we calculated the total energy for one or two antiferromagnetic configurations, σ, and the ferromagnetic configuration, yielding the exchange interaction energy,

$$\Delta E_{ex}(\sigma)=E(\sigma)-E(FM). \tag{17}$$

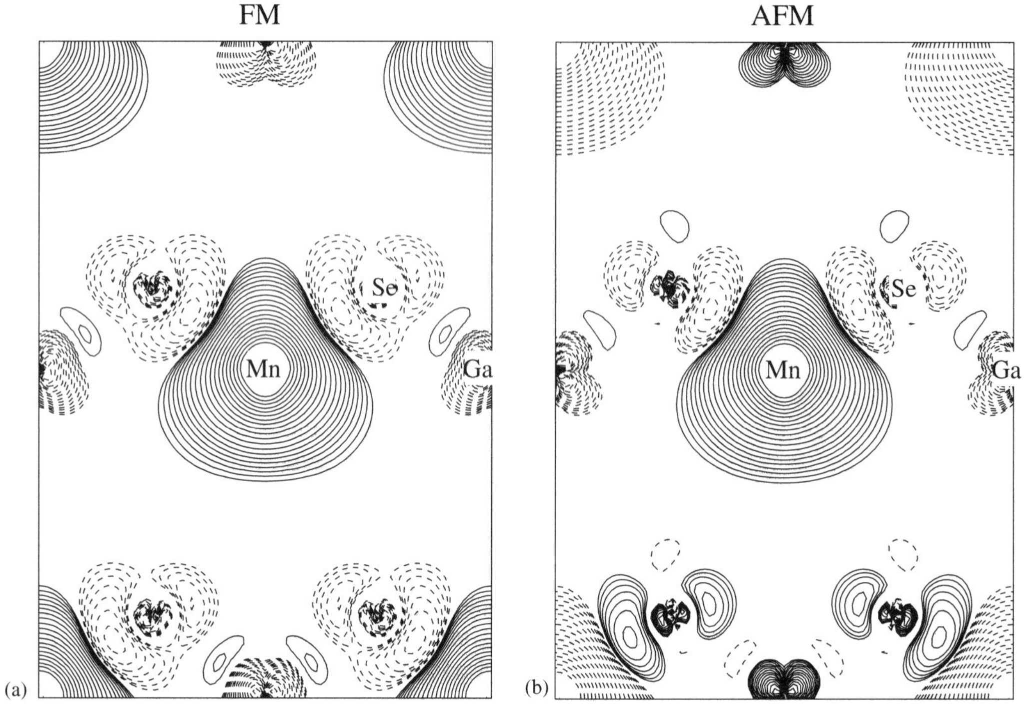

Fig. 7. Spin density of $CuGa_{1-x}Mn_xSe_2$ for $x = 0.50$ from FLAPW calculations, plotted in the (110) plane. Spin density contours start at $5 \times 10^{-4} e/a.u.^3$ and increase successively by a factor of $\sqrt{2}$. Positive spin density is represented by solid lines, while negative spin density is represented by dashed lines.

The spin configurations which we investigated can be described as a superlattice of period p and layer orientation **G**. Both the $DMol^3$ and FLAPW results show that the FM states are lower in energy than the AFM states for $x = 0.25$, 0.50 and 1.0. The paramagnetic state may not compete with the FM or AFM state because the $DMol^3$ calculated total energy of the paramagnetic state for $x = 0.25$ is 237 meV higher than the corresponding FM state. These first-principles studies strongly indicate that Mn doped $CuGaSe_2$ is a very stable ferromagnetic semiconductor.

The FLAPW calculated spin density of $CuGa_{1-x}Mn_xSe_2$ for $x = 0.50$ is shown in Fig. 7. Obviously, the spin moment is localized at the Mn site, and the Mn induced moment at the Se site is negative. The total spin moment of the system is 4.0 μ_B per Mn atom, with 3.94 μ_B inside the Mn muffin-tin sphere. The Mn induced magnetic moment at the Ga site is around 0.04 μ_B, while it is -0.06 μ_B at the Cu site. The AFM interaction between Mn 3d and Se 4p electrons is much stronger in the FM state than that in the AFM state. Furthermore, the bond length of Mn–Se in the FM state for $x = 0.50$ is 2.457 Å, which is slightly shorter than that in the AFM state due to the stronger magnetic interaction between Mn and Se atoms in the former case. This is in good agreement with our previous conclusion [117]—the AFM Mn 3d—anion p interaction in Mn doped semiconductors decreases the total energy and makes the system order ferromagnetically. Furthermore, the band structure for $CuGa_{1-x}Mn_xSe_2$ for $x = 0.50$, shown in Fig. 8, indicates that $CuGa_{1-x}Mn_xSe_2$ is half-metallic and that Mn doping provides holes as in $Ga_{1-x}Mn_xAs$ [118]. The states around the Fermi level come

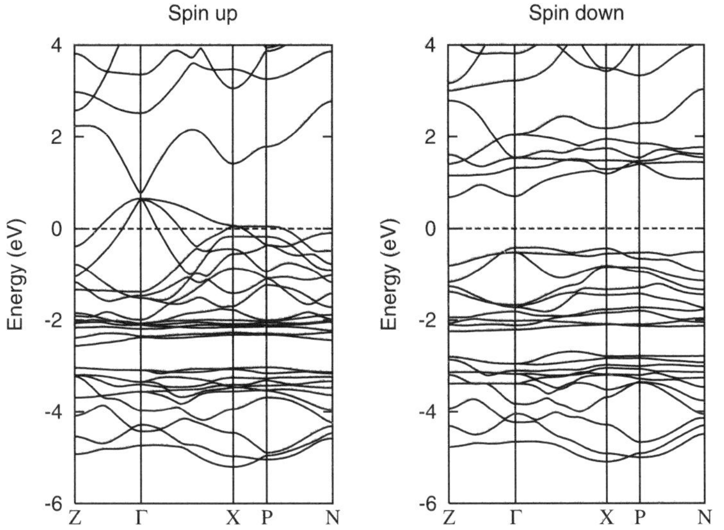

Fig. 8. Calculated FLAPW band structure of ferromagnetically coupled $CuGa_{1-x}Mn_xSe_2$ for $x = 0.50$. The Fermi level is set to zero.

mainly from the Mn 3d and Se 4p levels. These pictures are also valid for $CuGa_{1-x}Mn_xSe_2$ at $x = 0.25$ and 1.0.

The optimum way to estimate the Curie temperature is still an open question. One way to estimate the order-disorder transition temperature T_C from the calculated $J_1(x)$ is to make use of the Monte Carlo simulation for the nearest-neighbor classic Heisenberg Hamiltonian. Unfortunately, there is still no Monte Carlo simulation of a spin lattice in the chalcopyrite structure. However, the value of c/a in $CuGa_{1-x}Mn_xSe_2$ is very close to 2, which implies that its structure deformation from the zinc-blende is small. Therefore, in an attempt to give an estimate of T_C for Mn doped $CuGaSe_2$, we employ the Monte Carlo simulation results for $Cd_{1-x}Mn_xTe$, which has the zinc-blende structure, [101] namely, $T_C = 0.447|J_1|$. With the $J_1(x)$ obtained from FLAPW, the estimated T_C's are 114, 142, and 160 K for $x = 0.25$ and 0.50, and 1.0, respectively.

Finally, as pointed out by Dietl et al. [23], the Curie temperature is strongly dependent on the hole density, which is inversely proportional to the volume of the lattice. This suggests for this new class of FM semiconductors based on Mn doped I–III–VI_2 chalcopyrites, that higher Curie temperatures may be achieved by employing other materials with a smaller lattice constant such as $CuGaS_2$ and $CuAlS_2$. [1]

3.3. Exchange bias at interface between ferromagnetic NiFe and antiferromagnetic NiMn films: role of noncollinear magnetism and roughness

It is well-known that state-of-the-art ab initio density functional electronic structure calculations have achieved great success in the exciting field of thin film magnetism, in both explaining

[1] Preliminary results on Mn doped $CuGaS_2$ and $CuAlS_2$ just obtained confirm that the FM state is strongly preferred in both systems.

existing phenomena and, more importantly, in predicting the properties of new systems [113]. The prediction of enhanced magnetic moments with lowered coordination number at clean metal surfaces and interfaces has stimulated theoretical and experimental investigations for new magnetic systems and phenomena in man-made transition metal thin films, which has accompanied the renaissance of magnetism and led to a major impact on the magnetic recording industry. See review articles in [28].

Significant progress has been made for the treatment of the weak spin–orbit coupling (SOC) in magnetic transition metal systems, using modern first principles electronic structure methods, especially the highly precise full potential linearized augmented plane wave (FLAPW) approach [108,107]: Very reliable results for the magneto-crystalline anisotropy (MCA) energies, E_{MCA}, can now be obtained for most magnetic thin films (a few tenths meV/atom), and even for magnetic cubic bulk materials (a few μeV/atom). This enables us to solve a long-standing problem—the first principles determination of the magnetostrictive coefficients in bulk transition metals, and their alloys and compounds with rare-earth metals. Using linear response theory, magneto-optical properties such as the magneto-optical Kerr effect (MOKE) and soft X-ray magnetic circular dichroism (MCD) can now be determined. The results obtained for several selected systems indicate that high quality ab initio calculations of magnetic systems can achieve high accuracy and precision for a wide range of SOC induced magnetic properties of transition metal and rare-earth metal systems [113].

More recently, our attention has been directed to the challenging problem of unidirectional magnetic anisotropy—the so called exchange bias associated with anisotropy at an interface between ferromagnetic (FM) and antiferromagnetic (AFM) materials—holds promise as a useful phenomenon in controlling and designing magnetic applications such as storage and sensor devices. Beginning with the first discovery by Meiklejohn and Bean [68] in 1956, many experimental and theoretical studies have been performed in order to understand this phenomenon [71,31,96]. However, the theoretical understandings obtained, mostly based on a model Hamiltonian approach, have been revised several times over the past years. The first simple model [68] assumed an ideal FM/AFM interface in which the interfacial AFM spins are fully uncompensated and predicted the exchange bias to be two orders of magnitude larger than the experimentally observed values. To account for the discrepancies, different improvements have been suggested including domain wall formation, roughness and spin configurations such as a perpendicular coupling at the FM/AFM interface [65,61,53,89]. Although these give intuitive pictures, they are still far from providing quantitative understandings. Often, an approach applied for a specific system is found not valid for another system, due to a lack of quantitative predictions of the complexity of magnetic structures at FM/AFM interfaces.

Here we first determine magnetic structures, including noncollinear magnetism at the FM/AFM interface of NiFe/NiMn films [59,62,110,85,86] from first principles, which allow us to obtain (i) perpendicular coupling arising from biquadratic exchange and (ii) canting of magnetic moments at the FM/AFM interface that induces a ferromagnetic moment on the AFM spins. Based on these results, we discuss the coercivity and exchange bias through formation of a magnetic domain wall in the AFM layer, and determine the role of roughness at the FM/AFM interface as simulated with linear steps. In contrast to the qualitative understanding in previous studies [68,71,12,96,65,61,53,89], this work is the first quantitative prediction of the complex magnetism at the FM/AFM interface in realistic systems.

For these calculations, the noncollinear magnetism formalism [104,55] was incorporated into the FLAPW method with no shape approximation for the magnetization density, [72] i.e. the moment direction as well as the magnitude could vary continuously all over space.

We consider (001) and (111) interfaces between the FM NiFe and AFM NiMn of the exchange bias films; the (111) interface was observed in experiments [110,85]. As models, the $(NiFe)_2/(NiMn)_2$ superlattice structure with a (001) interface and the $(NiFe)_3/(NiMn)_3$ superlattice structure with a (111) interface were adopted, in which an $L1_0$ atomic ordering in the NiFe and NiMn layers was assumed. Note that ideally the (001) and (111) interfaces have no net magnetic moments on the AFM NiMn interface since the Mn moments align antiparallel to each other, i.e. compensated FM/AFM interfaces [71].

The LSDA self-consistent calculations were carried out without spin–orbit coupling. Next, we calculated the magnitude of the magneto-crystalline anisotropy (MCA) energy induced by the spin–orbit coupling—defined as the difference in total energies for the magnetic moments oriented along the in-plane and out of plane directions—to be of the order of 10^{-1} meV [70], and so is much smaller than the energies of interest here. Note that an in-plane orientation of the magnetic moments at the NiMn interface is favored, which are mainly dominated by the in-plane MCA from the NiMn layers. [2]

Figs. 9 and 10 show the calculated magnetization density on the Fe, Ni and Mn layers for the (001) interface, and the NiFe and NiMn layers for the (111) interface, respectively. Here the moments lie parallel to their interface. In both cases, the average FM Fe moments in Figs. 9(a) and 10(a) align perpendicular to the average AFM Mn moments in Figs. 9(c) and 10(b), respectively. The integrated Mn and Fe moments in the muffin-tin spheres at the (001) interface are 3.1 and 2.5 μ_B, and for the (111) interface, 3.0 and 2.5 μ_B, respectively. Those values are slightly smaller than those in their bulk [70]. The Fe and Mn moments cant slightly away from their FM or AFM axes. The canting of the Mn moments in the AFM interface for the (001) and (111) cases gives rise to a weak ferromagnetism with net moments of 0.07 and 0.20 μ_B, respectively.

In addition, the magnetization density at the Ni in Figs. 9(b) and 10(b) has a rather complicated pattern, with nonvanishing moments pointing perpendicular to the AFM Mn moments. Interestingly, the magnetization is also found to vary continuously on a smaller length scale inside the atom, i.e. *intra-atomic noncollinear magnetism exists*. As an example, for the (001) case, the moments on the right and left side portions of the Ni site in Fig. 9(b) cant slightly toward the [100] direction away from the average Ni moments, and so correlate with the neighboring Mn moments [cf., Fig. 9(c)] that point along the [100] direction; the moments on the top and bottom of the Ni site in Fig. 9(b) cant toward the $[\bar{1}00]$ direction, and correlate with the neighboring Mn moments [cf., Fig. 9(c)] pointing along the $[\bar{1}00]$ direction. Thus, the intra-atomic noncollinear magnetism is strongly affected by the coordination surrounding the atom. The intra-atomic noncollinear magnetism was found to be induced in the d_{xz} and d_{yz} orbitals, which are directly bonded to the neighboring FM Fe and AFM Mn moments.

[2] We tested the method on fcc Pu. Our results show that the easy magnetic axis is along [111] with a lower total energy of 6 meV compared to [001]. The total spin and orbital moments are 4.73 and 1.82 μ_B when the average magnetic moments points along [001], and 4.75 and 1.83 μ_B for [111]. These results are consistent with those in Ref. [13].

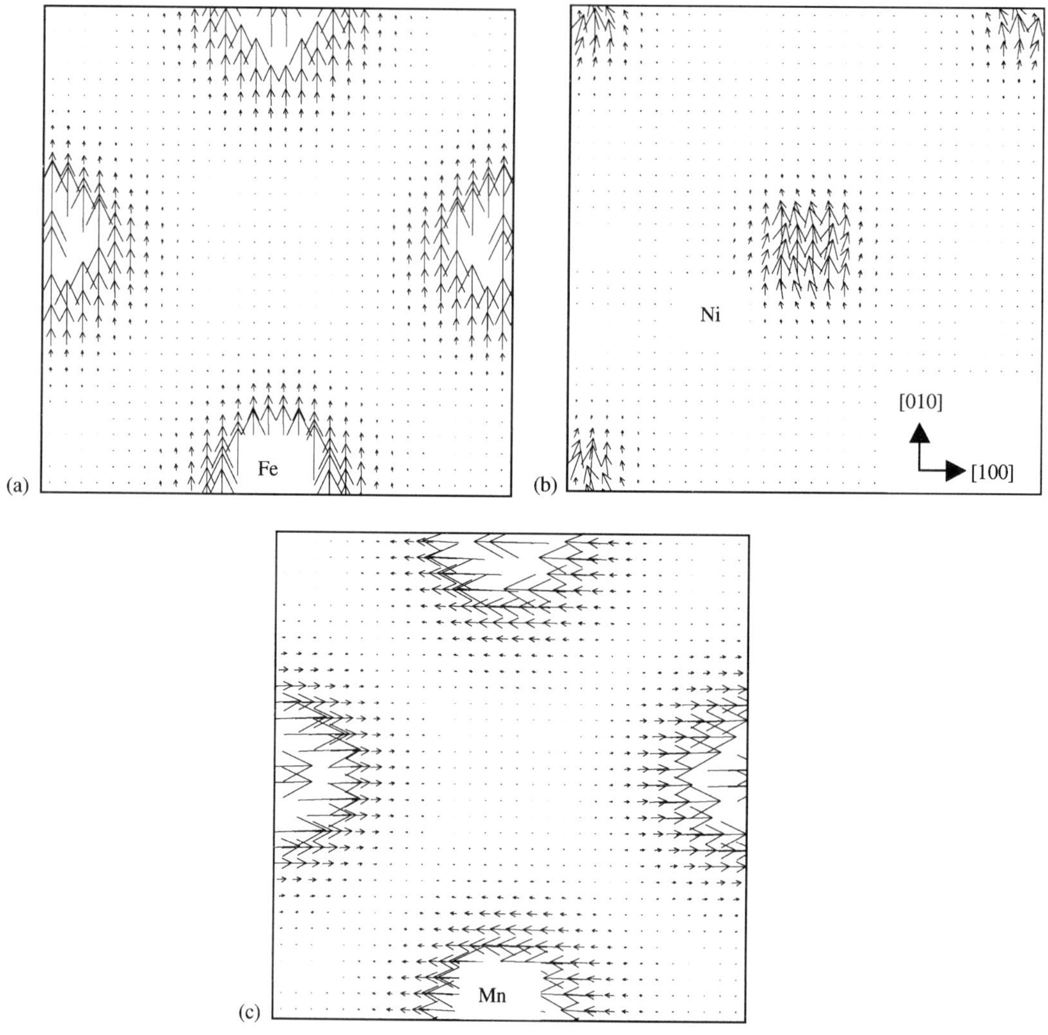

Fig. 9. Magnetization density on (a) FM Fe, (b) Ni and (c) AFM Mn layers at the (001) interface for the ($NiFe_2/NiMn_2$ superlattice structure. The moment direction and the magnitude are represented by arrows and their size.

The calculated total energies for both the (001) and (111) interfaces are lower than those in the collinear states by 13 and 24 meV/(Mn-atom-area), which correspond to 18 and 33 mJ/m^2 of biquadratic exchange energy (BEE). This causes a large coercivity in the exchange bias films [89]. In the (111) case, the noncollinear magnetic structure is expected to be stable above room temperature because of the large BEE, leading to a high blocking temperature as observed in experiments [62,86].

In order to discuss a mechanism of the exchange bias through magnetic domain wall formation in the AFM NiMn layers, we estimated the domain wall energy, given by $4\sqrt{A_{AFM}K_{AFM}}$ [17], where, K_{AFM} and A_{AFM} are the in-plane MCA energy of of the (111) plane in bulk AFM NiMn and exchange stiffness, respectively. Using the FLAPW method and the state tracking scheme [106], we

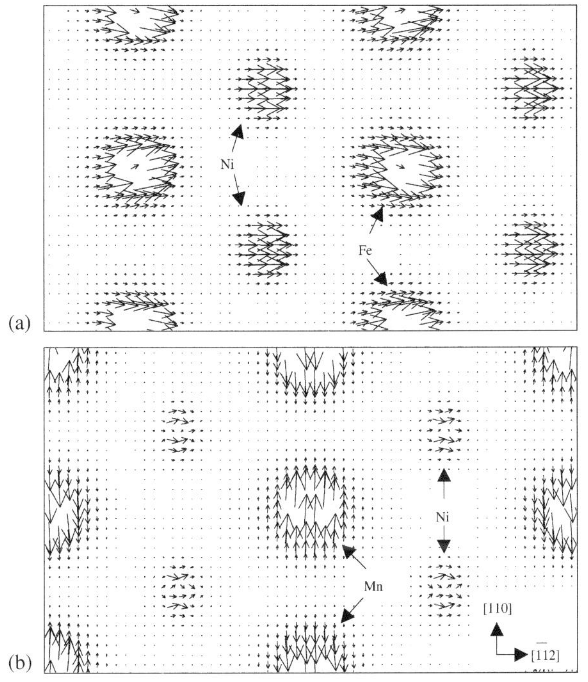

Fig. 10. Magnetization density on (a) FM NiFe, (b) AFM MnMn layers at the (111) interface for the $NiFe_3/NiMn_3$ superlattice structure. The moment direction and the magnitude are represented by arrows and their size.

obtained K_{AFM} of 0.17 meV/Mn-atom, which corresponds to 1.68×10^5 J/m^3. Also from the total energy difference between bulk FM and AFM NiMn, 0.43 eV/Mn-atom, the exchange constant in bulk AFM NiMn, J_{AFM}, is estimated by assuming the Heisenberg model to be 27 meV/Mn-atom, corresponding to an A_{AFM} of 2.0×10^{-11} J/m. The domain wall energy results in 7.4 mJ/m^2. Thus, the BEE is large enough to require the formation of the domain wall, and describes a strong coupling between the FM and AFM layers such that an extended influence would affect both together. Note that, using $\pi\sqrt{A_{AFM}K_{AFM}}$ [17], the wall width is roughly expected to be $\sim$ 370 Å, which corresponds to the AFM NiMn thickness needed to retain the large coercivity and/or the exchange bias as observed in experiment [86].

Finally, since roughness at the interface such as steps, islands or point defects plays a key role in the exchange bias, [61,99,89] we performed FLAPW calculations for the FM NiFe/AFM NiMn (001) interface including a line step defect along [010]. We employed a 12-layer superlattice structure assuming the bulk lattice parameters of AFM NiMn, as shown in Fig. 11. We found that a collinear antiferromagnetic coupling at the FM/AFM interface is highly favored over a ferromagnetic one with a 54 meV/(Mn-atom area) difference in their total energies. Thus, the step defect produces an uncompensated FM/AFM interface across the step, in which the Fe moments align antiparallel to the Mn moments. Further, since the energy difference is larger than the BEE, the introduction of step defects in the flat interface (where the FM Fe moments align perpendicular to the AFM Mn moment) induces a torque that causes the Fe moments to be antiparallel to the Mn moments. This gives rise to an unidirectional magnetic anisotropy of the exchange bias.

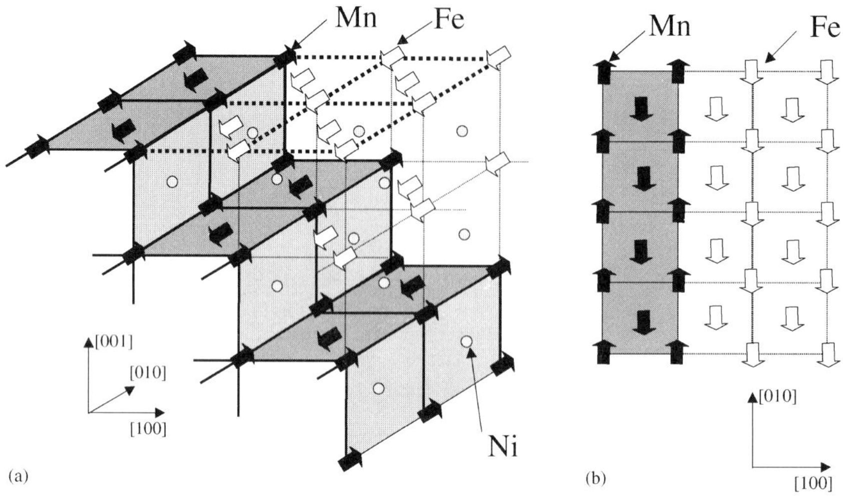

Fig. 11. (a) A model for the FM NiFe/AFM NiMn (001) interface including a line step defect along [010] and (b) the top view on the interface. Dark regions indicate the FM/AFM interface. Only Mn and Fe moments in (a) are given by arrows.

3.4. Influence of alloying additions on grain boundary cohesion of transition metals: first-principles determination and its phenomenological extension

Grain boundary cohesion is often the controlling factor limiting the ductility of high strength metallic alloys, see for example [14]. Thus, understanding the influence of transition metal alloying additions is of great importance in predicting and controlling grain boundary embrittlement (GBE) since the complexity of GBE behavior is often associated with the presence of substitutional alloying elements. Attempts at qualitatively explaining GBE on the electron–atom level go back to the 1960s [103]. Although predictions based on simple thermodynamic [24,98,56] and pair bonding models [90,114] have been proposed, the role of alloying elements on the cohesion of grain boundaries in alloys remains controversial. A thermodynamic theory by Rice and Wang [87] demonstrated that previous thermodynamic models are incorrect and identified the key surface-thermodynamic energy difference governing the quantitative role of a boundary segregant. In addition to the simple pair bonding approaches [90,114] which, like thermodynamic theories, employ energetics to describe the effect of segregants, electronic structure theorists have applied both empirical [20,66] and first-principles quantum-mechanical methods [79,25] in attempts to directly deduce the effect of segregants on grain boundary (GB) fracture resistance.

In contrast to these studies, our recent line of research [112,88,119,33,34] has applied rigorous first-principles methods to evaluate the surface-thermodynamic energy difference identified by Rice and Wang as directly connected to the ideal work of GB fracture. This thermodynamic theory describes the mechanism of metalloid induced intergranular embrittlement through the competition between plastic crack blunting and brittle boundary separation. Their most striking result is the prediction that the potency of a segregating solute in reducing the "Griffith work" of brittle boundary separation is a linear function of the difference in segregation energies for that solute at the GB, ΔE_b, and at the free surface (FS), ΔE_s. Simply put, a solute with positive energy difference $\Delta E_b - \Delta E_s$

(i.e., ΔE_b is less negative) will be a more potent embrittler, while a solute with a negative difference can enhance boundary cohesion.

Thus, a predictive theory for the intergranular cohesive properties can be established by directly determining both ΔE_b and ΔE_s, using present state-of-the-art first-principles local density electronic structure approaches. However, since $\Delta E_b - \Delta E_s$ is small (e.g., 0.4 eV per P atom for the P/Fe system), (i) the GB and FS systems must be treated precisely on an equal footing (uniquely possible with the thin film FLAPW method) and (ii) the atomic structures of both FS and GB should be well optimized for the cases with and without the impurity atoms.

In spite of the complexity of the mechanical behavior and GB atomic structures, general trends in certain mechanical properties can be correlated with specific features of the electronic structure [25,78]. First-principles computation has proved to be an accurate and powerful tool in attacking the problem of the mechanical properties of real materials such as GBE. An inherent advantage of the first-principles electronic theory is that it is independent of any adjustable parameters, and so, numerical results provide a solid basis or starting point for phenomenological theories. Without such a theory starting from first-principles, one has always to repeat the full procedure of calculations to predict quantitatively the effect of any substitutional element even on the same GB [119,34]. In this sense, the exact mechanism by which segregation elements cause embrittlement remains unclear.

For example, very recently [34], the effect of Mo and Pd substitutional segregation on the cohesion of the Fe $\Sigma 3(111)$ GB was investigated using the first-principles full-potential linearized augmented plane wave (FLAPW) total energy/atomic force method with the generalized gradient approximation [80]. Based on the Rice–Wang model [87], our total energy calculations show that Mo has a significant beneficial effect on the Fe GB cohesion, while Pd behaves as a weak embrittler. An analysis of the geometry optimization indicates that both Mo and Pd have a moderate atomic size to fit well in the GB substitutional site. The elastic energy associated with the Mo and Pd segregation was estimated with a rigid environment approximation. It was found that both Mo and Pd introduce a beneficial volume effect. Studies of the electronic structures show that the strong bonding capability associated with its half filled d bands makes Mo a cohesion enhancer (-0.90 eV) for the Fe $\Sigma 3(111)$ GB. By comparison, the weak bonding capability associated with its filled d bands leads Pd to be a weak embrittler ($+0.08$ eV).

These first-principles quantum-mechanical results support, in general, the main idea of the simple pair-bonding models [90,114] in that the elemental cohesive energy difference between the alloying element and the host element plays an important role in determining the effect of this alloying element on the host GB cohesion. They differ, however, quite significantly: For example, for Mo and Pd in the Fe GB, the embrittlement potency given by Seah's pair bonding model [90] is -1.5 and $+0.9$ eV, respectively. This significant difference with the values of -0.90 and $+0.08$ eV from our rigorous calculations indicates that a quantum mechanical treatment is necessary in an accurate study of the GBE. Also, the numerical results for Pd, which has a similar elemental cohesive energy to that of Fe, point to the importance of the role played by the GB free volume, which can be determined with high accuracy only by quantum mechanical calculation.

Based on the physical insight obtained from the results of our rigorous calculations, we have developed a phenomenological model by which the embrittlement potency of a substitutional atom in a given GB can be predicted without carrying out full first-principles calculations [35]. To be accurate and reliable, such a theory must use first-principles results, rather than adjustable parameters, as inputs. To be capable of prediction without solving the quantum mechanical Schrödinger equation,

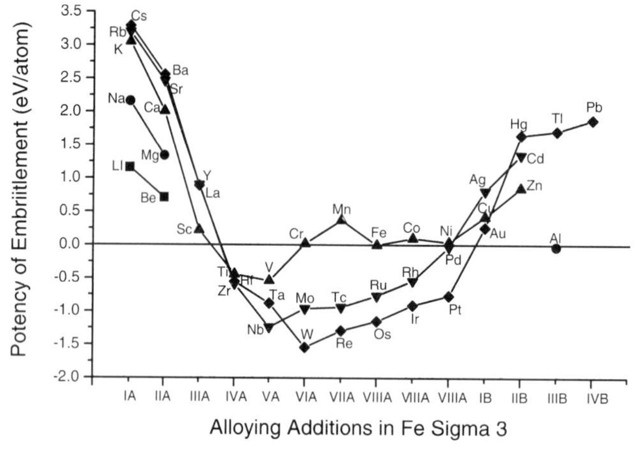

Fig. 12. Embrittlement potency of substitutional additions on the Fe $\Sigma 3(111)$ GB.

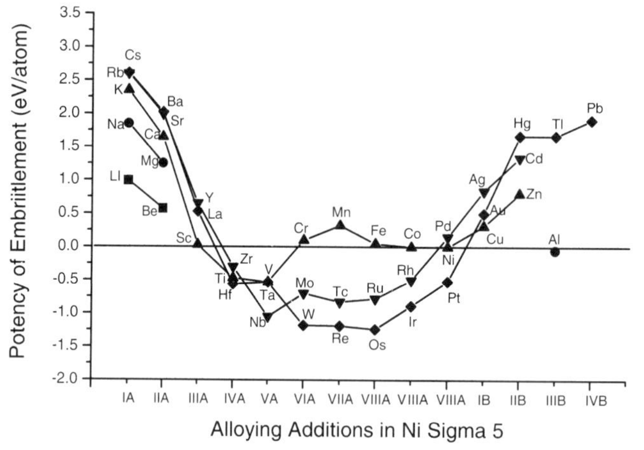

Fig. 13. Embrittlement potency of substitutional additions on the Ni $\Sigma 5(210)$ GB.

such a theory must also make use of atomic, or bulk quantities as inputs. This means that this model should have an electronic, rather than an atomistic or thermodynamic basis, in order to yield a quantum description of the mechanical behavior of a substitutional element at the GB.

Starting from first-principles, we formulated and developed, for the first time, an electronic level phenomenological theory to quantitatively predict the mechanical behavior of a substitutional metallic element in the grain boundary without carrying out full first-principles calculations, once the atomic structure of the clean grain boundary is determined. The model predicted embrittlement potencies of all the simple and transition metal elements as substitutional alloying additions to the Fe $\Sigma 3(111)$ and Ni $\Sigma 5(210)$ GB are plotted in Figs. 12 and 13, respectively. From our results, it is concluded that the strongest cohesion enhancer in the Fe $\Sigma 3(111)$ GB is W, followed by Re, Nb, and Os. The strongest cohesion enhancer in the Ni $\Sigma 5(210)$ GB is Os, followed by Re, W, and Nb. This model was tested and verified by detailed rigorous first-principles calculations on Fe and Ni based alloys.

Although we have focused on two specific GB types, i.e., bcc Fe $\Sigma3(111)$ and fcc Ni $\Sigma5(210)$, we expect our theory to be applicable to other high boundaries in general and instructive in the quantum design of ultrahigh-strength alloys through selection of alloying elements for GB cohesion enhancement. Further calculations and modelings of the segregation energy from crystalline solution to GB are underway to next control the segregation of desired components to the boundary.

4. Conclusions

Today's important applications of the powerful theories of physics owe much to recent developments in computational physics. Complementing the strides in numerical methods are tremendous advances in computer power. The availability of computers has made a tremendous difference in the kind of problems that can be attempted. The results of this basic research provide exciting insight to many applications. Today, we have found a key—computational physics/materials science on the computer—that is providing the answers to many questions; we are convinced that as computational physics and computers continue to evolve, the expansion of research on even more complex systems will become possible. The basic essential for success will continue to be the close collaboration with experimentalists, meaningful models to be pursued, and understanding the limitations of both theory and experiment.

The future looks very bright indeed! It is therefore even more fitting to recall, as I did at Hokkaido University at the Japan Physical Society meeting, October 1982, the parting words of Dr. Clarke (the founder of Hokkaido University) in 1877 "Boys, be ambitious" [27].

Acknowledgements

I am pleased to acknowledge and thank Ryoji Asahi, Alessandra Continenza, Wen-tong Geng, Miyoung Kim, Julia Medvedeva, Kohji Nakamura, Silvia Picozzi, Ruqian Wu, and Yu-Jun Zhao for collaboration and/or assistance with this manuscript. I also thank my other collaborators cited in the references.

Support was provided by the U.S. National Science Foundation supported Materials Research Center, Seagate Technology, L.L.C., and the Office of Naval Research (ONR) Grant No. N00014-94-1-018.

References

[1] B. Adolph, V.I. Gavrilenko, K. Tenelsen, F. Bechstedt, R. Del Sole, Phys. Rev. B 53 (1996) 9797.
[2] S. Albercht, L. Reining, R. Del Sole, G. Onida, Phys. Rev. Lett. 80 (1998) 4510.
[3] M. Alouani, L. Brey, N.E. Christernsen, Phys. Rev. B 37 (1988) 1167.
[4] O.K. Andersen, Phys. Rev. 12 (1975) 3060.
[5] R. Asahi, W. Mannstadt, A.J. Freeman, Phys. Rev. B 59 (1999) 7486.
[6] R. Asahi, W. Mannstadt, A.J. Freeman, Phys. Rev. B 62 (2000) 2552.
[7] D.E. Aspnes, A.A. Studna, Phys. Rev. B 27 (1983) 985.
[8] J. Baker, M. Muir, J. Andzelm, J. Chem. Phys. 102 (1995) 2063.
[9] G.A. Baraff, M. Schlüter, Phys. Rev. B 30 (1984) 3460.

[10] A.D. Becke, Phys. Rev. A 38 (1988) 3098;
J.P. Perdew, Phys. Rev. B 33 (1986) 8822;
J.P. Perdew, Phys. Rev. B 34 (1986) 7406.
[11] A.D. Becke, The challenge of d and f Electrons, in: D.R. Salahub, M.C. Zerne (Eds.), ACS Symposium series No. 394, American chemical Society, Washington, DC, 1989, p. 165.
[12] A.E. Berkowitz, K. Takano, J. Magn. Magn. Mater. 200 (1999) 552, and references therein.
[13] M. Born, R. Oppenheimer, Ann. Phys. Lpz. 84 (1927) 124.
[14] C.L. Briant, S.K. Banerji, in: C.L. Briant, S.K. Banerji (Eds.), Embrittlement of Engineering Alloys, Academic Press, New York, 1983, p. 21.
[15] B.M. Bylander, L. Kleinman, Phys. Rev. B 41 (1990) 7868.
[16] A. Canning, W. Mannstadt, A.J. Freeman, Comput. Phys. Comm. 130 (2000) 233.
[17] S. Chikazumi, Physics of Ferromagnetism, Wiley, New York, 1964, p. 186.
[18] S. Cho, et al., unpublished.
[19] A. Dal Corso, A. Pasquello, A. Baldereshi, R. Car, Phys. Rev. B 53 (1996) 1180.
[20] M.S. Daw, M.I. Baskes, Phys. Rev. Lett. 50 (1983) 1285.
[21] R. Del Sole, R. Girlanda, Phys. Rev. B 48 (1993) 11 789.
[22] B. Delley, J. Chem. Phys. 113 (2000) 7756.
[23] T. Dietl, H. Ohno, J. Cibert, D. Ferrand, Science 287 (2000) 1019.
[24] R.H. Doremus, J. Appl. Phys. 47 (1976) 1833.
[25] M.E. Eberhart, D.D. Vvedensky, Phys. Rev. Lett. 58 (1987) 61.
[26] H. Ehrenreich, M.H. Cohen, Phys. Rev. 155 (1959) 786.
[27] A.J. Freeman, Butsuri 38 (1983) 543.
[28] A.J. Freeman, S.D. Bader (Eds.), J. Magn. Magn. Mater 200 (1999).
[29] A.J. Freeman, E. Wimmer, Annu. Rev. Mater. Sci. 25 (1995) 7.
[30] M. Fuchs, M. Bockstedte, E. Pehlke, M. Scheffler, Phys. Rev. B 57 (1998) 2134–2145.
[31] A. Garcia, C. Elsaesser, J. Zhu, S. Louie, M.L. Cohen, Phys. Rev. B 46 (1992) 9829.
[32] C.B. Geller, J.E. Raynolds, W. Wolf, S. Picozzi, A. Continenza, R. Asahi, W. Mannstadt, E. Wimmer, Appl. Phys. Lett. 79 (2001) 368.
[33] W.T. Geng, A.J. Freeman, R. Wu, C.B. Geller, J.E. Raynolds, Phys. Rev. B 60 (1999) 7149.
[34] W.T. Geng, A.J. Freeman, R. Wu, G.B. Olson, Phys. Rev. B 63 (2001) 165 415.
[35] W.T. Geng, G.B. Olson, A.J. Freeman, Phys. Rev. B 63 (2001) 165 415
[36] G. Greenwood, Proc. Phys. Soc. (London) 71 (1958) 585.
[37] O. Gunnarsson, B.I. Lundqvist, S. Lundqvist, Solid State Comm. 11 (1972) 149.
[38] F. Gygi, A. Baldereschi, Phys. Rev. Lett. 62 (1989) 2160.
[39] D.R. Hamann, Phys. Rev. Lett. 76 (1996) 660.
[40] B. Hammer, K.W. Jacobsen, J.K. Norskov, Phys. Rev. Lett. 70 (1993) 3971.
[41] B. Hammer, M. Scheffler, K.W. Jarcobsen, J.K. Norskov, Phys. Rev. Lett. 73 (1994) 1400.
[42] L. Hedin, Phys. Rev. 139 (1965) A796.
[43] L. Hedin, B.I. Lundqvist, J. Phys. C 4 (1971) 2064.
[44] P. Hohenberg, W. Kohn, Phys. Rev. 136 (1964) B864.
[45] P. Hu, D.A. King, S. Crampin, M.-H. Lee, M.C. Payne, Chem. Phys. Lett. 230 (1994) 501.
[46] M.-Z. Huang, W.Y. Ching, Phys. Rev. B 47 (1993) 9449.
[47] M.S. Hybertsen, S.G. Louie, Phys. Rev. B 35 (1987) 5585.
[48] H.J.F. Jansen, A.J. Freeman, Phys. Rev. 30 (1984) 561.
[49] B.G. Johnson, P.M.W. Gill, J.A. Pople, J. Chem. Phys. 98 (1993) 5612.
[50] Y.-M. Juan, E. Kaxiras, R.G. Gordon, Phys. Rev. B 51 (1995) 9521.
[51] D.D. Koelling, G.O. Arbman, J. Phys. F 5 (1975) 2041.
[52] W. Kohn, L.J. Sham, Phys. Rev. 140 (1965) A1133.
[53] N.C. Koon, Phys. Rev. Lett. 78 (1997) 4865.
[54] M. Körling, J. Häglund, Phys. Rev. B 45 (1992) 13293.
[55] J. Kübler, K.-H. Höck, J. Sticht, A.R. Williams, J. Phys. F 18 (1988) 469.
[56] D.Y. Lee, E.V. Barrera, J.P. Stark, H.L. Marcus, Metall. Trans. A 15A (1984) 1415.

[57] T.C. Leung, C.T. Chan, B.N. Harmon, Phys. Rev. B 44 (1991) 2923.
[58] Z.H. Levine, D.C. Allan, Phys. Rev. B 43 (1991) 4187.
[59] T. Lin, D. Mauri, N. Staud, C. Hawng, J.K. Howard, G.L. Gorman, Appl. Phys. Lett. 65 (1994) 1183.
[60] T.L. Loucks, The Augmented-Plane-Wave Method, Benjamin, New York, 1967.
[61] A.P. Malozemoff, Phys. Rev. B 35 (1987) 3679.
[62] S. Mao, S. Gangopadhyay, N. Amin, E. Murdock, Appl. Phys. Lett. 69 (1996) 3539.
[63] S. Massidda, A. Continenza, M. Posternak, A. Baldereschi, Phys. Rev. Lett. 74 (1995) 2323.
[64] S. Massidda, A. Continenza, M. Posternak, A. Baldereschi, Phys. Rev. B 55 (1997) 13 494.
[65] D. Mauri, H.C. Siegmann, P.S. Bagus, E. Kay, J. Appl. Phys. 62 (1987) 3047.
[66] T. McMullan, M.J. Scott, E. Zaremba, Phys. Rev. B 35 (1987) 1076.
[67] G.A. Medvedkin, et al., Jpn. J. Appl. Phys. 39 (2000) L949.
[68] W.P. Meiklejohn, C.P. Bean, Phys. Rev. 102 (1956) 1413;
W.P. Meiklejohn, C.P. Bean, Phys. Rev. 105 (1957) 904.
[69] N. Moll, M. Bockstedte, M. Fuchs, E. Pehlke, M. Scheffler, Phys. Rev. B 52 (1995) 2550.
[70] K. Nakamura, M. Kim, A.J. Freeman, L. Zhong, J. Fernandez-de-Castro, IEEE Trans. Magn. 36 (2000) 3269.
[71] J. Nogués, I.K. Schuller, J. Magn. Magn. Mater. 192 (1999) 203.
[72] L. Nordström, D. Singh, Phys. Rev. Lett. 76 (1996) 4420.
[73] T. Oguchi, K. Terakura, A.R. Williams, Phys. Rev. B 28 (1983) 6443;
S.H. Liu, Phys. Rev. B 15 (1977) 4281.
[74] H. Ohno, Science 281 (1998) 951.
[75] H. Ohno, et al., Appl. Phys. Lett. 69 (3) (1996) 363.
[76] H. Ohno, et al., Nature (London) 408 (2000) 944.
[77] Y. Ohno, D.K. Young, B. Beschoten, F. Matsukura, H. Ohno, D.D. Awschalom, Nature (London) 402 (1999) 790.
[78] G.B. Olson, Science 277 (1997) 1237.
[79] G.S. Painter, F.W. Averill, Phys. Rev. Lett. 58 (1987) 234.
[80] J.P. Perdew, K. Burke, M. Ernzerhof, Phys. Rev. Lett. 77 (1996) 3865.
[81] J.P. Perdew, J.A. Chevay, S.H. Vosko, K.A. Jackson, M.R. Pederson, D.J. Singh, C. Fiolhais, Phys. Rev. B 46 (1992) 6671.
[82] J.P. Perdew, M. Levy, Phys. Rev. Lett. 51 (1983) 1884.
[83] S. Picozzi, A. Continenza, R. Asahi, W. Mannstadt, W. Wolf, E. Wimmer, C.B. Geller, Phys. Rev. B 61 (2000) 4677.
[84] D. Porezag, M.R. Pederson, J. Chem. Phys. 102 (1995) 9345.
[85] X. Portier, A.K. Petford-Long, T.C. Anthony, IEEE Trans. Magn. 33 (1997) 3679.
[86] Z. Qian, J.M. Sivertsen, J.H. Judy, B.A. Everitt, S. Mao, E.S. Murdock, J. Appl. Phys. 85 (1999) 6106.
[87] J.R. Rice, J.S. Wang, Mater. Sci. Eng. A 107 (1989) 23.
[88] L.P. Sagert, G.B. Olson, D.E. Ellis, Phil. Mag. B 77 (1997) 871;
L.P. Sagert, Ph.D. Thesis, Northwestern University, 1995.
[89] T.C. Schulthess, W.H. Butler, Phys. Rev. Lett. 81 (1998) 4516.
[90] M.P. Seah, Acta Metall. 28 (1979) 955.
[91] A. Seidl, A. Görling, P. Vogl, J.A. Majewski, M. Levy, Phys. Rev. B 53 (1996) 3764.
[92] L.J. Sham, M. Schlüter, Phys. Rev. Lett. 51 (1983) 1888;
L.J. Sham, M. Schlüter, Phys. Rev. B 32 (1985) 3883.
[93] D.J. Singh, Planewaves, Pseudopotentials and the LAPW Method, Kluwer Academic, Boston, 1994.
[94] J.C. Slater, Phys. Rev. 51 (1937) 846.
[95] J.M. Soler, A.R. Williams, Phys. Rev. B 40 (1989) 1560;
J.M. Soler, A.R. Williams, Phys. Rev. B 42 (1990) 9728.
[96] R.L. Stamps, J. Phys. D 33 (2000) R247.
[97] A.F. Starace, Phys. Rev. A 3 (1971) 1242.
[98] J.P. Stark, H.L. Marcus, Metall. Trans. A 8A (1977) 1423.
[99] K. Takano, R.H. Kodama, A.E. Berkowitz, W. Cao, G. Thomas, Phys. Rev. Lett. 79 (1997) 1130.
[100] P.H.T. Philipsen, G. te Velde, E.J. Baerends, Chem. Phys. Lett. 226 (1994) 583.
[101] H.T. Tiep, H. Kawamura, Phys. Rev. B 40 (1989) 7019.

[102] S.B. Trickey, F.R. Green Jr., F.W. Averill, Phys. Rev. B 8 (1973) 4822.
[103] See, e.g., A.R. Troiano, Trans. Am. Soc. Met. 52 (1960) 54.
[104] J. von Barth, L. Hedin, J. Phys. C 5 (1972) 1629.
[105] C.S. Wang, B.M. Klein, Phys. Rev. B 24 (1981) 3417.
[106] D.S. Wang, R. Wu, A.J. Freeman, Phys. Rev. Lett. 70 (1993) 869.
[107] M. Weinert, E. Wimmer, A.J. Freeman, Phys. Rev. B 26 (1982) 4571.
[108] E. Wimmer, H. Krakauer, M. Weinert, A.J. Freeman, Phys. Rev. B 24 (1981) 864, and references therein.
[109] W. Wolf, E. Wimmer, S. Massidda, M. Posternak, C.B. Geller, Bull. APS 43 (1998) 797.
[110] B.Y. Wong, C. Mitumata, S. Prakash, D.E. Langhlin, T. Kobayashi, IEEE Trans. Magn. 32 (1996) 3425.
[111] R. Wu, A.J. Freeman, Comput. Phys. Comm. 76 (1993) 58.
[112] R. Wu, A.J. Freeman, G.B. Olson, Science 265 (1994) 376.
[113] R.Q. Wu, A.J. Freeman, J. Magn. Magn. Mater. 200 (1999) 498;
R.Q. Wu, A.J. Freeman, J. Magn. Magn. Mater. 100 (1991) 497.
[114] P. Wynblatt, R.C. Ku, in: W.C. Johnson, J.M. Blakely (Eds.), Interface Segregation, ACM, New York, 1977.
[115] R. Yu, D. Singh, H. Krakauer, Phys. Rev. B 43 (1991) 6411.
[116] R. Yu, H. Krakauer, D. Singh, Phys. Rev. B 45 (1992) 8671.
[117] Yu-Jun Zhao, W.T. Geng, A.J. Freeman, T. Oguchi, Rapid Comm. Phys. Rev. B 63 (2001) 201202.
[118] Yu-Jun Zhao, W.T. Geng, K.T. Park, A.J. Freeman, Phys. Rev. B 64 (2001) 035207.
[119] L.P. Zhong, R. Wu, A.J. Freeman, G.B. Olson, Phys. Rev. B 55 (1997) 11133.

ELSEVIER

Journal of Computational and Applied Mathematics 149 (2002) 57–69

JOURNAL OF COMPUTATIONAL AND APPLIED MATHEMATICS

www.elsevier.com/locate/cam

A coherent analysis of Stokes flows under boundary conditions of friction type

Hiroshi Fujita

Research Institute of Educational Development, Tokai University, 2-28-4 Tomigaya, Shibuya-ku, 151-0063 Tokyo, Japan

Received 28 October 2001; received in revised form 14 January 2002

Abstract

This paper is concerned with the stationary and nonstationary flow of viscous incompressible fluid under boundary conditions of friction type, which are certain nonlinear boundary conditions similar to the so-called Signorini boundary condition in elasticity. We assume that the flow is governed by the linear Stokes equation, while the boundary condition is nonlinear. From the methodological viewpoint, the analysis is carried out in a coherent way, starting from study of the related boundary value problems for the stationary flow by means of the theory of variational inequalities, and getting to wellposedness of the initial boundary value problems for the nonstationary flow by means of the nonlinear semigroup theory. From the viewpoint of applications, we mention original motivations and include some new generalizations like the cases of anisotropic friction and inhomogeneous boundary value.

Keywords: Nonlinear partial differential equation; Navier–Stokes equation; Variational inequality; Nonlinear boundary condition; Nonlinear semigroup

1. Introduction and summary

The purpose of this paper is to report on the results of our recent study of certain nonlinear boundary conditions for motions of a viscous incompressible fluid, which we call boundary conditions of friction type (BCF in abbr.). Although most of the content has been adapted from the author's previous presentations [4 8,10] of which the first onc is his scrics of lccturcs at Collègc dc France delivered in October of 1993 on an invitation by the late Prof. J.L. Lions, the present paper contains some new generalizations, particularly in the last section. Also we have tried to organize the paper in a coherent way so that it can clarify the methodology of analysis.

E-mail address: hfujita@yoyogi.ycc.u-tokai.ac.jp (H. Fujita).

PII: S0377-0427(02)00520-4

In his actual talk at the conference, the author expressed his view concerning the trends and issues in scientific and engineering computing in Japan, which was done in accordance with the session theme where his talk was given. However, this part of the author's is not included in the present paper and is expected to be published elsewhere.

1.1. Motivations and objectives

The motivation of our study on BCF comes from practical intention to model mathematically certain flow phenomena in applications, while the study of BCF deserves theoretical interests for its own sake because of the coherent use of celebrated methods in applied functional analysis, i.e., the variational inequalities (see, e.g., [3,11]) and the nonlinear semigroup theory (see, e.g., [1,15]).

In fact, there exist such flow phenomena, like flow through a drain with its bottom covered by sherbet of mud or like flow through a certain filter, that their mathematical model seems to require boundary conditions allowing a slip or leak. Furthermore, in some cases we observe that the slip or leak does not take place if the flow is gentle, namely, as long as the magnitude of stress on the boundary due to the fluid motion is less than a threshold. Our BCF have been introduced to meet these situations. As an successful example in this direction, we can refer to interesting works of simulation of certain environmental and ecological phenomena by Kawarada and his collaborators [13,14], where they have made an extensive use of BCF and interface conditions of similar nature.

1.2. Description of BCF

While there are various kinds of BCF, we describe here in this section the leak boundary condition which is most typical and convincing in order to fix the idea.

The motion of our incompressible fluid is described by velocity u and pressure p. We consider the flow in a bounded domain Ω in $\mathbb{R}^3$, whose boundary Γ is smooth. And we assume that Γ consists of the two components Γ_0 (say, the inner wall) and S (the outer wall). Furthermore, for the sake of simplicity in writing the solvability conditions of the boundary value problems which we are going to deal with, we impose the homogeneous Dirichlet condition

$$u = 0 \quad \text{on } \Gamma_0. \tag{1.1}$$

On S, we impose the leak BCF of our main concern, but again for simplicity, we assume that it is complemented by the nonslip condition

$$u_t = 0 \quad \text{on } S, \tag{1.2}$$

where u_t denotes the tangential component of u. Thus, our Leak-BCF actually means a kind of boundary conditions of nonslip and possible leak. Mathematically, the leak condition is formulated in terms of a given positive function g on S, which is called the barrier function and will play the role to control the threshold for occurrence of the leak. It is written as

$$-\sigma_n \in g\partial|u_n| \quad \text{on } S. \tag{1.3}$$

Here $\sigma_n = \sigma_n(u, p)$ is the normal component of the stress on S. u_n stands for the normal component of u there. We recall that

$$\sigma_n = \sigma(u, p)_n = -p + 2\nu n\, e(u)n, \tag{1.4}$$

where ν is the (kinematic) viscosity and n the outer unit normal to the boundary, while $e(u)$ stands for the strain rate tensor $e(u) = (e_{ij}(u))$:

$$e_{ij} = e_{ij}(u) = \frac{1}{2}\left(\frac{\partial u_j}{\partial x_i} + \frac{\partial u_i}{\partial x_j}\right).$$

In (1.3), $\partial|\cdot|$ means the sub-differential of the real absolute value $|\cdot|$. Without using the notation of $\partial|\cdot|$, the leak condition (1.3) can be re-written as follows (see Appendix A):

$$|\sigma_n(u,p)| \leqslant g \quad \text{on } S \tag{1.5}$$

and

$$\begin{cases} |\sigma_n| < g \Rightarrow u_n = 0, \\ |\sigma_n| = g \Rightarrow \begin{cases} u_n = 0 \text{ or } u_n \neq 0, \\ u_n \neq 0 \Rightarrow -\sigma_n = g\dfrac{u_n}{|u_n|}. \end{cases} \end{cases} \tag{1.6}$$

1.3. Target problems and the results

The boundary value problem, *Leak-BVP*, is to find $\{u, p\}$ which satisfies the steady Stokes equation

$$-\nu\Delta u + \nabla p = f, \quad \operatorname{div} u = 0 \tag{1.7}$$

and the above-mentioned boundary conditions on Γ_0 and on S.

In (1.7), f stands for the external force, while for simplicity the density ρ of the fluid has been taken to be equal to 1. After some preparation concerning weak solutions of (1.7), which is made in Section 2, we shall show the solvability of a weak version of Leak-BVP by means of variational inequalities, Leak-VI in Section 3. In Section 5 we proceed to the initial value problem for the nonstationary flow of the fluid under the Leak-BCF, assuming $f \equiv 0$. Then the flow is governed by the time-dependent Stokes equation.

$$\frac{\partial u}{\partial t} = \nu\Delta u - \nabla p, \quad \operatorname{div} u = 0. \tag{1.8}$$

The initial condition is given by

$$u(0,\cdot) = a \tag{1.9}$$

Leak-IVP (the initial value problem with Leak-BCF).

For a given initial value $a = a(x)$, find $u = u(t,x)$ and $p = p(t,x)$ which satisfy the Stokes equation (1.8) in $[0,+\infty) \times \Omega$, the initial condition (1.9) at $t = 0$, the Dirichlet boundary condition (1.1) on Γ_0 and Leak-BCF (1.2), (1.3) on S.

The solvability of Leak-IVP will be shown in a certain abstract sense by means of the nonlinear semigroup theory in Hilbert space, which originated from the pioneer work by Kōmura [15] in 1967. Section 5 is devoted to brief comments on various directions of generalization of our formulation and analysis.

2. Weak solutions of Stokes equation and of Leak-BVP

We recall some fundamental facts in regard to the weak formulation (variational formulation) of the steady Stokes equation (1.7). The following subspaces of $H^1(\Omega)$ of vector functions are needed here and later:

$$
\begin{aligned}
&V = \{u \in H^1(\Omega); \quad u = 0 \text{ on } \Gamma_0\},\\
&H_0^1(\Omega) = \{u \in H^1(\Omega); \quad u = 0 \text{ on } \Gamma\},\\
&H_\sigma^1(\Omega) = \{u \in H^1(\Omega); \quad \operatorname{div} u = 0\},\\
&V_\sigma = \{u \in H_\sigma^1(\Omega); \quad u = 0 \text{ on } \Gamma_0\},\\
&H_{0,\sigma}^1(\Omega) = \{u \in H_0^1(\Omega); \quad \operatorname{div} u = 0\}.
\end{aligned} \tag{2.1}
$$

We introduce a basic quadratic form (the dissipation form) on V by

$$a(u,v) = 2\nu \int_\Omega \sum_{i,j=1}^{3} e_{ij}(u) e_{ij}(v)\,\mathrm{d}x \quad (u,v \in V). \tag{2.2}$$

This quadratic form is continuous over V in the $H^1(\Omega)$-topology. And by virtue of the Dirichlet boundary conditions on Γ_0, we have Korn's inequality (see e.g., [2,12]):

Lemma 2.1. *There exits positive (domain) constants c_0, c_1 such that*

$$c_0 \|u\|_{H^1(\Omega)}^2 \leqslant a(u,u) \leqslant c_1 \|u\|_{H^1(\Omega)}^2 \quad (\forall u \in V). \tag{2.3}$$

If $u \in V_\sigma$ and $\{u, p\}$ is smooth, we obtain by integration by parts

$$a(u,v) - (p, \operatorname{div} v) = (-\nu\Delta u + \nabla p, v) + \int_S \sigma \cdot v \,\mathrm{d}\Gamma \quad (u \in V_\sigma \cap H^2(\Omega),\ p \in H^1(\Omega),\ v \in V), \tag{2.4}$$

where the L^2-inner product is denoted by $(\cdot,\cdot)$ for the scalar case as well as for the vector case, and where $\sigma = \sigma(u, p)$ stands for the stress vector which is expressed in terms of the normal vector n as follows (in the summation convention):

$$\sigma = (\sigma_i) = (-pn_i + 2\nu e_{i,j}(u) n_j) \quad (i = 1,2,3). \tag{2.5}$$

Definition 2.1 (*Weak solutions of the Stokes equation*). The pair of $u \in H_\sigma^1(\Omega)$ and $p \in L^2(\Omega)$ is a weak solution of the steady Stokes equation (1.7) if

$$a(u,\varphi) - (p, \operatorname{div} \varphi) = (f, \varphi) \quad (\forall \varphi \in H_0^1(\Omega)) \tag{2.6}$$

p is called the pressure associated with u and is unique except for an additive constant when u is fixed.

Condition (2.6) is called the weak (steady) Stokes equation. In the case that the weak solution $\{u, p\}$ happens to be smooth, $\{u, p\}$ satisfies the steady Stokes equation in the classical sense. If φ

in (2.6) is divergence-free, then we have

$$a(u,\varphi) = (f,\varphi) \quad (\forall \varphi \in H^1_{0,\sigma}(\Omega)) \tag{2.7}$$

Condition (2.7) might be called the solenoidal weak Stokes equation. The following lemma holds true because Ω is bounded:

Lemma 2.2. *Let* $u \in H^1_\sigma(\Omega)$ *be a solution of the solenoidal weak Stokes equation* (2.7). *Then there exists* $p \in L^2(\Omega)$ *such that* $\{u, p\}$ *is a weak solution of the* (*steady*) *Stokes equation* (2.6).

Due to this lemma, any solution $u \in H^1_{0,\sigma}$ of (2.7) can be called a weak solution of the steady Stokes equation.

2.1. Stress components of weak solutions

For a weak solution $\{u, p\}$, we can define its stress vector σ as a distribution belonging to $H^{-1/2}(\Gamma)$, through the argument below. Before doing so, we introduce some subclasses of V and V_σ:

$$K = \{u \in V;\ u_{\mathrm{t}} = 0 \text{ on } S\},$$

$$K_\sigma = \{u \in V_\sigma;\ u_{\mathrm{t}} = 0 \text{ on } \Gamma\}. \tag{2.8}$$

Note that if u is a solution of Leak-BVP together with its associated pressure, then $u \in K_\sigma$. Actually, we are concerned with the normal stress σ_n on S of a weak solution $\{u, p\}$, u belonging to K_σ. Furthermore, we need the classes of (scalar and vector) functions on S listed below, in order to define σ_n in the distribution sense and to make later arguments:

$$Y = \text{the scalar } H^{1/2}(\Gamma), \quad Y_0 = \left\{\eta \in Y;\ \int_\Gamma \eta \,\mathrm{d}\Gamma = 0\right\}. \tag{2.9}$$

Now take an $\eta \in Y$ and consider any $\varphi_\eta \in K$ such that $(\varphi_\eta)_t = 0$ and $(\varphi_\eta)_n = \eta$ on S. Then, if $\{u, p\}$ is a smooth solution of the Stokes equation, we have by (2.4)

$$\int_S \sigma_n \eta \,\mathrm{d}\Gamma = a(u,\varphi_\eta) - (p, \operatorname{div}\varphi_\eta) - (f,\varphi_\eta). \tag{2.10}$$

The right-hand side of (2.10) does not depend on the manner of extension from η to φ_η since $\{u, p\}$ satisfies the weak Stokes equation, while it depends on linearly and continuously on η in $H^{1/2}(S)$-topology. Consequently, through (2.10) we can define a bounded linear functional Σ_n on Y, which can be viewed as a distribution in $H^{-1/2}(S)$. This Σ_n is again denoted by σ_n and we write formally

$$\int_S \sigma_n \varphi_n \,\mathrm{d}\Gamma = a(u,\varphi) - (p, \operatorname{div}\varphi) - (f,\varphi) \quad (\forall \varphi \in K). \tag{2.11}$$

Definition 2.2 (*Weak solution of Leak-BVP*). $\{u, p\}$ is a weak solution of Leak-BVP if the following three conditions hold true:

(i) $u \in K_\sigma$ and $p \in L^2(\Omega)$;

(ii) $\{u, p\}$ satisfies the weak Stokes equation (2.6);
(iii) $\{u, p\}$ satisfies the leak condition (1.3) in the a.e. sense. Namely

$$-\sigma(u, p)_n \in g\partial|u_n| \quad \text{(a.e. on } S). \tag{2.12}$$

3. Variational inequalities for Leak-BVP

The variational inequalities for analysis of Leak-BVP is now formulated:
Leak-VI: Find $u \in K_\sigma$ and $p \in L^2(\Omega)$ such that

$$a(u, v-u) - (p, \operatorname{div}(v-u)) + j(v) - j(u) \geqslant (f, v-u) \quad (\forall v \in K), \tag{3.1}$$

where

$$j(v) = \int_S g|v_n| \, \mathrm{d}\Gamma \quad (\forall v \in K). \tag{3.2}$$

If $\{u, p\}$ is a solution of Leak-VI, then it satisfies the weak Stokes equation (2.6), which can be seen by taking an arbitrary $\varphi \in H_0^1(\Omega)$ and substituting $v = u \pm \varphi$ into (3.1).

Since we can use (2.11) for weak solutions of the Stokes equation, we can reduce (3.1) to

$$\int_S \sigma_n (v-u)_n \, \mathrm{d}\Gamma + j(v) - j(u) \geqslant 0 \quad (\forall v \in K). \tag{3.3}$$

3.1. Theorems for Leak-VI

We claim

Theorem 3.1. *Leak-VI and the weak Leak-BVP are equivalent in the sense that $\{u, p\}$ is a solution of Leak-VI if and only if it is a weak solution of Leak-BVP.*

Proof. Suppose that $\{u, p\}$ is a weak solution of Leak-BVP. We have only to prove the inequality (3.3). In view of (A.1), we have from (1.3)

$$g|v_n| - g|u_n| \geqslant -\sigma_n (v-u)_n \quad \text{a.e. on } S \ (\forall v \in K).$$

Integrating the inequality above, we get to

$$j(v) - j(u) \geqslant -\int_S \sigma_n (v-u)_n \, \mathrm{d}\Gamma,$$

which is nothing but (3.3).

Conversely, suppose that $\{u, p\}$ is a solution of Leak-VI. Already we have seen that $\{u, p\}$ is a solution of the weak Stokes equation (2.6). It remains to prove the leak condition (1.3). From (3.3) follows:

$$-\int_S \sigma_n (v-u)_n \, \mathrm{d}\Gamma \leqslant j(v) - j(u) \leqslant \int_S g|(v-u)_n| \, \mathrm{d}\Gamma \quad (\forall v \in K). \tag{3.4}$$

Namely, we have

$$-\int_S \sigma_n \eta \, \mathrm{d}\Gamma \leqslant \int_S g|\eta| \, \mathrm{d}\Gamma \quad (\forall \eta \in Y). \tag{3.5}$$

Since this inequality holds true if we replace η by $-\eta$, we have

$$\left|\int_S \sigma_n \eta \, \mathrm{d}\Gamma\right| \leqslant \int_S g|\eta| \, \mathrm{d}\Gamma \quad (\forall \eta \in Y). \tag{3.6}$$

Here we make a duality argument. Actually, let us consider the Banach space M of L^1-type of functions defined on S with the weighted measure $g\,\mathrm{d}\Gamma$, i.e., with the norm

$$\|\eta\|_M = \int_S g|\eta| \, \mathrm{d}\Gamma. \tag{3.7}$$

(3.6) means that σ_n defines a linear functional on $Y \subset M$ with its functional norm bounded by 1. Since Y is dense in M, σ_n can be viewed as an element in the dual space M^* of M. As a matter of fact, M^* is an L^∞-type space with its norm given by

$$\|\eta\|_{M^*} = \operatorname*{ess\,sup}_{s \in S} \frac{|\eta(s)|}{g(s)}. \tag{3.8}$$

Therefore, σ_n turns out to be a bounded function on S subject to (1.5). We are now going to show the second equality in (A.5). Coming back to (3.4), we put $v = 0$ there, obtaining

$$-\int_S \sigma_n u_n \, \mathrm{d}\Gamma - \int_S g|u_n| \, \mathrm{d}\Gamma \geqslant 0,$$

which leads to

$$\int_S (\sigma_n u_n + g|u_n|) \, \mathrm{d}\Gamma = 0$$

with the aid of (1.5), and leads further to the second equality of (A.5) in the a.e. sense on S. Thus $\{u, p\}$ is a weak solution of Leak-BVP. □

We proceed to one of our main theorems:

Theorem 3.2. *Leak-VI has a solution $\{u, p\}$, of which u is unique and p is unique except for an additive constant. The range of the additive constant to p is limited to $\{0\}$ or to a finite closed interval. This assertion is true for Leak-BVP.*

Proof. *Uniqueness*: Let $\{u_i, p_i\}$ be solutions of Leak-VI ($i = 1, 2$). Then we have

$$a(u_1, u_2 - u_1) + j(u_2) - j(u_1) \geqslant (f, u_2 - u_1),$$

$$a(u_2, u_1 - u_2) + j(u_1) - j(u_2) \geqslant (f, u_1 - u_2),$$

since $\operatorname{div} u_1 = 0$, $\operatorname{div} u_2 = 0$. Adding these two inequalities, we have $a(u_2 - u_1, u_2 - u_1) \leqslant 0$, which gives $u_2 - u_1 = 0$ by virtue of Korn's inequality. After obtaining the uniqueness of u, it is easy to examine the uniqueness of p in $L^2(\Omega)/\mathbb{R}$. Then the range of the additive constant to p is seen through (1.3).

Existence: We start from variational inequalities within solenoidal functions.

Leak-VI$_\sigma$: Find $u \in K_\sigma$ such that

$$a(u, v - u) + j(v) - j(u) \geqslant (f, v - u) \quad (\forall v \in K_\sigma). \tag{3.9}$$

The existence of the solution u of Leak-VI$_\sigma$ can be shown by a standard argument in the theory of variational inequalities (e.g., see Glowinski's book [11]). Then in the same way as before, we can verify that u is a solution of (2.7) and see that there exists an associated pressure p. Let us fix this p. $\{u, p\}$ may not satisfy (1.3) but we can use (2.11) for $\sigma_n(u, p)$. If $v \in K_\sigma$, then by means of (2.11) and (3.9) we have

$$\int_S \sigma_n(v-u)_n \, d\Gamma = a(u, v-u) - (p, \operatorname{div}(v-u)) - (f, v-u)$$
$$= a(u, v-u) - (f, v-u) \geqslant -j(v) + j(u). \tag{3.10}$$

Hence we have

$$\int_S \sigma_n(v-u)_n \, d\Gamma + j(v) - j(u) \geqslant 0 \quad (\forall v \in K_\sigma). \tag{3.11}$$

Partly repeating the argument in the proof of the preceding theorem, we deduce

$$\left| \int_S \sigma_n \eta \, d\Gamma \right| \leqslant \int_S g|\eta| \, d\Gamma \quad (\forall \eta \in Y_0) \tag{3.12}$$

in consideration that $(v-u)_n$ ranges over Y_0 as v varies over K_σ. Note here that Y_0 is not dense in the L^1-type Banach space M introduced in the proof of the previous theorem. We can, however, regard σ_n as a linear functional defined on Y_0 which is a subspace of M, whose functional norm is bounded by 1. At this point, we apply the Hahn–Banach theorem and see that there exist an element λ^* in the dual space M^* such that

$$\langle \lambda^*, \eta \rangle = \langle \sigma_n, \eta \rangle \quad (\forall \eta \in Y_0) \tag{3.13}$$

and

$$\|\lambda^*\|_{M^*} \leqslant 1. \tag{3.14}$$

From (3.14), we see that λ^* is a bounded function on S and is subject to

$$|\lambda^*| \leqslant g \quad \text{a.e. on } S. \tag{3.15}$$

On the other hand, (3.13) implies

$$\lambda - \sigma_n = -k^*$$

for some constant k^*. Let us put $p^* = p + k^*$. Then

$$\lambda^* = \sigma_n(u, p) - k^* = \sigma_n(u, p^*)$$

and from (3.15) it holds true that

$$|\sigma_n(u, p^*)| \leqslant g \quad \text{a.e. on } S. \tag{3.16}$$

Furthermore, we may write (3.11) for $\{u, p^*\}$ as

$$\int_S \sigma_n^*(v-u)_n \, d\Gamma + j(v) - j(u) \geqslant 0 \quad (\forall v \in K_\sigma). \tag{3.17}$$

From (3.16) and (3.17) with $v = 0$, we can deduce

$$\sigma_n^* u_n + g|u_n| = 0 \quad \text{(a.e. on } S)$$

in a parallel way as in the proof of preceding theorem.

Thus we have shown that $\{u, p^*\}$ satisfies (A.5) and is a solution of Leak-VI and hence of the weak Leak-BVP. □

4. Nonlinear semigroups theory and Leak-IVP

The crucial step in applying the nonlinear semigroup theory to our nonstationary problems, Leak-IVP, is to define the Stokes operator A in accordance with the linear and nonlinear boundary conditions in consideration, so that the multi-valued A is *maximally monotone* (*m-accretive*) in the Hilbert space $X = L^2(\Omega)$. Here the maximal monotonicity of A is equivalent (see, for instance [1]) to

$$(f_1 - f_2, u_1 - u_2) \geqslant 0 \quad (\forall u_1, u_2 \in D(A),\ \forall f_1 \in Au_1, \forall f_2 \in Au_2) \tag{4.1}$$

and

$$\text{Range of } (I + A) = X. \tag{4.2}$$

In fact, we define the Stokes operator with Leak-BCF by putting

$$D(A) = \{u \in K_\sigma;\ u \text{ is a solution of Leak-VI for some } f \text{ (and for some } p)\} \tag{4.3}$$

and defining Au for each fixed $u \in D(A)$ as

$$f \in Au \Leftrightarrow u \text{ is the solution of Leak}-\text{VI for } f \quad \text{(and for some } p). \tag{4.4}$$

Then we have

Theorem 4.1. *The Stokes operator A with Leak-BCF is a maximal monotone operator.*

In fact, the monotonicity (4.1) of A can be seen as follows: let $\{u_i, p_i\}$ be the solution of Leak-VI for f_i, $(i = 1, 2)$. Then we have

$$a(u_1, u_2 - u_1) + j(u_2) - j(u_1) \geqslant (f_1, u_2 - u_1),$$

$$a(u_2, u_1 - u_2) + j(u_1) - j(u_2) \geqslant (f_2, u_1 - u_2),$$

since $\operatorname{div} u_1 = 0$, $\operatorname{div} u_2 = 0$. Adding these two inequalities, we have $a(u_1 - u_2, u_2 - u_1) \geqslant (f_1 - f_2, u_2 - u_1)$, which gives (4.1), since $a(u, u)$ is positive definite. On the other hand, equality (4.2) is shown by repeating the previous argument for the solvability of Leak-BVP with $a(u, v)$ replaced by $a(u, v) + (u, v)$, or by making use of a theorem which can be applied to a monotone operator A with the property "Range of $A = X$" (see [1]).

In this way, the generation theorem in the nonlinear semigroup theory can be applied and we have

Theorem 4.2. *If $a \in D(A)$, then Leak-IVP is solvable uniquely and strongly in the following sense: there exists uniquely an absolutely continuous function $u = u(t)$ defined on $[0, +\infty)$ and with values in $X = L^2(\Omega)$ such that*

(i) *$u(t) \in D(A)$ for every t,*
(ii) *it holds true that $u(0) = a$ and*

$$\frac{\mathrm{d}u}{\mathrm{d}t} \in -Au(t) \quad \text{(a.e. } t\text{)}. \tag{4.5}$$

Moreover, the right derivative of u satisfies

$$\frac{\mathrm{d}^+u}{\mathrm{d}t} = -A^0u(t) \quad (\forall t \in [0, +\infty)), \tag{4.6}$$

where A^0 is the canonical restriction of A, which assigns to each $u \in D(A)$ the unique element A^0u which is the element with minimum norm within the closed convex set Au.

Remark. As a matter of fact, if we make use of those theorems in the NSG theory which are concerned with generators of the sub-differential type, then we can relax the condition on the initial value a so that $a \in K_\sigma$ is sufficient for the strong solvability of the abstract Leak-IVP (see [1,9]). On the other hand, as mentioned in [6], some direct characterization of the operator A is possible by virtue of a regularity theorem due to Saito [16,17], which can be applied if the data are slightly more regular.

5. Some generalizations

Here we state brief remarks on some generalizations of our study.

5.1. Slip-BCF

Similarly to Leak-BCF, we can formulate the slip boundary condition of friction type (Slip-BCF) by replacing (1.3) by

$$-\sigma_t \in g\partial|u_t| \quad \text{on } S, \tag{5.1}$$

where $\sigma_t = \sigma(u)_t$ is the tangential component of the stress. Additionally, we impose the non-leak boundary condition, $u_n = 0$ on S. Then our analysis goes in parallel to the previous one (or more simply), with the barrier functional and admissible function classes renewed as

$$\begin{aligned} j(v) &= \int_S g|u_t|\,\mathrm{d}\Gamma \quad (v \in V), \\ K &= \{v \in V;\quad u_n = 0 \text{ on } S\}, \\ K_\sigma &= \{v \in V_\sigma;\quad u_n = 0 \text{ on } S\}. \end{aligned} \tag{5.2}$$

for technical details, we refer to a forthcoming paper [9].

5.2. Anisotropic friction

In some applications, it would be practical to introduce anisotropic boundary conditions of friction type, where the threshold for occurrence of leak or slip depends on the direction. One model to meet such a phenomena can be created by adding a linear perturbation term to the variational inequalities. For instance, suppose that we are concerned with such a situation of slip type. Then we could try a slip condition of the following form:

$$-\sigma_t + k(s) \in g\partial|u_t| \quad \text{on } S, \tag{5.3}$$

where $k = k(s)$ is a tangential vector function given on S. k could be singular to some extent (e.g., in $H^{-1/2}$-class). All of our previous arguments are valid, if we add the term

$$-\int_S k(v-u)_t \,\mathrm{d}\Gamma, \quad (\forall v \in K),$$

to the left-hand side of the basic variational inequalities. Another model can be obtained introducing a smooth two-dimensional matrix function $m = m(s)$ on S which is uniformly invertible. The slip condition of the form

$$-\sigma_t \in g\,\partial|m\,u_t| \quad \text{on } S$$

can be dealt with as before, if we use the barrier functional

$$\int_S g\,|m\,u_t|d\Gamma \quad (v \in V).$$

5.3. Slip/leak from inhomogeneous boundary value

Sometimes, it would be necessary to consider leak or slip from a nonhomogeneous boundary value. For instance, if S is an axially symmetric surface and actually stands for a wall rotating around the axis, then it is natural to regard that the original boundary value, from which the slip takes place, is a nontrivial tangential vector function β determined by the rotation. In such a case, we introduce the slip condition of the form

$$-\sigma_t \in g\partial|u_t - \beta| \quad \text{on } S.$$

Then the variational analysis can be carried out successfully with the barrier functional defined as

$$j(v) = \int_S g|u_t - \beta|\,\mathrm{d}\Gamma.$$

5.4. Other generalizations

(1) The formulation and method of analysis can be applied to the case where Ω is an exterior domain and $\Gamma = S$ is compact, as long as the initial boundary value is concerned. As a matter of fact, we must make a more delicate argument since the associated pressure does belong to $L^2_{\mathrm{loc}}(\Omega)$ but not to $L^2(\Omega)$ and since the original BVP may not be uniquely solvable. See [7] for some detail.

(2) We can deal with flow problems with an interface condition of friction type (ICF). The formulation and method of analysis are parallel to the case of BCF. See [8] for some detail.

Appendix A. Elementary remarks on subdifferentials

Let $\psi = \psi(x)$ be a lower semi-continuous convex function on $\mathbb{R}^1$. The subdifferential of ψ at a point x_0, which is denoted by $\partial\psi(x_0)$, is by definition the set of all ξ which satisfies

$$\psi(x) - \psi(x_0) \geqslant \xi(x - x_0) \quad (\forall x \in \mathbb{R}^1). \tag{A.1}$$

If ψ is differentiable at x_0, then $\partial\psi(x_0) = \{\psi'(x_0)\}$. Generally, $\partial\psi$ is a multi-valued function on $\mathbb{R}^1$. The definition above of the subdifferential is naturally extended to convex functions on $\mathbb{R}^n$ and even to convex functions on a Hilbert space X. In the latter case, we have for each $x_0 \in X$

$$\xi \in \partial\psi(x_0) \quad \Leftrightarrow \quad \psi(x) - \psi(x_0) \geqslant (\xi, x - x_0) \quad (\forall x \in X). \tag{A.2}$$

It is easy to check the case of $\psi(x) = |x|$ and obtain

$$\partial|x| = \begin{cases} \text{the closed interval } [-1, 1], & (x = 0), \\ 1, & (x > 0), \\ -1, & (x < 0). \end{cases} \tag{A.3}$$

Similarly, when $|x|$ denote the norm (= the length) of real 2-vectors, we have

$$\partial|x| = \begin{cases} \text{the closed unit disk } |x| \leqslant 1, & (x = 0), \\ \dfrac{x}{|x|}, & (x \neq 0). \end{cases} \tag{A.4}$$

Lemma A.1. *The leak condition* (1.3) *is equivalent to the following set of conditions on* S:

$$|\sigma_n| \leqslant g, \quad \sigma_n u_n + g|u_n| = 0 \quad \text{on } S. \tag{A.5}$$

Proof. From (1.3) follows (A.5) immediately by virtue of (A.3). Conversely, by (A.5) we can calculate for any real number z as follows:

$$\begin{aligned} &g|z| - g|u_n| + \sigma_n(z - u_n) \\ &\quad = g|z| + \sigma_n z - (g|u_n| + \sigma_n u_n) \\ &\quad = g|z| + \sigma_n z \geqslant 0, \end{aligned} \tag{A.6}$$

which implies (1.3) in virtue of (A.1). □

References

[1] H. Brezis, Opérateurs Maximaux Monotone et Semi-groupes de Contractions dans les Espaces de Hilbert, Mathematical Studies, Vol. 5, North-Holland, Amsterdam, 1973.

[2] P.G. Ciarlet, Mathematical Elasticity, Theory of Plates, Vol. II, Studies in Mathematics and its Applications, Vol. 27, North-Holland, Amsterdam, 1997.

[3] G. Duvaut, J.L. Lions, Les Inéquations en Méchanique et en Physique, Dunod, Paris, 1972 (English version, Springer, Berlin, 1976).

[4] H. Fujita, Flow Problems with Unilateral Boundary Conditions, Leçons, Collège de France, October 1993.

[5] H. Fujita, A mathematical analysis of motions of viscous incompressible fluid under leak or slip boundary conditions, Res. Inst. Math. Sci. Kōkyūroku 888 (1994) 199–216.
[6] H. Fujita, Non-stationary Stokes flows under leak boundary conditions of friction type, J. Comput. Math. 19 (2001) 1–8.
[7] H. Fujita, Variational inequalities and nonlinear semi-groups applied to certain nonlinear problems for the Stokes equation, in: Proceedings of International Conference: Tosio Kato's Method and Principle for Evolution Equations in Mathematical Physics, Sapporo, July 27–29, 2001, RIMS Kyoto University, 2002, pp. 70–85.
[8] H. Fujita, Remarks on the Stokes flow under slip and leak boundary conditions of friction type, in: Topics in Mathematical Fluid Mechanics, Vol. A (celebrating J. Heywood 60th birthday), Aracne, Rome, 2002, to appear.
[9] H. Fujita, On the Stokes equation with the slip and leak boundary conditions of friction type: existence of stationary solutions, in preparation.
[10] H. Fujita, H. Kawarada, Variational inequalities for the Stokes equation with the boundary conditions of friction type, Recent Development in Domain Decomposition Methods and Flow Problems, Gakkotosho, 1998, pp. 15–33.
[11] R. Glowinski, Numerical Methods for Nonlinear Variational Problems, Springer Series in Computational Physics, Springer, Berlin, 1984.
[12] C.O. Horgan, Korn's inequalities and their applications in continuum mechanics, SIAM Rev. 37 (1995) 491–511.
[13] H. Kawarada, H. Fujita, H. Suito, Wave motion breaking upon the shore, Recent Developments in Domain Decomposition Methods and Flow Problems, Gakuto International Series, Mathematical Sciences and Applications, Vol. 11, Gakkotosho, 1998, pp. 145–159.
[14] H. Kawarada, H. Suito, Multi-phase flow with reaction, Lecture Note in Computational Science and Engineering, vol. 19, Mathematical Modeling and Numerical Simulation in Continuum Mechanics, Springer, 2001, pp. 39–56.
[15] Y. Kōmura, Nonlinear semigroups in Hilbert space, J. Math. Soc. Japan 19 (1967) 493–507.
[16] N. Saito, On the Stokes equation with the slip and leak boundary conditions of friction type: regularity of solutions, in preparation.
[17] N. Saito, H. Fujita, Regularity of solutions to the Stokes equations under a certain nonlinear boundary conditions, Navier–Stokes Equations, Theory and Numerical Methods, Marcel-Dekker, New York, 2001, pp. 73–86.

ELSEVIER Journal of Computational and Applied Mathematics 149 (2002) 71–82

JOURNAL OF COMPUTATIONAL AND APPLIED MATHEMATICS

www.elsevier.com/locate/cam

Direct simulation of the motion of a settling ellipsoid in Newtonian fluid

Tsorng-Whay Pan[a], Roland Glowinski[a,*], Giovanni P. Galdi[b]

[a] *Department of Mathematics, University of Houston, Houston, 4800 Calhouwd Road, 77204 Houston, TX, USA*
[b] *Department of Mechanical Engineering, University of Pittsburg, Pittsburg, PA 15261, USA*

Received 24 October 2001; received in revised form 18 December 2001

Abstract

In this paper, we discuss the generalization of a Lagrange multiplier-based fictitious domain method to the simulation of the motion of general shape particles in Newtonian fluid. Preliminary numerical results of a settling ellipsoid in a narrow channel filled with a Newtonian fluid are presented. As expected, the ellipsoid turns its broadside to the stream in the simulations.

1. Introduction

The orientation of symmetric long body (loosely, a long body is a body where one dimension is much prevailing upon the other two) in liquids of different nature is a fundamental issue in many problems of practical interest (see [15], and references therein). In this article we first discuss the generalization of a Lagrange multiplier based fictitious domain method [8,10] to the simulation of the motion of particles of general shape in a Newtonian fluid. Unlike the cases where the particles are spheres, we attach two points to each particle of general shape and move them according to the rigid-body motion of the particle in order to track the motion of the particle. The equations describing the motion of these two points are solved by a distance preserving scheme so that rigidity can be maintained. We then apply it to simulate the settling of an ellipsoid in a narrow channel filled with a Newtonian fluid. In the simulation, the ellipsoid turns its broadside to the stream as expected. Preliminary numerical results are presented.

* Corresponding author.
E-mail address: roland@math.uh.edu (R. Glowinski).

PII: S0377-0427(02)00521-6

2. A model problem and fictitious domain formulation for three-dimensional particulate flow

To perform the *direct numerical simulation* of the interaction between particles and fluid, we have developed a methodology which is a combination of a distributed Lagrange multiplier-based fictitious domain (also called domain embedding) method and operator splitting methods [6–10], this approach (or closely related ones derived from it) has become the method of choice for other investigators around the world (Refs., [2,23]). We are going to recall the ideas at the basis of the above methodology by considering the motion of a single particle in a Newtonian viscous incompressible fluid (of density ρ_f and viscosity ν_f) under the effect of gravity; actually, the generalization to a thousand of spherical particles in 3D and 10 000 circular particles in 2D is possible as shown in [10,18,19]. For the situation depicted in Fig. 1 (for nonspherical particle cases), the flow is modeled by the Navier–Stokes equations, namely, (with obvious notation)

$$\rho_f \left[\frac{\partial \mathbf{u}}{\partial t} + (\mathbf{u} \cdot \nabla)\mathbf{u} \right] - \nu_f \Delta \mathbf{u} + \nabla p = \rho_f \mathbf{g} \quad \text{in } \Omega \backslash \bar{B}(t), (0, T), \tag{1}$$

$$\nabla \cdot \mathbf{u} = 0 \quad \text{in } \Omega \backslash \bar{B}(t), (0, T), \tag{2}$$

$$\mathbf{u}(0) = \mathbf{u}_0(\mathbf{x}) \quad (\text{with } \nabla \cdot \mathbf{u}_0 = 0), \tag{3}$$

$$\mathbf{u} = \mathbf{g}_0 \quad \text{on } \Gamma \times (0, T) \quad \text{with} \int_\Gamma \mathbf{g}_0 \cdot \mathbf{n} \, d\Gamma = 0; \tag{4}$$

where $\Gamma = \partial\Omega$, $\mathbf{g}$ is gravity and $\mathbf{n}$ is the unit normal vector pointing outward to the flow region. We assume a *no-slip condition* on $\gamma(=\partial B)$. The motion of particle B satisfies the Euler–Newton's equations, namely

$$\mathbf{v}(\mathbf{x}, t) = \mathbf{V}(t) + \overrightarrow{\omega}(t) \times \overrightarrow{\mathbf{G}(t)\mathbf{x}} \quad \forall \{\mathbf{x}, t\} \in \bar{B}(t) \times (0, T), \tag{5}$$

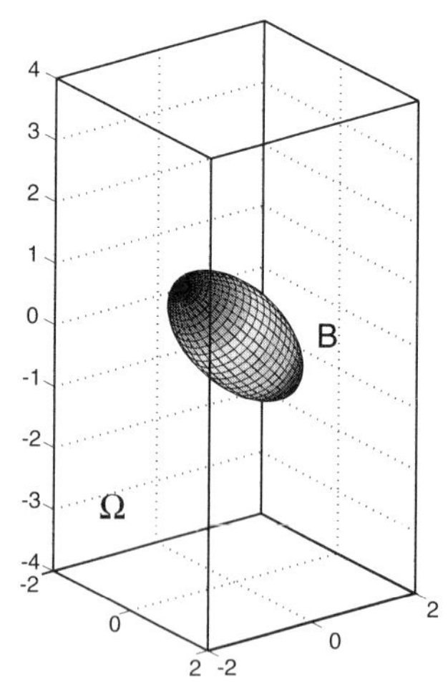

Fig. 1. The flow region with one particle.

$$\frac{\mathrm{d}\mathbf{G}}{\mathrm{d}t} = \mathbf{V}, \tag{6}$$

$$M_\mathrm{p}\frac{\mathrm{d}\mathbf{V}}{\mathrm{d}t} = M_\mathrm{p}\mathbf{g} + \mathbf{F}_\mathrm{H} + \mathbf{F}^r, \tag{7}$$

$$\frac{\mathrm{d}(\mathbf{I}_\mathrm{p}\boldsymbol{\omega})}{\mathrm{d}t} = \mathbf{T}_\mathrm{H} + \overrightarrow{\mathbf{Gx}_r} \times \mathbf{F}^r \tag{8}$$

with hydrodynamical forces and torques

$$\mathbf{F}_\mathrm{H} = -\int_\gamma \boldsymbol{\sigma}\mathbf{n}\,\mathrm{d}\gamma, \quad \mathbf{T}_\mathrm{H} = -\left(\int_\gamma \overrightarrow{\mathbf{Gx}} \times \boldsymbol{\sigma}\mathbf{n}\,\mathrm{d}\gamma\right), \tag{9}$$

completed by the following initial conditions:

$$\mathbf{G}(0) = \mathbf{G}_0, \quad \mathbf{V}(0) = \mathbf{V}_0, \quad \boldsymbol{\omega}(0) = \boldsymbol{\omega}_0. \tag{10}$$

Above, M_p, $\mathbf{I}_\mathrm{p}$, $\mathbf{G}$, $\mathbf{V}$ and $\boldsymbol{\omega}$ are the mass, inertia, center of mass, velocity of the center of mass and angular velocity of particle B, respectively. In (8) we found preferable to deal with the *kinematic angular momentum* $\mathbf{I}_\mathrm{p}\boldsymbol{\omega}$ making the formulation more conservative. In order to avoid particle–particle and particle–wall penetration which can happen in the numerical simulation, we have introduced a artificial force $\mathbf{F}^r$ in (7) (for more details, see, e.g., [8] and [10]) and then a torque in (8) acting on the point $\mathbf{x}_r$ where $\mathbf{F}^r$ applies on B.

To solve system (1)–(10) we can use, for example, *Arbitrary Lagrange–Euler* (ALE) methods as in [12,14,17], or *fictitious domain methods*, which allow the flow calculation on a fixed grid, as in [6–10]. The fictitious domain methods that we advocate have some common features with the *immersed boundary method* of Ch. Peskin (see, e.g., Refs. [20–22]) but also some significant differences in the sense that we take systematically advantage of *distributed Lagrange multipliers* to force the rigid-body motion inside the particle, which seems still to be a relatively novel approach in this context, and whose possibilities have not been fully explored yet. As with the methods in [20–22], our approach takes advantage of the fact that the flow can be computed on a grid which does not have to vary in time, a substantial simplification indeed.

The principle of fictitious domain methods is simple. It consists of

- Filling the particle with a fluid having the same density and viscosity as the surrounding one.
- Compensating the above step by introducing, in some sense, an *anti-particle* of mass $(-1)M_\mathrm{p}\rho_\mathrm{f}/\rho_\mathrm{s}$ and inertia $(-1)\mathbf{I}_\mathrm{p}\rho_\mathrm{f}/\rho_\mathrm{s}$, taking into account the fact that any rigid-body motion $\mathbf{v}(\mathbf{x},t)$ verifies $\nabla\cdot\mathbf{v} = 0$ and $\mathbf{D}(\mathbf{v}) = \mathbf{0}$ (ρ_s: particle density).
- Finally, imposing the rigid-body velocity on $\bar{B}(t)$, namely

  $$\mathbf{v}(\mathbf{x},t) = \mathbf{V}(t) + \vec{\omega}(t) \times \overrightarrow{\mathbf{G}(t)\mathbf{x}} \quad \forall \mathbf{x}\in\bar{B}(t)\ \forall t\in(0,T), \tag{11}$$

  via a Lagrange multiplier $\boldsymbol{\lambda}$ supported by $\bar{B}(t)$. Vector $\boldsymbol{\lambda}$ forces rigidity in $B(t)$ in the same way that ∇p forces $\nabla\cdot\mathbf{v} = 0$ for incompressible fluids.

We obtain then an equivalent formulation of (1)–(10) defined on the whole domain, namely
For a.e. $t > 0$, find $\{\mathbf{u}(t), p(t), \mathbf{V}(t), \mathbf{G}(t), \boldsymbol{\omega}(t), \boldsymbol{\lambda}(t)\}$

such that

$$\mathbf{u}(t) \in \mathbf{W}_{g_0}(t),\ p(t) \in L_0^2(\Omega),\ \mathbf{V}(t) \in \mathbb{R}^3,\ \mathbf{G}(t) \in \mathbb{R}^3, \boldsymbol{\omega}(t) \in \mathbb{R}^3,\ \boldsymbol{\lambda}(t) \in \Lambda(t) \tag{12}$$

and

$$\begin{aligned} &\rho_f \int_\Omega \left[\frac{\partial \mathbf{u}}{\partial t} + (\mathbf{u}\cdot\boldsymbol{\nabla})\mathbf{u}\right]\cdot \mathbf{v}\,d\mathbf{x} - \int_\Omega p\boldsymbol{\nabla}\cdot\mathbf{v}\,d\mathbf{x} \\ &+ \nu_f \int_\Omega \boldsymbol{\nabla}\mathbf{u} : \boldsymbol{\nabla}\mathbf{v}\,d\mathbf{x} - \langle \boldsymbol{\lambda}, \mathbf{v} - \mathbf{Y} - \boldsymbol{\theta}\times\mathbf{Gx}\rangle_{\Lambda(t)} \\ &+ \left(1 - \frac{\rho_f}{\rho_s}\right)\left[M_p \frac{d\mathbf{V}}{dt}\cdot\mathbf{Y} + \frac{d(\mathbf{I}_p\boldsymbol{\omega})}{dt}\cdot\boldsymbol{\theta}\right] - \mathbf{F}^r\cdot\mathbf{Y} - \overrightarrow{\mathbf{Gx}_r}\times\mathbf{F}^r\cdot\boldsymbol{\theta} \\ &= \left(1 - \frac{\rho_f}{\rho_s}\right) M_p \mathbf{g}\cdot\mathbf{Y} + \rho_f \int_\Omega \mathbf{g}\cdot\mathbf{v}\,d\mathbf{x} \quad \forall \mathbf{v} \in (H_0^1(\Omega))^3,\ \forall \mathbf{Y} \in \mathbb{R}^3,\ \forall \boldsymbol{\theta} \in \mathbb{R}^3, \end{aligned} \tag{13}$$

$$\int_\Omega q\boldsymbol{\nabla}\cdot\mathbf{u}(t)\,d\mathbf{x} = 0 \quad \forall q \in L^2(\Omega), \tag{14}$$

$$\frac{d\mathbf{G}}{dt} = \mathbf{V}, \tag{15}$$

$$\langle \boldsymbol{\mu}, \mathbf{u}(t) - \mathbf{V}(t) - \boldsymbol{\omega}(t)\times\mathbf{G}(t)\mathbf{x}\rangle_{\Lambda(t)} = 0 \quad \forall \boldsymbol{\mu} \in \Lambda(t), \tag{16}$$

$$\mathbf{V}(0) = \mathbf{V}_0, \quad \boldsymbol{\omega}(0) = \boldsymbol{\omega}_0, \quad \mathbf{G}(0) = \mathbf{G}_0, \tag{17}$$

$$\mathbf{u}(\mathbf{x},0) = \tilde{\mathbf{u}}_0(\mathbf{x}) = \begin{cases} \mathbf{u}_0(\mathbf{x}) & \forall \mathbf{x} \in \Omega\backslash\overline{B(0)}, \\ \mathbf{V}_0 + \boldsymbol{\omega}_0\times\mathbf{G}_0\mathbf{x} & \forall \mathbf{x} \in \overline{B(0)} \end{cases} \tag{18}$$

with the following functional spaces:

$$\mathbf{W}_{g_0}(t) = \{\mathbf{v} \mid \mathbf{v} \in (H^1(\Omega))^3, \mathbf{v} = \mathbf{g}_0(t) \text{ on } \Gamma\},$$

$$L_0^2(\Omega) = \left\{q \,\middle|\, q \in L^2(\Omega), \int_\Omega q\,d\mathbf{x} = 0\right\}, \quad \Lambda(t) = (H^1(B(t)))^3.$$

In (12)–(18), only the center of mass, the translation velocity of the center of mass and the angular velocity of the particle are considered. Knowing these two velocities and the center of mass of the particle, one is able to translate and rotate the particle in space by tracking two extra points $\mathbf{x}_1$ and $\mathbf{x}_2$ in each particle, which follow the rigid-body motion:

$$\frac{d\mathbf{x}_i}{dt} = \mathbf{V}(t) + \overrightarrow{\boldsymbol{\omega}}(t)\times\overrightarrow{\mathbf{G}(t)\mathbf{x}_i}, \quad \mathbf{x}_i(0) = \mathbf{x}_{i,0}, \quad i = 1,2. \tag{19}$$

In practice we shall track two orthogonal normalized vectors rigidly attached to the body B and originating from the center of mass $\mathbf{G}$.

3. Time and space discretization

For simplicity, we assume that $\Omega \subset \mathbb{R}^3$ and is a rectangular parallelepiped. Concerning the *space approximation* of problem (12)–(19) by a *finite element method*, we have

$$\mathbf{W}_h = \{\mathbf{v}_h \mid \mathbf{v}_h \in (C^0(\bar{\Omega}))^3, \mathbf{v}_h|_T \in (P_1)^3 \quad \forall T \in \mathscr{T}_h\}, \tag{20}$$

$$\mathbf{W}_{0h} = \{\mathbf{v}_h \mid \mathbf{v}_h \in \mathbf{W}_h,\ \mathbf{v}_h = \mathbf{0} \text{ on } \Gamma\}, \tag{21}$$

$$L_h^2 = \{q_h \mid q_h \in C^0(\bar{\Omega}),\ q_h|_T \in P_1 \quad \forall T \in \mathscr{T}_{2h}\}, \quad L_{0h}^2 = \left\{q_h \mid q_h \in L_h^2, \int_\Omega q_h \,\mathrm{d}\mathbf{x} = 0\right\}, \tag{22}$$

where $\mathscr{T}_h$ is a tetrahedrization of Ω, $\mathscr{T}_{2h}$ is twice coarser than $\mathscr{T}_h$, and P_1 is the space of the polynomials in three variables of degree $\leqslant 1$. A finite-dimensional space approximating $\Lambda(t)$ is as follows: let $\{\xi_i\}_{i=1}^N$ be a set of points form $\overline{B(t)}$ which cover $\overline{B(t)}$ (uniformly, for example); we define then

$$\Lambda_h(t) = \left\{\boldsymbol{\mu}_h \mid \boldsymbol{\mu}_h = \sum_{i=1}^N \boldsymbol{\mu}_i \delta(\mathbf{x} - \xi_i),\ \boldsymbol{\mu}_i \in \mathbb{R}^3,\ \forall i = 1,\dots,N\right\}, \tag{23}$$

where $\delta(\cdot)$ is the Dirac measure at $\mathbf{x} = \mathbf{0}$. Then we shall use $\langle\cdot,\cdot\rangle_h$ defined by

$$\langle \boldsymbol{\mu}_h, \mathbf{v}_h\rangle_h = \sum_{i=1}^N \boldsymbol{\mu}_i \cdot \mathbf{v}_h(\xi_i) \quad \forall \boldsymbol{\mu}_h \text{ in } \Lambda_h(t),\ \mathbf{v}_h \in \mathbf{W}_h. \tag{24}$$

A typical choice of points for defining (23) is a collection of grid points for velocity field covered by the interior of the particle $B(t)$ and selected points from the boundary of $B(t)$.

Using the above finite-dimensional spaces leads to the following approximation for problem (12)–(19):

For a.e. $t > 0$, find $\{\mathbf{u}_h(t), p_h(t), \mathbf{V}(t), \mathbf{G}(t), \boldsymbol{\omega}(t), \boldsymbol{\lambda}_h(t)\}$ such that

$$\mathbf{u}(t) \in \mathbf{W}_h(t),\ p_h(t) \in L_{0h}^2(\Omega),\ \mathbf{V}(t) \in \mathbb{R}^3, \mathbf{G}(t) \in \mathbb{R}^3, \boldsymbol{\omega}(t) \in \mathbb{R}^3, \boldsymbol{\lambda}_h(t) \in \Lambda_h(t) \tag{25}$$

and

$$\begin{aligned}
&\rho_\mathrm{f} \int_\Omega \left[\frac{\partial \mathbf{u}_h}{\partial t} + (\mathbf{u}_h \cdot \boldsymbol{\nabla})\mathbf{u}_h\right] \cdot \mathbf{v}\,\mathrm{d}\mathbf{x} - \int_\Omega p_h \boldsymbol{\nabla} \cdot \mathbf{v}\,\mathrm{d}\mathbf{x} \\
&\quad + \nu_\mathrm{f} \int_\Omega \boldsymbol{\nabla}\mathbf{u}_h : \boldsymbol{\nabla}\mathbf{v}\,\mathrm{d}\mathbf{x} - \langle \boldsymbol{\lambda}_h, \mathbf{v} - \mathbf{Y} - \boldsymbol{\theta} \times \mathbf{G}\mathbf{x}\rangle_h \\
&\quad + \left(1 - \frac{\rho_\mathrm{f}}{\rho_\mathrm{s}}\right) \left[M_\mathrm{p} \frac{\mathrm{d}\mathbf{V}}{\mathrm{d}t} \cdot \mathbf{Y} + \frac{\mathrm{d}(\mathbf{I}_\mathrm{p}\boldsymbol{\omega})}{\mathrm{d}t} \cdot \boldsymbol{\theta}\right] - \mathbf{F}^r \cdot \mathbf{Y} - \overrightarrow{\mathbf{G}\mathbf{x}_r} \times \mathbf{F}^r \cdot \boldsymbol{\theta} \\
&= \left(1 - \frac{\rho_\mathrm{f}}{\rho_\mathrm{s}}\right) M_\mathrm{p}\mathbf{g} \cdot \mathbf{Y} + \rho_\mathrm{f} \int_\Omega \mathbf{g} \cdot \mathbf{v}\,\mathrm{d}\mathbf{x} \quad \forall \mathbf{v} \in \mathbf{W}_{0h},\ \forall \mathbf{Y} \in \mathbb{R}^3,\ \forall \boldsymbol{\theta} \in \mathbb{R}^3,
\end{aligned} \tag{26}$$

$$\int_\Omega q \boldsymbol{\nabla} \cdot \mathbf{u}_h(t)\,\mathrm{d}\mathbf{x} = 0 \quad \forall q \in L_h^2, \tag{27}$$

$$\mathbf{u}_h = \mathbf{g}_{0h} \quad \text{on } \Gamma, \tag{28}$$

$$\frac{\mathrm{d}\mathbf{G}}{\mathrm{d}t} = \mathbf{V}, \tag{29}$$

$$\frac{\mathrm{d}\mathbf{x}_i}{\mathrm{d}t} = \mathbf{V}(t) + \vec{\omega}(t) \times \overrightarrow{\mathbf{G}(t)\mathbf{x}_i}, \quad \mathbf{x}_i(0) = \mathbf{x}_{i,0}, \quad i = 1,2, \tag{30}$$

$$\langle \boldsymbol{\mu}, \mathbf{u}_h(t) - \mathbf{V}(t) - \omega(t) \times \mathbf{G}(t)\mathbf{x} \rangle_h = 0 \quad \forall \boldsymbol{\mu} \in \Lambda(t), \tag{31}$$

$$\mathbf{V}(0) = \mathbf{V}_0, \quad \omega(0) = \omega_0, \quad \mathbf{G}(0) = \mathbf{G}_0, \tag{32}$$

$$\mathbf{u}(\mathbf{x}, 0) = \tilde{\mathbf{u}}_{0h}(\mathbf{x}). \tag{33}$$

In (28), $\mathbf{g}_{0h}$ is an approximation of $\mathbf{g}_0$ belonging to $\gamma\mathbf{W}_h = \{\mathbf{z}_h \mid \mathbf{z}_h \in (C^0(\Gamma))^3,\ \mathbf{z}_h = \tilde{\mathbf{z}}_h|_\Gamma \text{ with } \tilde{\mathbf{z}}_h \in \mathbf{W}_h\}$ and verifying $\int_\Gamma \mathbf{g}_{0h} \cdot \mathbf{n}\,\mathrm{d}\Gamma = 0$.

3.1. An operator-splitting scheme à la Marchuk–Yanenko

Many operator-splitting schemes can be used to time-discretize (25)–(33). One of the advantage of operator-splitting schemes is that we can decouple difficulties like (i) the incompressibility condition, (ii) the nonlinear advection term, and (iii) a rigid-body motion projection, so that each one of them can be handled separately, and in principle optimally. Let Δt be a time discretization step and $t^{n+s} = (n+s)\Delta t$. By an operator-splitting scheme à la Marchuk–Yanenko in [16], we have the following scheme after dropping some of the subscripts h (similar ones are discussed in [6–10]):

$$\mathbf{u}^0 = \tilde{\mathbf{u}}_0, \quad \mathbf{G}^0 = \mathbf{G}_0, \quad \mathbf{V}^0 = \mathbf{V}_0, \quad \omega^0 = \omega_0, \quad \mathbf{x}_1^0 = \mathbf{x}_{1,0}, \quad \mathbf{x}_2^0 = \mathbf{x}_{2,0} \quad \text{given}; \tag{34}$$

for $n \geqslant 0$, $\mathbf{u}^n (\simeq \mathbf{u}(t^n))$, $\mathbf{G}^n$, $\mathbf{V}^n$, ω^n, $\mathbf{x}_1^n$ and $\mathbf{x}_2^n$ being known, we compute $\mathbf{u}^{n+1/5}$, $p^{n+1/5}$ via the solution of

$$\begin{aligned}
&\rho_\mathrm{f} \int_\Omega \frac{\mathbf{u}^{n+1/5} - \mathbf{u}^n}{\Delta t} \cdot \mathbf{v}\,\mathrm{d}\mathbf{x} - \int_\Omega \mathrm{p}^{n+1/5} \boldsymbol{\nabla} \cdot \mathbf{v}\,\mathrm{d}\mathbf{x} = 0 \quad \forall \mathbf{v} \in \mathbf{W}_{0h},\\
&\int_\Omega q \boldsymbol{\nabla} \cdot \mathbf{u}^{n+1/5}\,\mathrm{d}\mathbf{x} = 0 \quad \forall q \in L_h^2,\\
&\mathbf{u}^{n+1/5} \in \mathbf{W}_h, \quad \mathbf{u}^{n+1/5} = \mathbf{g}_{0h}^{n+1} \quad \text{on } \Gamma, \quad p^{n+1/5} \in L_{0h}^2.
\end{aligned} \tag{35}$$

Next, compute $\mathbf{u}^{n+2/5}$ via the solution of

$$\begin{aligned}
&\int_\Omega \frac{\partial \mathbf{u}}{\partial t} \cdot \mathbf{v}\,\mathrm{d}\mathbf{x} + \int_\Omega (\mathbf{u}^{n+1/5} \cdot \boldsymbol{\nabla})\mathbf{u} \cdot \mathbf{v}\,\mathrm{d}\mathbf{x} = 0 \quad \forall \mathbf{v} \in \mathbf{W}_{0h}^{n+1,-}, \quad \text{a.e. on } (t^n, t^{n+1}),\\
&\mathbf{u}(t^n) = \mathbf{u}^{n+1/5},\\
&\mathbf{u}(t) \in \mathbf{W}_h, \quad \mathbf{u}(t) = \mathbf{g}_{0h}^{n+1} \quad \text{on } \boldsymbol{\Gamma}_-^{n+1} \times (t^n, t^{n+1}),
\end{aligned} \tag{36}$$

and set $\mathbf{u}^{n+2/5} = \mathbf{u}(t^{n+1})$.

Then, compute $\mathbf{u}^{n+3/5}$ via the solution of

$$\rho_{\mathrm{f}}\int_{\Omega}\frac{\mathbf{u}^{n+3/5}-\mathbf{u}^{n+2/5}}{\Delta t}\cdot\mathbf{v}\,\mathrm{d}\mathbf{x}+\alpha\nu_{\mathrm{f}}\int_{\Omega}\nabla\mathbf{u}^{n+3/5}:\nabla\mathbf{v}\,\mathrm{d}\mathbf{x}=\rho_{\mathrm{f}}\int_{\Omega}\mathbf{g}\cdot\mathbf{v}\,\mathrm{d}\mathbf{x}$$

$$\forall\mathbf{v}\in\mathbf{W}_{0h};\ \mathbf{u}^{n+3/5}\in\mathbf{W}_h,\quad \mathbf{u}^{n+3/5}=\mathbf{g}_{0h}^{n+1}\quad\text{on }\Gamma. \tag{37}$$

Now predict the motion of the center of mass and the angular velocity of the particle via

$$\frac{\mathrm{d}\mathbf{G}}{\mathrm{d}t}=\mathbf{V}(t), \tag{38}$$

$$\frac{\mathrm{d}\mathbf{x}_i}{\mathrm{d}t}=\mathbf{V}(t)+\overrightarrow{\omega}(t)\times\overrightarrow{\mathbf{G}(t)\mathbf{x}_i},\quad\text{for } i=1,2, \tag{39}$$

$$(1-\rho_{\mathrm{f}}/\rho_{\mathrm{s}})M_{\mathrm{p}}\frac{\mathrm{d}\mathbf{V}}{\mathrm{d}t}=(1-\rho_{\mathrm{f}}/\rho_{\mathrm{s}})M_{\mathrm{p}}\mathbf{g}+\mathbf{F}^{r}, \tag{40}$$

$$(1-\rho_{\mathrm{f}}/\rho_{\mathrm{s}})\frac{\mathrm{d}(\mathbf{I}_{\mathrm{p}}\boldsymbol{\omega})}{\mathrm{d}t}=\overrightarrow{\mathbf{G}\mathbf{x}_r}\times\mathbf{F}^{r}, \tag{41}$$

$$\mathbf{G}(t^n)=\mathbf{G}^n,\quad \mathbf{V}(t^n)=\mathbf{V}^n,\quad (\mathbf{I}_{\mathrm{p}}\boldsymbol{\omega})^n=(\mathbf{I}_{\mathrm{p}}\boldsymbol{\omega})(t^n),\quad \mathbf{x}_1(t^n)=\mathbf{x}_1^n,\quad \mathbf{x}_2(t^n)=\mathbf{x}_2^n, \tag{42}$$

for $t^n<t<t^{n+1}$. Then set $\mathbf{G}^{n+4/5}=\mathbf{G}(t^{n+1})$, $\mathbf{V}^{n+4/5}=\mathbf{V}(t^{n+1})$, $(\mathbf{I}_{\mathrm{p}}\boldsymbol{\omega})^{n+4/5}=(\mathbf{I}_{\mathrm{p}}\boldsymbol{\omega})(t^{n+1}), \mathbf{x}_1^{n+4/5}=\mathbf{x}_1(t^{n+1})$, and $\mathbf{x}_2^{n+4/5}=\mathbf{x}_2(t^{n+1})$.

With the center $\mathbf{G}^{n+4/5}$, $\mathbf{x}_1^{n+4/5}$ and $\mathbf{x}_2^{n+4/5}$ obtained at the above step, we enforce the rigid-body motion in the region $B(t^{n+4/5})$ occupied by the particle

$$\rho_{\mathrm{f}}\int_{\Omega}\frac{\mathbf{u}^{n+1}-\mathbf{u}^{n+4/5}}{\Delta t}\cdot\mathbf{v}\,dx+\beta\nu_{\mathrm{f}}\int_{\Omega}\nabla\mathbf{u}^{n+1}:\nabla\mathbf{v}\,\mathrm{d}\mathbf{x}$$

$$+(1-\rho_{\mathrm{f}}/\rho_{\mathrm{s}})M_{\mathrm{p}}\frac{\mathbf{V}^{n+1}-\mathbf{V}^{n+4/5}}{\Delta t}\cdot\mathbf{Y}+(1-\rho_{\mathrm{f}}/\rho_{\mathrm{s}})\frac{(\mathbf{I}_{\mathrm{p}}\boldsymbol{\omega})^{n+1}-(\mathbf{I}_{\mathrm{p}}\boldsymbol{\omega})^{n+4/5}}{\Delta t}\cdot\boldsymbol{\theta}$$

$$=\langle\boldsymbol{\lambda}^{n+4/5},\mathbf{v}-\mathbf{Y}-\boldsymbol{\theta}\times\overrightarrow{\mathbf{G}^{n+4/5}\mathbf{x}}\rangle_h\quad\forall\mathbf{v}\in\mathbf{W}_{0h},\ \mathbf{Y}\in\mathbb{R}^3,\ \boldsymbol{\theta}\in\mathbb{R}^3,$$

$$\mathbf{u}^{n+1}\in\mathbf{W}_h,\quad \mathbf{u}^{n+1}=\mathbf{g}_{0h}^{n+1}\quad\text{on }\Gamma,\quad \boldsymbol{\lambda}^{n+4/5}\in\Lambda_h^{n+4/5},\quad \mathbf{V}^{n+1}\in\mathbb{R}^3,\quad \boldsymbol{\omega}^{n+1}\in\mathbb{R}^3, \tag{43}$$

$$\langle\boldsymbol{\mu},\mathbf{u}^{n+1}-\mathbf{V}^{n+1}-\boldsymbol{\omega}^{n+1}\times\overrightarrow{\mathbf{G}_j^{n+4/5}\mathbf{x}}\rangle_h=0\quad\forall\boldsymbol{\mu}\in\Lambda_h^{n+4/5}. \tag{44}$$

In (34)–(44), $\Gamma_-^{n+1}=\{\mathbf{x}\,|\,\mathbf{x}\in\Gamma,\ \mathbf{g}_{0h}^{n+1}(\mathbf{x})\cdot\mathbf{n}(\mathbf{x})<0\}$ and $\mathbf{W}_{0h}^{n+1,-}=\{\mathbf{v}\,|\,\mathbf{v}\in\mathbf{W}_h,\mathbf{v}=0\text{ on }\Gamma_-^{n+1}\}$, $\Lambda_h^{n+s}=\Lambda_h(t^{n+s})$, and $\alpha+\beta=1$. But in the numerical simulation, we usually choose $\alpha=1$ and $\beta=0$.

3.2. On the solution of subproblems (35), (36), (37), (38)–(42), and (43)–(44)

The degenerated quasi-Stokes problem (35) is solved by an Uzawa/preconditioned conjugate gradient algorithm as in [11], where the discrete elliptic problems from the preconditioning are solved by a matrix-free fast solver from FISHPAK due to Adams et al. [1]. The advection problem (36) for the velocity field is solved by a wave-like equation method as in [4,5]. Problem (37) is a classical discrete elliptic problem which can be solved by the same matrix-free fast solver.

System (38)–(42) is a system of ordinary differential equations thanks to operator splitting. For its solution one can choose a time step smaller than Δt (i.e., we can divide Δt into smaller steps) to predict the translation velocity of the center of mass, the angular velocity of the particle, the position of the center of mass and the regions occupied by each particle so that the repulsion forces can be effective to prevent particle–particle and particle-wall overlapping. At each subcycling time step, keeping the distance as constant between the pair of points $\mathbf{x}_1$ and $\mathbf{x}_2$ in each particle is important since we are dealing with rigid particles. We have applied the following approach to satisfy the above constraint.

- Translate $\mathbf{x}_1$ and $\mathbf{x}_2$ according to the new position of the mass center at each subcyling time step.
- Rotate $\mathbf{Gx}_1$ and $\mathbf{Gx}_2$, the relative positions of $\mathbf{x}_1$ and $\mathbf{x}_2$ to the center of mass $\mathbf{G}$, by the following Crank–Nicolson scheme (a Runge–Kutta scheme of order 2 in fact):

$$\frac{\mathbf{Gx}_i^{\text{new}} - \mathbf{Gx}_i^{\text{old}}}{\tau} = \boldsymbol{\omega} \times \frac{\mathbf{Gx}_i^{\text{new}} + \mathbf{Gx}_i^{\text{old}}}{2} \tag{45}$$

for $i = 1,2$ with τ as a subcyling time step. By (45), we have $|\mathbf{Gx}_i^{\text{new}}|^2 = |\mathbf{Gx}_i^{\text{old}}|^2$ for $i = 1,2$ and $|\mathbf{Gx}_2^{\text{new}} - \mathbf{Gx}_1^{\text{new}}|^2 = |\mathbf{Gx}_2^{\text{old}} - \mathbf{Gx}_1^{\text{old}}|^2$ (i.e., scheme (45) is distance and in fact shape preserving).

After solving (38)–(42), the rigid-body motion is enforced in $B(t^{n+4/5})$, via Eq. (44). At the same time those hydrodynamical forces acting on the particles are also taken into account in order to update the translation and angular velocities of the particles. To solve (43)–(44), we use a conjugate gradient algorithm as discussed in [8]. Since we take $\beta = 0$ in (43) for the simulation, we actually do not need to solve any nontrivial linear systems for the velocity field; this saves a lot of computing time. To get the angular velocity $\boldsymbol{\omega}^{n+1}$, computed via

$$\boldsymbol{\omega}^{n+1} = (\mathbf{I}_{\text{p}}^{n+4/5})^{-1}(\mathbf{I}_{\text{p}}\boldsymbol{\omega})^{n+1}, \tag{46}$$

we need to have $\mathbf{I}_{\text{p}}^{n+4/5}$, the inertia of the particle $B(t^{n+4/5})$. We first compute the inertia $\mathbf{I}_0$ in the coordinate system attached to the particle. Then via the center of mass $\mathbf{G}^{n+4/5}$ and points $\mathbf{x}_1^{n+4/5}$ and $\mathbf{x}_2^{n+4/5}$, we have the rotation transformation $\mathbf{Q}$ ($\mathbf{Q}\mathbf{Q}^{\text{T}} = \mathbf{Q}^{\text{T}}\mathbf{Q} = \mathbf{I}_d$, $\det \mathbf{Q} = 1$) which transforms vectors expressed in the particle frame to vectors in the flow domain coordinate system and $\mathbf{I}_{\text{p}}^{n+s} = \mathbf{Q}\mathbf{I}_0\mathbf{Q}^{\text{T}}$. Actually in order to update matrix $\mathbf{Q}$, we can also use *quaternion* techniques, as shown, in the review paper [3].

4. Numerical experiments

In the test cases, we consider the simulation of the motion of a settling ellipsoid in a narrow channel of infinity length filled with a Newtonian fluid. The computational domain in $\Omega = (0,1) \times (0,0.25) \times (0.4)$ initially, then it moves down with the center of the ellipsoid (see, e.g., [13] for adjusting the computational domain according to the position of the particle). The fluid density is $\rho_{\text{f}} = 1$ and the fluid viscosity is $\nu_{\text{f}} = 0.01$. The flow field initial condition is $\mathbf{u} = \mathbf{0}$. The three semi-axes of the ellipsoid are 0.2, 0.1 and 0.1. The initial velocity and angular velocity of the ellipsoid are $\mathbf{0}$.

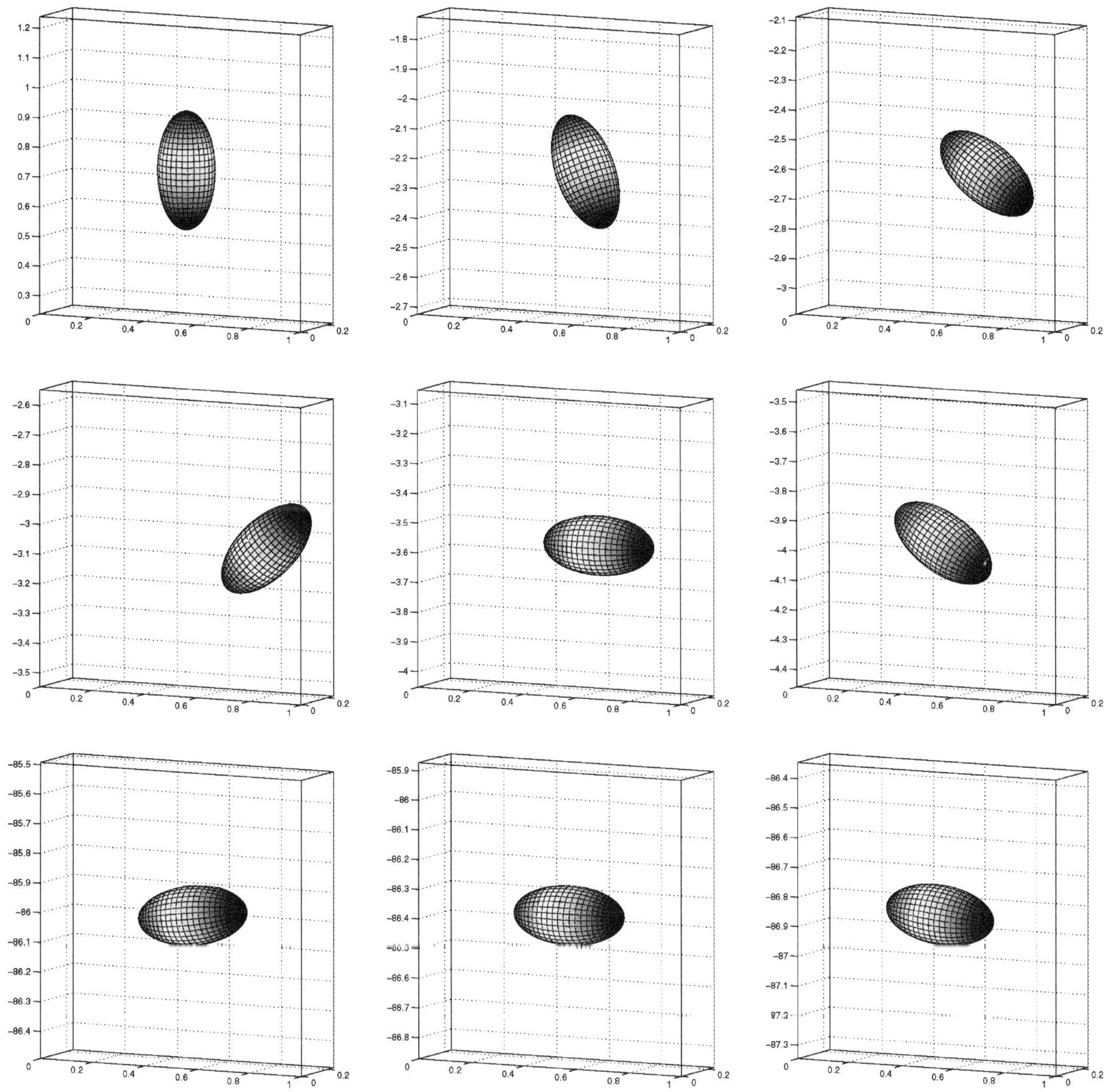

Fig. 2. Position of the ellipsoid at $t=0$, 0.41, 0.46, 0.56, 0.66, 0.75, 18.1, 18.18, and 18.28 (from left to right and from top to bottom). The density of the ellipsoid is $\rho_s = 1.25$.

In the first case, the density of the ellipsoid is $\rho_s = 1.25$. Its vertical axis is the longest semi-axis (see Fig. 2). The mesh size for the velocity field (resp., pressure) is $h_v = 1/80$ (resp., $h_p = 2h_v$). The time step is $\Delta t = 0.001$. In Fig. 2, the positions of the ellipsoid at different times in the channel are shown. The motion of the ellipsoid is very violent at the beginning, it moves very close to the side wall after release from its initial position (see Fig. 2). Later on the motion becomes periodic (see Figs. 2 and 3). As expected, the ellipsoid turns its broadside to the stream while oscillating as

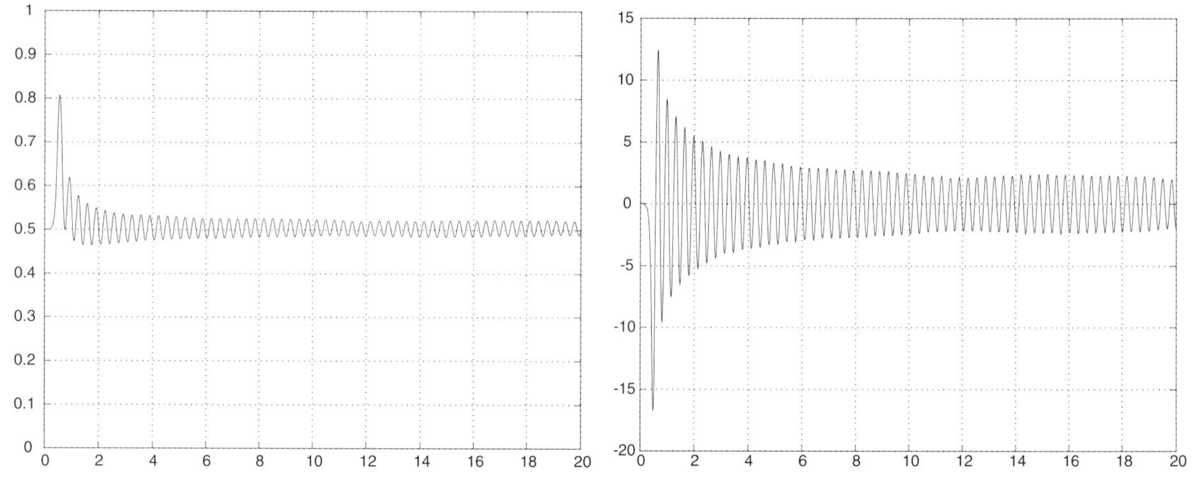

Fig. 3. Histories of the x-coordinate of the center (left) and the y-component of the angular velocity (right) of the ellipsoid of the density $\rho_s = 1.25$.

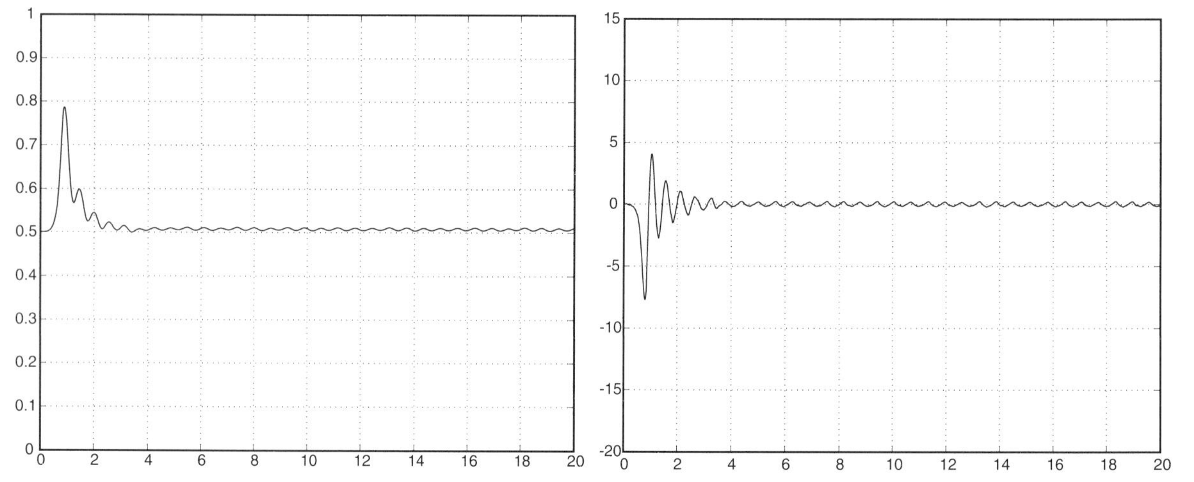

Fig. 4. Histories of the x-coordinate of the center (left) and the y-component of the angular velocity (right) of the ellipsoid of the density $\rho_s = 1.1$.

shown in the last three snapshots of Fig. 2. The averaged particle speed at the end of the simulation is about 4.256 so the particle Reynolds number with the long axis as characteristic length is 170.24. In the second case, the density of the ellipsoid is $\rho_s = 1.1$. The initial position is the same as the previous one. The mesh size for the velocity field (resp., pressure) is $h_v = 1/64$ (resp., $h_p = 2h_v$). The time step is $\Delta t = 0.001$. The averaged particle speed at the end of simulation is about 2.246. The angular velocity with respect to the y-axis is relative small compared to the one in the previous case (see Fig. 4) and the ellipsoid turns its broadside to the stream and oscillates much less.

Acknowledgements

We acknowledge the helpful comments and suggestions of E.J. Dean, J. He, H.H. Hu, P.Y. Huang, D.D. Joseph, Y. Kuznetsov and J. Periaux. We acknowledge also the support of NSF (grants ECS-9527123, CTS-9873236, DMS-9973318, and CCR-9902035), Texas Board of Higher Education (ARP grant 003652-0383-1999), and DOE/LASCI (grant R71700K-292-000-99).

References

[1] J. Adams, P. Swarztrauber, R. Sweet, FISHPAK: a package of Fortran subprograms for the solution of separable elliptic partial differential equations, The National Center for Atmospheric Research, Boulder, CO, 1980.
[2] F.P.T. Baaijens, A fictitious domain/mortar element method for fluid-structure interaction, Internat. J. Numer. Methods Fluids 35 (2001) 743–761.
[3] J.C.K. Chou, Quaternion kinematic and dynamic differential equations, IEEE Trans. Robot. Automat. 8 (1992) 53–64.
[4] E.J. Dean, R. Glowinski, A wave equation approach to the numerical solution of the Navier–Stokes equations for incompressible viscous flow, C.R. Acad. Sci. Paris, Sér. 1 325 (1997) 783–791.
[5] E.J. Dean, R. Glowinski, T.-W. Pan, A wave equation approach to the numerical simulation of incompressible viscous fluid flow modeled by the Navier–Stokes equations, in: J.A. De Santo (Ed.), Mathematical and Numerical Aspects of Wave Propagation, SIAM, Philadelphia, PA, 1998, pp. 65–74.
[6] R. Glowinski, T. Hesla, D.D. Joseph, T.-W. Pan, J. Périaux, Distributed Lagrange multiplier methods for particulate flows, in: M.O. Bristeau, G. Etgen, W. Fitzgibbon, J.L. Lions, J. Périaux, M.F. Wheeler (Eds.), Computational Science for the 21st Century, Wiley, Chichester, UK, 1997, pp. 270–279.
[7] R. Glowinski, T.-W. Pan, T. Hesla, D.D. Joseph, J. Périaux, A fictitious domain method with distributed Lagrange multipliers for the numerical simulation of particulate flow, in: J. Mandel, C. Farhat, X.C. Cai (Eds.), Domain Decomposition Methods 10, American Mathematical Society, Providence, RI, 1998, pp. 121–137.
[8] R. Glowinksi, T.-W. Pan, T. Hesla, D.D. Joseph, A distributed Lagrange multiplier/fictitious method for flows around moving rigid bodies: application to particulate flows, Int. J. Multiphase Flow 25 (1999) 755–794.
[9] R. Glowinski, T.-W. Pan, T. Hesla, D.D Joseph, J. Périaux, A distributed Lagrange multiplier/fictitious domain method for flow around moving rigid bodies: application to particulate flow, Internat. J. Numer. Methods Fluids 30 (1999) 1043–1066.
[10] R. Glowinski, T.-W. Pan, T.I. Hesla, D.D. Joseph, J. Périaux, A fictitious domain approach to the direct numerical simulation of incompressible viscous flow past moving rigid bodies: application to particulate flow, J. Comput. Phys. 169 (2001) 363–426.
[11] R. Glowinski, T.W. Pan, J. Périaux, Distributed Lagrange multiplier methods for incompressible flow around moving rigid bodies, Comput. Methods Appl. Mech. Eng. 151 (1998) 181–194.
[12] H.H. Hu, Direct simulation of flows of solid–liquid mixtures, Int. J. Multiphase Flow 22 (1996) 335–352.
[13] H.H. Hu, D.D. Joseph, M.J. Crochet, Direct simulation of fluid particle motions, Theoret. Comput. Fluid Dynamics 3 (1992) 285–306.
[14] A. Johnson, T. Tezduyar, 3-D simulation of fluid-rigid body interactions with the number of rigid bodies reaching 100, Comput. Methods Appl. Mech. Eng. 145 (1997) 301–321.
[15] Y.J. Liu, D.D. Joseph, Sedimentation of particles in polymer solutions, J. Fluid Mech. 255 (1993) 565–595.
[16] G.I. Marchuk, Splitting and alternating direction methods, in: P.G. Ciarlet, J.L. Lions (Eds.), Handbook of Numerical Analysis, North-Holland, Amsterdam, 1990, pp. 197–462.
[17] B. Maury, R. Glowinski, Fluid particle flow: a symmetric formulation, C.R. Acad. Sci. Paris, Sér. 1 324 (1997) 1079–1084.
[18] T.-W. Pan, D.D Joseph, R. Bai, R. Glowinski, V. Sarin, Fluidization of 1204 spheres: simulation and experiments, J. Fluid Mech. 451 (2002) 169–191.
[19] T.-W. Pan, D.D. Joseph, R. Glowinski, Modeling Rayleigh–Taylor instability of a sedimenting suspension of several thousand circular particles in a direct numerical simulation, J. Fluid Mech. 434 (2001) 23–37.
[20] C.S. Peskin, Numerical analysis of blood flow in the heart, J. Comput. Phys. 25 (1977) 220–252.

[21] C.S. Peskin, Lectures on Mathematical Aspects of Physiology, Lectures in Applied Mathematics, Vol. 19, 1981, pp. 69–107.
[22] C.S. Peskin, D.M. McQueen, Modeling prosthetic heart valves for numerical analysis of blood flow in the heart, J. Comput. Phys. 37 (1980) 113–132.
[23] G.J. Wagner, N. Moes, W.K. Liu, T. Belytschko, The extended finite element method for rigid particles in Stokes flow, Internat. J. Numer. Methods. Eng. 51 (2001) 293–313.

ELSEVIER

Journal of Computational and Applied Mathematics 149 (2002) 83–100

JOURNAL OF COMPUTATIONAL AND APPLIED MATHEMATICS

www.elsevier.com/locate/cam

GasTurbnLab: a multidisciplinary problem solving environment for gas turbine engine design on a network of nonhomogeneous machines☆

E.N. Houstis[a,b,c,*], A.C. Catlin[a], P. Tsompanopoulou[d], D. Gottfried[a], G. Balakrishnan[a], K. Su[a], J.R. Rice[a]

[a]*Department of Computer Sciences, Purdue University, West Lafayette, IN 47807, USA*
[b]*Department of Computer Engineering and Informatics, University of Patras, Patras, Greece*
[c]*Department of Computer Engineering and Communications, University of Thessaly, Volos, Greece*
[d]*Department of Informatics, University of Oslo, Oslo, Norway*

Received 31 October 2001; received in revised form 31 January 2002

Abstract

Gas turbine engines are very complex (with 20–40,000 parts) and have extreme operating conditions. The important physical phenomena take place on scales from 10–100 microns to meters. A complete and accurate dynamic simulation of an entire engine is enormously demanding. Designing a complex system, like a gas turbine engine, will require fast, accurate simulations of computational models from multiple engineering disciplines along with sophisticated optimization techniques to help guide the design process. In this paper, we describe the architecture of an agent based software framework for the simulation of various aspects of a gas turbine engine, utilizing a "network" of collaborating numerical objects through a set of interfaces among the engine parts. Moreover, we present its implementation using the Grasshopper agent middleware and provide simulation results that show the feasibility of the computational paradigm implemented.

Keywords: Multidisciplinary problem solving environments; Mobile agents; Mathematical network model; Interface relaxation; Domain decomposition

☆ This work was partially supported by the US Department of Energy ASCI program contract LG-6982 and NSF grants 0075506-EIA, DMI-0122214.

* Corresponding author. Department of Computer Sciences, Purdue University, 47807 West Lafayette, IN, USA. Tel.: +1-765-494-6181.

E-mail address: enh@cs.purdue.edu (E.N. Houstis).

PII: S0377-0427(02)00522-8

1. Introduction

A physical system in the real world normally consists of a large number of components that have different shapes, obey different physical laws and manufacturing/design constraints, and interact through geometric and physical interfaces. Mathematically, the physical behavior of each component is modeled by a partial or ordinary differential (PDE or ODE) system with various formulations for the geometry, PDE, ODE, interface/boundary/linkage and constraint conditions in many different geometric regions [6,23]. It is difficult to imagine creating a monolithic software system to accurately model such a real problem with complicated artifacts such as the turbo engine, which has literally hundreds of odd shaped parts and a dozen physical phenomena. Therefore, one needs a new computational paradigm which is applicable to a wide variety of practical problems, allows for software reuse in order to achieve lower costs and high quality, and is suitable for some reasonably fast numerical methods [9]. Most physical systems and manufactured artifacts can be modeled as a mathematical network whose nodes represent the physical components in a system or artifact. Each node has a mathematical model of the physics of the component it represents and a solver agent for its analysis. Individual components are chosen so that each node corresponds to a simple PDE or ODE problem defined on a regular geometry [2,12,15].

There exist many standard, reliable PDE/ODE solvers that can be applied to these local node problems. In addition, there are nodes that correspond to interfaces (e.g. ODEs, objective functions, relations, common parameters and their constraints) that model the collaborating parts in the global model. Moreover, the analysis of an artifact changes through time, thus some of the interfaces appear and disappear during the analysis session. To solve the global problem, we let these local solvers collaborate with each other to relax (i.e., resolve) the interface conditions. An interface controller or mediator agent collects boundary values, dynamic/shape coordinates, and parameters/constraints from neighboring subdomains and adjusts boundary values and dynamic/shape coordinates to better satisfy the interface conditions. Therefore, the network abstraction of a physical system or artifact allows us to build a software system that is a network of collaborating well-defined numerical objects through a set of interfaces. Some of the theoretical issues of this methodology have been addressed for the case of collaborating PDE models in [16–19,21,22]. The results obtained so far verify the feasibility and potential of network-based prototyping. Throughout this paper, we refer to the framework and software kernel for supporting the above computational scenario as the multidisciplinary problem solving environment (MPSE) for simulating multidisciplinary applications [4,9,11].

We believe that the software architecture of an MPSE can be effectively supported by the collaborating PDE solvers simulation approach [1,3,14,24] and be realized by an agent-oriented [5] paradigm. MPSEs must exploit and build on the new technologies of computing. This paper describes the implementation of an MPSE framework based on a commercial agent-based system for the gas turbine engine design which has all the characteristics of a multidisciplinary application. The paper is organized as follows. Section 2 describes the gas turbine engine as a multi-physics application. Section 3 presents the design and application infrastructure of the agent-based MPSE framework. Section 4 describes the implementation of the GasTurbnLab MPSE using this framework while in Section 5, we present results obtained by GasTurbnLab for four engine configurations and discuss computational issues such as convergence and time stepping selection, and quality of simulations. We conclude our discussion in Section 6 with an analysis of the overall MPSE framework architecture and the major challenges in validating this architecture and its principle objectives through the implementation of GasTurbnLab.

2. The gas turbine engine multidisciplinary application

The gas turbine engine is an engineering triumph. It has more than 1300 parts with rotational speeds to 16,000 rpm for axial and 50,000 rpm for radial flow components. For aircraft applications, it operates with maneuver loads of up to 10 g, with flow path pressures and temperatures to 40 atmospheres and 1400 F. The important physical phenomena take place on scales from 10–1000 microns to meters. A complete and accurate simulation of an entire engine is enormously demanding; it is unlikely that the required computing power, simulation technology or software systems will be available in the next decade. In this paper, we consider simulating the compressor–combustor–turbine coupling in a gas turbine engine [3]. For this, we plan to design and implement an agent-based MPSE, referred to as GasTurbnLab, to study complex physical phenomena such as stall, surge and turbine blade fatigue. The software infrastructure supporting CFD and combustion simulations in the prototype GasTurbnLab MPSE consists of ALE3D and KIVA-3V legacy FORTRAN codes. The hardware infrastructure for these simulations and the implementation of GasTurbnLab MPSE is a grid 96 heterogeneous CPUs organized in three homogeneous Intel PC clusters running Solaris. For the gas turbine engine design simulations, we use a collaborating PDE solvers paradigm [2,8,10] implemented by the Grasshopper Agent Platform [7] which is mobile agent system interoperability facilities specification (MASIF). It can be easily extended to any grid environment.

3. Framework for an agent-based MPSE

In this section we describe the design of a general MPSE framework that can be used to simulate complex multi-physics phenomena governed by PDE models including gas turbine engine. A network of distributed machines is the hardware infrastructure. Since PDE simulations are often defined on geometric domains, natural or artificial geometric boundaries can be used to split the problem and the underlying simulation into many smaller sub-problems. Each sub-problem can then be solved independently, with mediator interactions along the boundaries for interface relaxation [16,18]. Thus, the MPSE framework for composite PDE simulations is based on domain decomposition with geometric objects, using a network of PDE solvers and interface mediators executing on a network of distributed machines.

3.1. Functional specifications of the MPSE

The design of the MPSE framework is driven by the underlying geometric modularity of the problem. The geometry is assumed to have a root node for the target physical object (domain), and the user is allowed to subdivide it in multiple ways. The resulting network of geometric objects defines a network of PDE models, where each object (subdomain) is modeled by a PDE. Each subdomain has some neighbors and possibly some fixed boundaries. If each neighboring connection is represented by an arrow, we get an abstraction of a network of PDE models. Since the PDEs

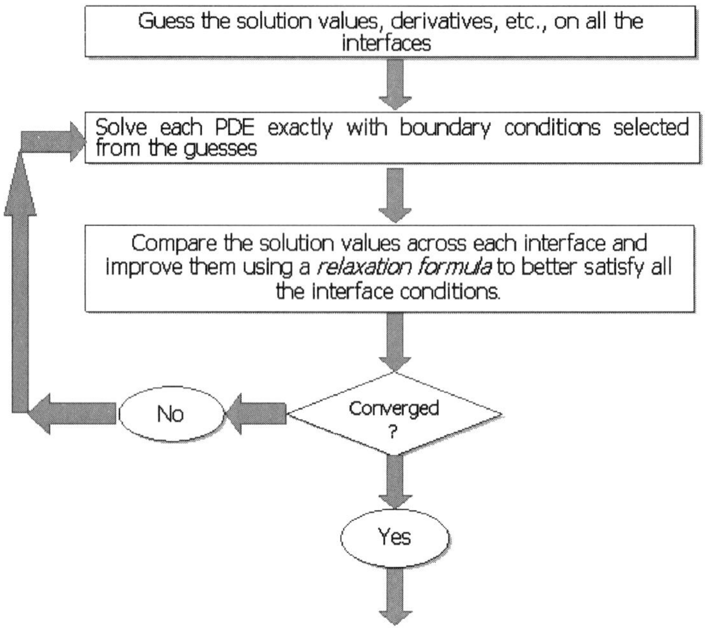

Fig. 1. The interface relaxation iteration.

on each subdomain are usually not the same, this represents a composite PDE problem. The MPSE framework maps the network of PDE models resulting from a user-specified partitioning onto a set of computational solvers running on a network of machines. This resource allocation will be done in an optimal manner to minimize the communication overheard between computational agents of neighboring subdomains. Under the assumption that any single PDE problem of the composite problem can be solved exactly, the interface relaxation technique will be used to solve the composite PDE problem. The interface relaxation algorithm is based on the iteration shown in Fig. 1. This MPSE supports user specification of a set of geometric objects that partition a composite physical domain, with a corresponding network of PDE solvers that collaborate via mediation to find a solution for the composite problem. The MPSE framework relies on an agent-based infrastructure. All system components, including the user interface, the legacy computational PDE solvers and the mediators, operate as agents. These agents are implemented using the Grasshopper Agent Platform [7]. A brief discussion of Grasshopper is given first, since a description of its functionality is essential for describing the MPSE agent-based framework.

3.2. The Grasshopper agent system

Grasshopper is an agent development platform that supports distributed agent-based applications. The platform provides a base for communication services, mobile computing power and dynamic information retrieval. It is essentially a mobile agent platform that is built on top of a distributed processing environment, integrating the traditional client-server paradigm and mobile agent technology. The primary feature of the Grasshopper platform is its location independent computing, driven

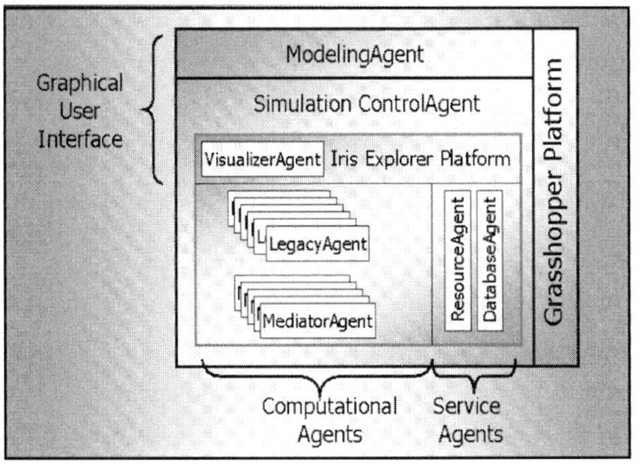

Fig. 2. The agent-based design of the MPSE. The SimulationControlAgent is the client which controls the entire simulation process. Legacy and mediator codes (generally Fortran or C programs) are transformed into server agents.

by the ability to move agents between different systems. It is a powerful environment that facilitates the creation of agents, transparently locating them and controlling their execution. These agent-based applications are interoperable with other agent systems that are MASIF compliant. The Grasshopper distributed agent environment is composed of region, agencies and agents. At the top of the hierarchy is the region that manages the distributed components in the Grasshopper environment. Agencies and agents are associated with a particular region. Each region has a registry that maintains information about all components associated with it. An agency is the runtime environment for mobile and stationary agents, providing the functionality to support agent execution. The agency is responsible for a number of services, including: (1) communication services for all remote interactions that take place between Grasshopper components, their movements and transport, (2) registration service to track all currently hosted agents, (3) management services that monitor external control of agents, (4) transport services for migration of agents from one agency to another, (5) security services for protecting remote interactions and agency resources and (6) persistence services for enabling the storage of agents for possible recovery. Agents are computer programs characterized by a set of attributes. They can be mobile or stationary. Mobile agents move from one location to another within a region to take advantage of local interactions, and thus are capable of reducing network load by migrating. Stationary agents are associated with a particular location only, and are incapable of migration.

3.3. The MPSE agent-based design

The Grasshopper environment supports the traditional client-server structure. MPSE clients and servers are Grasshopper agents which provide modeling (specification of the physical object to be simulated), simulation control and computational service. The MPSE framework consists of 6 classes of agents (Fig. 2): the ModelingAgent, the SimulationControlAgent (SCA), the VisualizerAgent (VA), the LegacyCodeAgents (LCA), the MediatorAgents (MA) and the ServiceAgents. The ModelingAgent and SCA are client agents; the remaining classes of agents are servers. The

ModelingAgent, SCA and VA have graphical user interfaces which support user interaction. The ServiceAgents provide services exclusively for the SCA, such as resource identification and database access for agent information retrieval. The ModelingAgent is the top level user interface. It allows users to define the target simulation object (domain) and its geometric decomposition into subdomains. Users can then select the configuration of subdomains to be simulated. These subdomains are passed to grid or meshing tools (such as TrueGrid [20]), which are incorporated into the ModelingAgent as self-contained systems. The meshing tools are used to generate the grid and define the boundary conditions for the user-selected subdomains. The ModelingAgent assists users to identify the interfaces between the domains, to define data exchange information for each interface, and to select solvers for each subdomain. This information becomes the startup data for the SCA.

The SCA requests and manages the services of all server agents. It controls the entire simulation process: launching/terminating the computational servers, managing their interactions, and facilitating the cross-network asynchronous communication of computational and control data. The VA is a server agent that receives solution data and renders it graphically using the IRIS Explorer data visualization system [13]. The LCAs are the computational agents that simulate the subdomains of the decomposed physical object; each LCA encapsulates an established legacy code targeted for a specific physical sub-object. The MPSE iteration algorithm requires an LCA to communicate its boundary data (via the SCA) to one or more MAs. The MAs encapsulate the mediation codes that are responsible for adjusting and resolving the interfaces (represented by the boundary data) between neighboring subdomains, each of which is simulated by one LCA. The SCA can handle any number of subdomains; i.e., it can control any number and type of communicating legacy code and mediator agents. In the following sections, we describe a template procedure for creating agents from legacy and mediator codes, and then discuss the generic operation of the SCA.

3.4. Implementation of a LegacyCodeAgent

The initial challenge was to create a template procedure for embedding legacy Fortran code into the server agent structure. The resulting LCA could then exercise control over the Fortran code and enable the data flow by (1) starting up the legacy code, (2) pausing after each Fortran iteration to communicate the required boundary data to the SCA, (3) receiving the mediated boundary data in return, (4) continuing with the next iteration, and (5) signaling to the Fortran code that the process should terminate so that the final solution output files can be generated. The following template procedure was developed to transform legacy code, which generally starts out as a Fortran executable, into a C-wrapped legacy code library as follows (Fig. 3):

- Change the Fortran "main" to a subroutine. Command line arguments are passed as parameters from the C-wrapper,
- Start the Fortran subroutine as a thread from the C-wrapper routine,
- Define C structures to hold the data representing boundary information,
- Write a C data transfer routine to copy boundary data back and forth between the C and Fortran data structures,
- Insert a call to the C data transfer routine in the Fortran iteration loop, and
- Insert control code in the C-wrapper to sleep/wake the C data transfer routine, thus effectively controlling the pause and restart of the Fortran iterations.

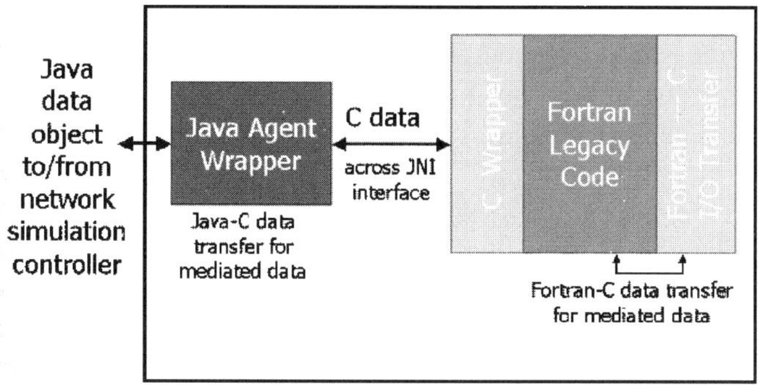

Fig. 3. Generic template procedure for embedding legacy Fortran code into a Java Grasshopper agent. This procedure transforms a Fortran executable into an LCA legacy library, which can be invoked and controlled by the LCA.

The C code for copying the data (both in the wrapper for the conversion from Java object to C, and in the data transfer routine for conversion from C to Fortran) requires a number of source code lines corresponding to the amount of data to copy. Less than 20 lines of C code are required for starting the thread and controlling the start/stop of the Fortran iterations. The changes to the Fortran code are minimal. The C-wrapper is defined with a JNI (Java Native Interface) interface to the Java agent code, and the C boundary data is passed up through the JNI interface as function parameters to the Java agent. The modified Fortran source code and the two new C routines (the wrapper and the data transfer routine) are compiled together as a legacy library, which is loaded into the Java LCA server when it is instantiated. The Java legacy agent starts the C-wrapper and waits for the JNI object containing the boundary data. When the data object is received, the agent serializes it and communicates it to the SCA. The SCA passes the object to the mediator, and returns the mediated data object to the agent, which copies it into a JNI object and passes it down to the C-wrapper. The SCA is also responsible for passing the LCA the number of iterations when it is instantiated, and the LCA passes this information to the legacy code so that the iteration process can terminate properly.

3.5. Implementation of a MediatorAgent

Mediators are customized Java or C programs that are easily encapsulated as Java agents. The mediation code is wrapped by agent code that handles the boundary data communication between the MediatorAgent (MA) and the SCA. In general, the mediator code is heavily dependent on the legacy codes and the data types that are exchanged between the two legacy codes. Each shared boundary (interface) between two subdomains requires an MA to mediate the boundary data. This mediation code must take into account the computational requirements of both solvers as well as the geometry (grid) of the domains on both sides of the boundary. The mediator must interpolate the grid points for the data exchange, since the grid points on the interface boundary for the two domains are not the same.

3.6. The SCA and simulation control

The SCA is responsible for managing the entire simulation in accordance with the specifications provided by the user through the ModelingAgent. Users select the subdomain (physical parts) to be simulated, generate geometry and parameter files for these parts, identify the length of the simulation process (in time units), and specify the output requirements. The SCA determines the software and hardware components to be used in simulating the specified configuration, that is, the SCA selects the LCAs and MAs to be instantiated and chooses the machine resources on which they will run. The SCA also translates the simulation time specification into iteration specifications for each LCA, computing an iteration correspondence factor to ensure that the simulation time when boundary information is exchanged will match for solvers on both sides of the interface.

Since the SCA is the intermediary for data exchange, it must contain code that is customized for the data structures involved for a particular MPSE. The data exchange routines, however, are the only part of the SCA which is MPSE specific. The remaining code is the standard asynchronous client agent sample code provided with the Grasshopper system. The SCA client agent invokes the LCAs and MAs, receives data from the LCAs, coordinates exchanges for each interface (between two LCAs and their corresponding MA), returns mediated data to the LCAs, and sends data to the VA when requested to do so. The entire simulation process begins and ends with the SCA.

4. GasTurbnLab: an agent-based MPSE for gas turbine engine design

We now present the implementation of GasTurbnLab assuming the framework described in the previous section and a simplified view of the turbine engine as it is illustrated in Fig. 4. We identify two existing legacy code PDE solvers to be encapsulated as LCA, and we write the customized interface mediation code for two MA to support the exchange of interface boundary data between the LCAs.

4.1. The GasTurbnLab LegacyCodeAgents

Two legacy codes have been encapsulated as LCAs: ALE3D, an advanced CFD code for simulating turbines, and KIVA, an advanced combustion simulation code. The ALE3D solver models the stator

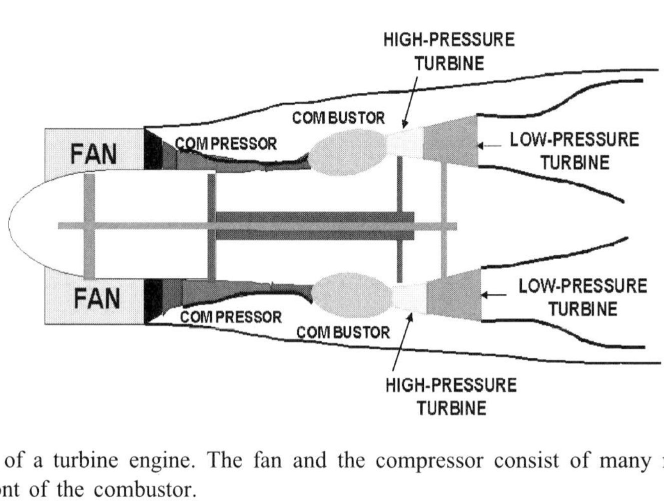

Fig. 4. A simplified view of a turbine engine. The fan and the compressor consist of many rows of stators and rotors, ending with a stator in front of the combustor.

and rotor parts of the engine compressor, and the KIVA solver modules the engine combustor. The two agents created by encapsulating these solvers, the AleAgent and the KivaAgent, have been used successfully to simulate several engine configurations for stator rows, rotor rows and combustors. The solutions generated by the simulations have been validated, and the results are described in Section 5. Simulations for more complicated configurations that answer questions about stall, surge and fatigue are also described; these simulations are currently underway.

4.2. The GasTurbnLab MediatorAgents

Mediation code is MPSE specific code that handles the exchange of boundary data on the interface between two domains. Two MAs, a basic mediator (BasicMA) and an advanced mediator (AdvMA), were developed for mediating the boundary data for the GasTurbnLab domain interfaces. Both MAs are capable of mediating inflow or outflow interfaces for any combination of ALE3D and KIVA engine parts (domains). Either mediator can be used to obtain accurate solution results for problems with no unsteadiness (i.e., steady-state problems). The difference between the mediator codes is in the selection of boundary data used to mediate the interfaces; for unsteady problems, the AdvMA provides a more accurate solution than the BasicMA. The BasicMA was created as a simple solution to the boundary interface problem. Based on this experience, the AdvMA was created to provide a more accurate transfer of unsteady fluid mechanic phenomena across the interface.

The BasicMA treats the boundary between domains in the same way inflow/outflow boundaries are treated when open to the atmosphere, i.e., when either inflow or outflow boundary is open to the atmosphere, while the atmospheric conditions are specified at the beginning of the simulation. The BasicMA simply updates these conditions after each time step based on the aerodynamic data from the adjacent domain. Except for the data being passed to them from the adjacent domain, the inflow/outflow boundary algorithms are the same whether the boundary is connected to the open atmosphere or to another ALE3D or KIVA domain. The drawback of this technique is that the inflow/outflow boundary condition algorithms are inherently less accurate than the interior algorithms. This mediator, however, has been shown to produce acceptable results on ALE3D-KIVA interfaces.

The AdvMA calculates elemental and nodal quantities at the interface between domains in nearly the same way as the interior elemental and nodal quantities are calculated. When the AdvMA is used, the inflow/outflow algorithms are bypassed and more accurate algorithms are employed. These algorithms are nearly identical to the interior algorithms. Specifically, there are two areas in which the AdvMA provides more accurate values at the interface between the two domains. First, ALE3D and KIVA codes both use a second-order upwinding scheme for the calculation of density and energy fluxes of interior faces during their advection steps. The AdvMA maintains second-order upwinding at the interface between the legacy codes while the BasicMA does not. Second, the AdvMA more accurately handles the exit boundary of a domain during the Lagrange calculation when it is connected to another domain. Unlike the BasicMA, the AdvMA provides forces at the exit so that the acceleration of exit nodes are computed the same way as interior node accelerations.

4.3. The GasTurbnLab SCA

When a user invokes the GasTurbnLab SCA from the Grasshopper graphical interface, the modeling specifications from the ModelingAgent are passed as input. The specifications describe the configuration of combustors, stator rows and rotor rows, listing the interfaces between these engine parts which are to be mediated. In addition, the geometry and solver parameters for each engine part, as well as the total simulation time are identified. Each engine part is simulated by one LCA, and each interface between two engine parts is mediated by one MA. After verifying that the required number of LCA and MA servers are available, the SCA determines which servers (i.e., which agents on which machines) are to be instantiated.

The SCA creates a KivaAgent proxy for each combustor, an AleAgent proxy for each stator and an AleAgent proxy for each rotor. The SCA then determines which MAs are required to mediate the interface boundaries identified by the user. The BasicMA mediates any interface involving a KivaAgent; the AdvMA mediates interfaces where two AleAgents exchange boundary data. The required number of AdvMAs and BasicMAs proxies are created, one per identified interface. The geometry and parameter files for the solver startups are passed to the LCAs via mobile FileAgents. The SCA uses the specified simulation time in ms and the time step information from all solver parameter files to compute the number of iterations required for each LCA. Since the KivaAgent time steps are generally much larger than the AleAgent time steps, the SCA determines a coordinating factor that allows the ALE3D iterations and the KIVA iterations to exchange data when the simulation time values match. For a given ALE3D–KIVA interface, the AleAgent may direct the ALE3D legacy code to iterate 10, 20 or even 60 times before sending interface data, while the KivaAgent directs the KIVA code to iterate once. The simulation time at the exchange is then equal for each code. The SCA is responsible for computing and communicating these factors to all LCAs so that the exchanges occur properly as the simulation time advances.

The SCA continues to facilitate the data exchange during the entire simulation process, receiving data from the LCAs and MAs, and sending them to their proper destinations. When the final simulation time step is finished, each LCA terminates its legacy code (ensuring that the proper exit processing occurs) and sends a termination signal to the SCA. The SCA shuts down the entire simulation when all LCAs have terminated. It then allows the user to access the VA for data visualization.

5. GasTurbnLab: simulation results

In this section, we present some numerical results to indicate (a) the user specified parameters involved, (b) the convergence criteria of the *interface relaxation method* used, and (c) the quality of results obtained. The results are based on four configurations of engine parts consisting of stators, rotors and combustors that have been simulated. Results on more complex turbine engine configurations and the performance evaluation of the GasTurbnLab MPSE will be presented elsewhere. The engine configurations considered here explore the result of connecting the inflow or outflow domain boundary of one engine part to the boundary of another engine part for data exchange via mediation. The mediation includes resolution of nonmatching discrete interfaces using interpolation and implementation of the relaxation method specified in Section 3. The time stepping of the

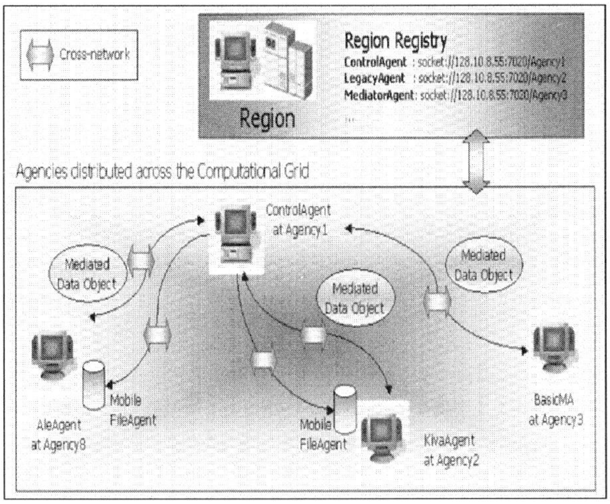

Fig. 5. Sample simulation scenario from the GasTurbnLab MPSE. The scenario shows a stator–combustor two domain simulation, mediated at the outflow boundary of the stator and the inflow boundary of the combustor by the BasicMA mediator agent. They communicate boundary interface information using the SCA as an intermediary. Users interact primarily with the Modeling and VisualizerAgents. All agents are built and executed within the Grasshopper environment.

ALE3D- and KIVA-based agents must be specified by the user so that these two codes communicate at the same simulation time. The nature of the computations involved in these two codes requires multiple steps of ALE3D for a single KIVA step. The quality of convergence for each engine configuration is assessed by criteria associated with each legacy code. In the case of ALE3D, we monitor the absolute change of the mass flow rate averaged over a period. For the KIVA code, we monitor the relative change of velocity, pressure, kinetic energy, dissipation rate, and chemical species. In all computations presented here, each agent code was executed on a single CPU. It's worth noticing that GasTurbnLab can run on a network of heterogeneous processors that support the Grasshopper middleware.

The first configuration (case 1) connects an upstream stator (AleAgent) to a downstream combustor (KivaAgent), using the BasicMA to mediate the data along the interface for stator outflow and combustor inflow (see the agent diagram in Fig. 5). This simulation was run for several variations of geometry, initial pressure conditions and number of iterations. Results of the stator–combustor simulations are shown in Figs. 6 and 7. Fig. 6 shows the mass flow rate at the inflow to the ALE3D domain. This mass flow rate has reached a steady-state value, indicating a converged aerodynamic solution. Fig. 7 shows the temperature profile within the combustor, and reveals the presence of a flame.

The second configuration (case 2) connects an upstream stator to a downstream rotor and is simulated using two AleAgents. The interface boundary was mediated by the AdvMA, and involves the exchange of force, flux and velocity data, with three mediator exchanges per time step. This simulation ran for 16,000 iterations with a time step of 0.5 μs. The resulting mass flow rate, Fig. 8, has a periodic character due to the relative motion of the rotor and stator for this highly loaded configuration. The measure of convergence in this case is the mass flow rate averaged over a period of unsteadiness. The time-averaged mass flow rate changes by less one-tenth of a percent for the

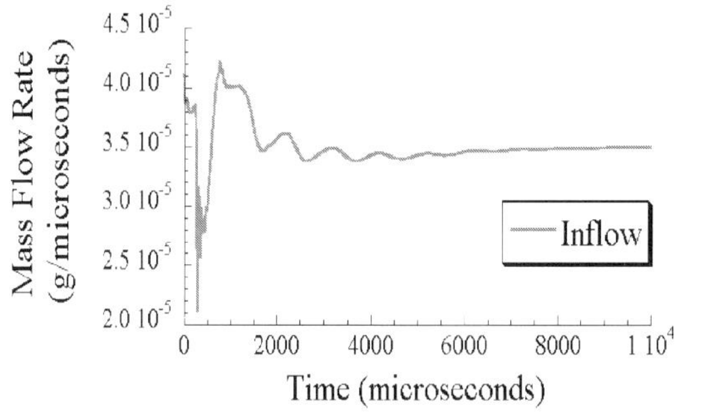

Fig. 6. The mass flow rate at the stator inflow moves towards a steady-state value after 1000 KIVA time steps (10,000 corresponding ALE3D time steps) in the stator–combustor simulation (case 1).

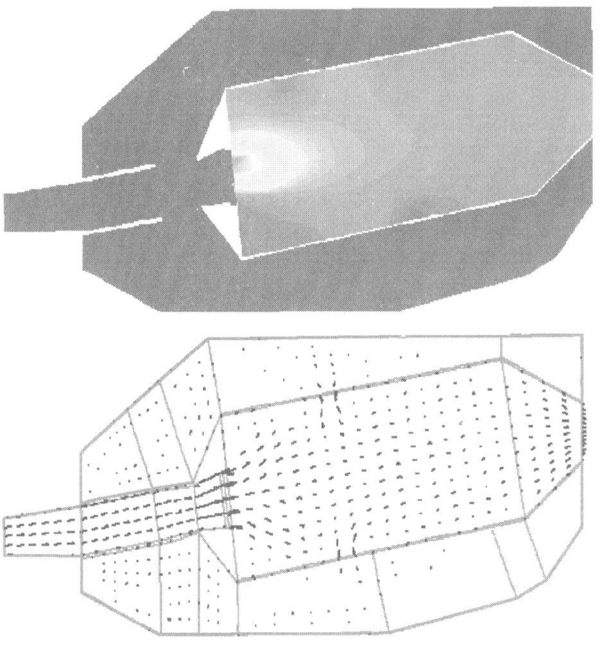

Fig. 7. Combustor plots which show the temperature contour indicating an ignition process in the beginning of combustion (top) and the velocity vector field (bottom). The results are from a case 1.

last two cycles, indicating a converged solution. The final mass flow was 5.69e-5 g/μs at stator inflow and 5.71e-5 g/μs at the rotor outflow. The total pressure ratio was 1.587. The axial velocity is depicted by the color contour plot in Fig. 10. It shows a shock emanating from the rotor and impacting the upstream stator. The presence of the shock indicates the flow field is transonic in the rotor frame of reference.

The engine configuration of case 3 is depicted in Fig. 9. It contains three AleAgents, simulating the stator, rotor, and stator passages upstream of the combustor, and one KivaAgent simulating the

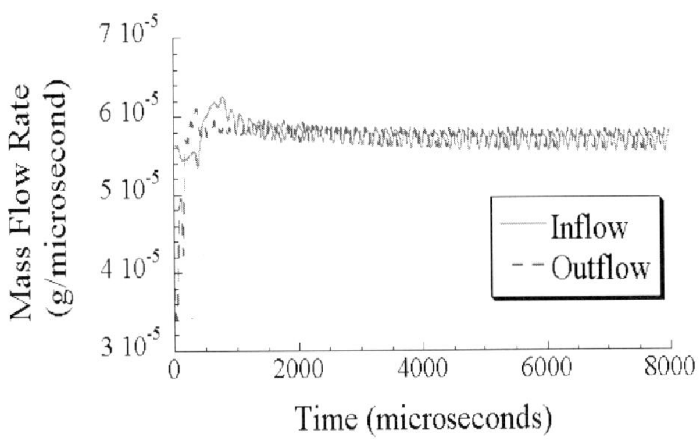

Fig. 8. Mass flow rate time history for the inflow of the stator and outflow of the rotor for case 2.

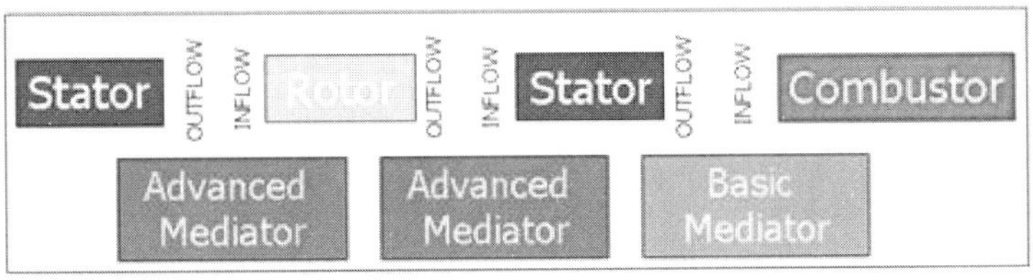

Fig. 9. A block diagram of case 3 engine configuration.

Fig. 10. Axial velocity plot for case 2. Maximum velocity (red) is 180 m/s; minimum velocity (blue) is 0 m/s.

combustor flow field. The interface boundary between the first stator (also called the IGV) and the rotor and between the rotor and the downstream stator are mediated by the AdvMA. The interface boundary between the downstream stator and combustor is mediated by the BasicMA. This case serves as the first test case for a rotating ALE3D domain upstream of a stator, but also as the first case using the two types of mediator in one simulation. Fig. 11 shows the mass flow rate time history for the inflow of the first ALE3D domain and the outflow of the third ALE3D domain. After

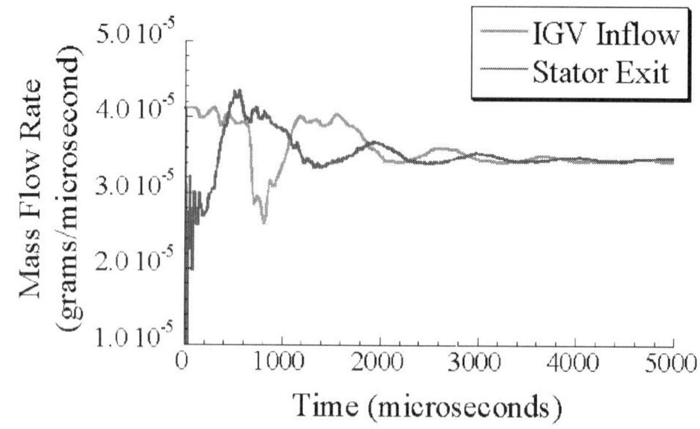

Fig. 11. Mass flow rate time history for the inflow of the IGV and outflow of the stator for case 3.

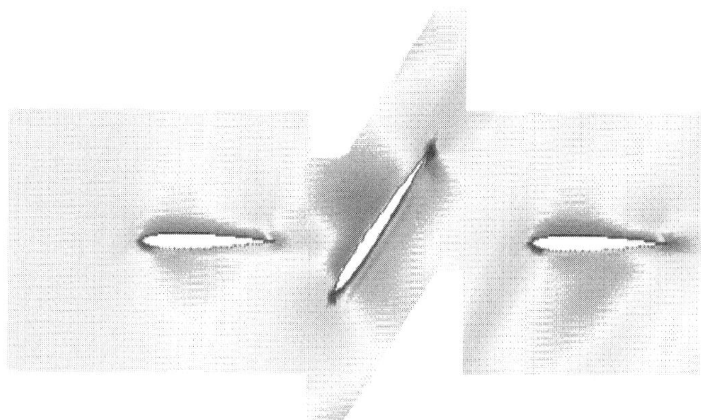

Fig. 12. Axial velocity plot for case 3. Maximum velocity (red) is 85 m/s; minimum velocity (blue) is 30 m/s.

5000 μs, the mass flow rate has reached a nearly constant value, indicating sufficient convergence. The final mass flow rate is 3.42e-5 g/μs. This case does not display a periodic unsteadiness in the mass flow rate because the blades were designed to have zero loading. Fig. 12 shows the axial velocity field for all three ALE3D domains. As evident in the figure, the airfoils are zero-camber airfoils having a NACA0008 profile. The flow conditions were specified such that each airfoil has zero incidence angle, and thus experiences zero loading. The rotor is at 45° stagger and is rotating at 5511 rpm. No shock waves are evident in this figure since the entire flow field is subsonic. As expected with the zero loading design, the total pressure ratio across the three ALE3D domains is 1.00. There appear to be wakes emanating from the aft portion of the airfoils. However, these are not true viscous wakes because viscous effects are not modeled in the ALE3D domains. These apparent wakes are due to small numerical losses from the airfoil surface boundary conditions being propagated downstream as velocity deficits. Fig. 13 shows the velocity vectors for the combustor flow field. Due to low activation energy and low stagnation pressure in the combustor, the flame did not start for this simulation.

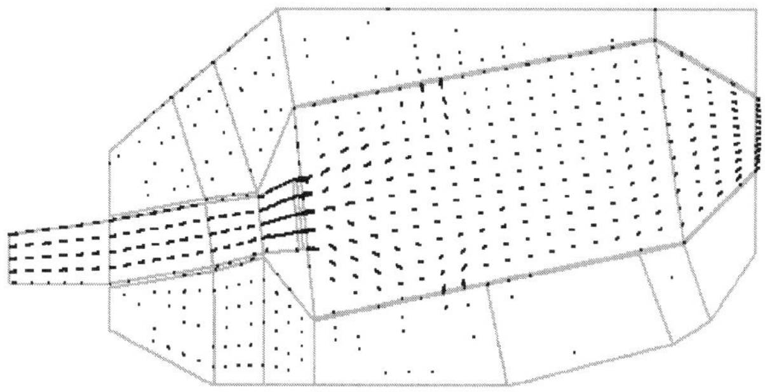

Fig. 13. Velocity vector plot for combustor flow field for case 3.

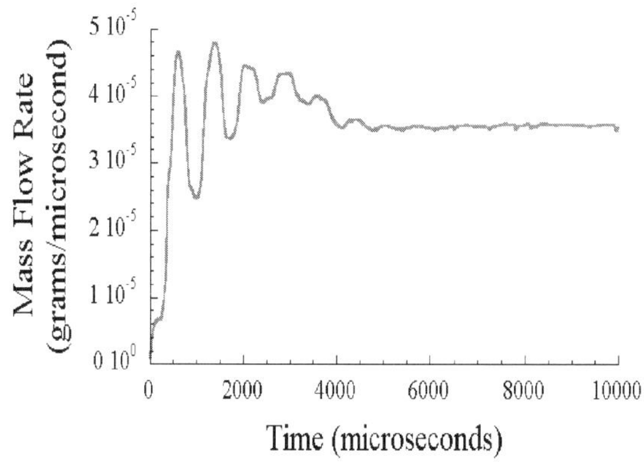

Fig. 14. Mass flow rate at outflow of ALE3D domain for case 4.

Case 4 has one ALE3D domain and one KIVA domain. The ALE3D domain is a stator downstream of the combustor. This is the first test case of the BasicMA with the ALE3D domain downstream. The ALE3D stator profile is again a zero-camber NACA0008 blade designed to have zero loading. Fig. 14 shows that after 10 ms of simulation the mass flow rate has reached a constant value of 4.00e-5 g/μs. Fig. 15 shows the axial velocity color contour plot of the ALE3D domain. Fig. 16 shows the temperature contours for the combustor. Ignition has just begun in this simulation, so the flame has not yet reached its steady-state temperature.

These four test cases demonstrate that the various interface boundary conditions, including the two kinds of mediators, are working properly.

6. Conclusion

In summary, we have described an agent-based framework for building multidisciplinary problem solving environments for complex, large-scale simulations of physical artifacts consisting of numerous parts with physical and geometric interfaces. The design is based on a "network" of

Fig. 15. Axial velocity plot for case 4. Maximum velocity (red) is 114 m/s; minimum velocity (blue) is −40 m/s.

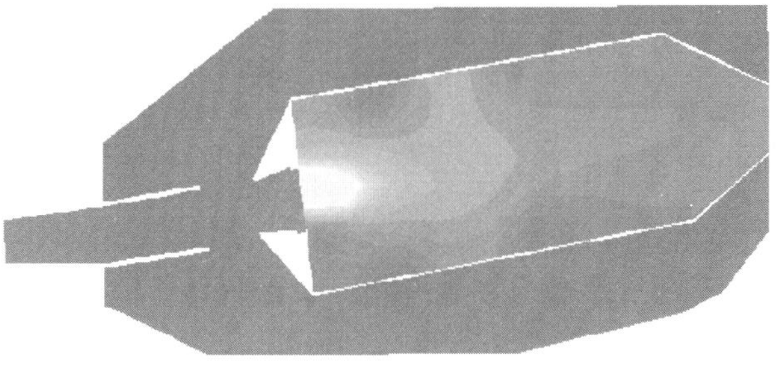

Fig. 16. Temperature contour plot, for case 4, showing ignition is just beginning.

collaborating numerical objects through the interfaces among the artifact parts. The MPSE framework uses the Grasshopper Agent Platform for its agent-based infrastructure, making heavy use of the flexible and robust agent creation and distributed execution features of the Grasshopper environment. The MPSE framework can be extended at the user interface level by adding additional modules that support user interaction for model specification or data visualization. Furthermore, the framework can be extended at the enabling services and computational levels by creating new agents which perform additional services or computations. To facilitate legacy code incorporation, we have proposed a simple and very general wrapper method for encapsulating Fortran or C code as an agent library. The GasTurbnLab MPSE is an implementation of the agent-based MPSE framework for the simulation of gas turbines. The large body of legacy code needed for this simulation can be easily incorporated within the MPSE framework using the technique outlined in Section 4. A suitable load balancing algorithm can be implemented within the SCA for better distributed performance of the highly compute intensive simulations utilizing the mobile functionality of the grasshopper agents. The ModelingAgent can be tailored appropriately with suitable problem specification modules that include tools such as TrueGrid and MeshTV. The GasTurbnLab MPSE implementation may contain a library of Explorer modules for such problem specification tools, including additional modules for different solution visualization tools. This would enable the scientist to customize GasTurbnLab with suitable pre- and post-processing modules for each target gas turbine simulation problem. The proposed MPSE framework architecture is scalable, enabling the implementation of very large scale, distributed problem solving environments for scientific simulations. It is also versatile and simple enough to be used to build prototype problem solving environments to analyze and validate mathematical techniques for interface relaxation. Thus, it is a useful environment that advances the state of the art in simulating complex physical phenomena.

References

[1] L. Boloni, D.C. Marinescu, J.R. Rice, P. Tsompanopoulou, E.A. Vavalis, Agent based scientific simulation and modeling, Concurrency: Practice Experience 12 (2000) 845–861.

[2] T.T. Drashansky, E.N. Houstis, N. Ramakrishnan, J.R. Rice, Networked agents for scientific computing, Assoc. Comput. Mach. 42 (3) (1999) 48–54.

[3] S. Fleeter, E. Houstis, J. Rice, C. Zhou, A. Catlin, GasTurbnLab: a problem solving environment for simulating gas turbines, Proceedings of the 16th IMACS World Congress, August 2000, No. 104-5, 5pp.

[4] E. Gallopoulos, E.N. Houstis, J.R. Rice, Computer as thinker/doer: problem solving environments for computational science, IEEE Comput. Sci. Eng. 1 (1994) 11–23.

[5] M.R. Genesereth, An agent-based framework for interoperability, in: J.M. Bradshaw (Ed.), Software Agents, MIT Press, Cambridge, MA, 1997, pp. 317–345.

[6] D.A. Gottfried, Simulation of fluid-structure interaction in turbomachines, Ph.D. Thesis, Mechanical Engineering Department of Purdue University, May 2000.

[7] The Grasshopper Agent Platform, IKV++ GmbH, Kurfurstendamm 173–174, D-10707 Berlin, Germany, http://www.ikv.de.

[8] E.N. Houstis, G. Balakrishnan, A. Catlin, N. Dhanjani, S. Lalis, M. Stamatogiannakis, J.R. Rice, C. Houstis, Network-based scientific computing, in: R.F. Boisvert, Ping Tak Peter Tang (Eds.), The Architecture of Scientific Software, Kluwer Academic Publishers, Norwell, MA, 2001, pp. 3–28.

[9] E.N. Houstis, A. Joshi, J.R. Rice, T. Drashansky, S. Weerawarana, Towards multidisciplinary problem solving environments, HPCU J 1 (1) (1997), http://www.hpcu.org/hpcunews/vol1/issue1.

[10] E.N. Houstis, J.R. Rice, Future problem solving environments for computational science, Math. Comput. Simulation 54 (4–5) (2000) 243–257.

[11] E.N. Houstis, J.R. Rice, S. Markus, S. Weerawarana, Network based scientific problem solving environments, HPCU J. 1 (1) (1997), http://www.hpcu.org/hpcunews/vol1/issue1.

[12] E.N. Houstis, J.R. Rice, N. Ramakrishnan, T. Drashansky, S. Weerawarana, A. Joshi, C.E. Houstis, Agent based systems to support multidisciplinary problem solving environments for computational science, in: M. Zelkowitz (Ed.), Advances in Computers, 1998.

[13] IRIS Explorer Toolkit, IRIS Explorer Center (North America), 1400 Opus Place, Suite 200, Downers Grove, IL 60515-5702, USA, http://www.nag.com/IEC.

[14] S. Markus, E.N. Houstis, A.C. Catlin, J.R. Rice, P. Tsompanopoulou, E. Vavalis, D. Gottfried, Ke Su, G. Balakrishnan, An agent-based netcentric framework for multidisciplinary problem solving environments (MPSE), Internat. J. Comput. Eng. Sci. 1 (2000) 33–60.

[15] H.S. McFaddin, J.R. Rice, Collaborating PDE solvers, Appl. Numer. Math. 10 (1992) 279–295.

[16] H.S. McFaddin, J.R. Rice, RELAX: a platform for software relaxation, in: E.N. Houstis, J.R. Rice, R. Vichnevetsky (Eds.) Expert Systems for Scientific Computing, North-Holland, Amsterdam, 1992, pp. 175–194.

[17] M. Mu, J.R. Rice, Modeling with collaborating PDE solvers—theory and practice, Contemp. Math. 180 (1994) 427–438.

[18] J.R. Rice, P. Tsompanopoulou, E.A. Vavalis, Interface relaxation methods for elliptic differential equations, Appl. Numer. Math. 32 (1999) 219–245.

[19] J.R. Rice, P. Tsompanopoulou, E.A. Vavalis, Fine tunning interface relaxation methods for elliptic differential equations, Appl. Numer. Math., to appear.

[20] TrueGrid, XYZ Scientific Applications Inc., USA, http://www.truegrid.com.

[21] P. Tsompanopoulou, Collaborative PDE solvers: theory and practice, Ph.D. Thesis, Mathematics Department, University of Crete, Greece, 2000.

[22] P. Tsompanopoulou, L. Boloni, D. Marinescu, J. Rice, The design of software agents for a network of PDE solvers, Proceedings of the Agents'99: Third International Conference on Autonomous Agents, ACM Press, New York, 1999.

[23] R.G. Voigt, Requirements for multidisciplinary design of aerospace vehicles on high performance computers, ICASE Report No. 89-70, September 1989, 9pp.

[24] P. Wu, E.N. Houstis, EPPOD: a problem solving environment for parallel electronic prototyping of physical object design, J. Parallel Distributed Comput. 42 (1997) 157–172.

ELSEVIER

Journal of Computational and Applied Mathematics 149 (2002) 101–118

JOURNAL OF COMPUTATIONAL AND APPLIED MATHEMATICS

www.elsevier.com/locate/cam

Meso-scale fusion: a method for molecular electronic state calculation in inhomogeneous materials

Shi-aki Hyodo

Frontier Research Group V, Toyota Central R&D Labs., Inc., Nagakute, Aichi 480-1192, Japan

Received 28 October 2001; received in revised form 28 January 2002

Abstract

In order to obtain a prediction about materials at the molecular level in a large-scale inhomogeneous environment, some discussions were made based on realizing an electronic state calculation under such a condition. The process of defining the electronic state in an inhomogeneous environment was discussed as the concept of "*meso-scale fusion*". This process is likely to *fuse* the information in the macroscopic condition and the electronic structure in inhomogeneous materials. A delicate problem which arises in the connection of a quantum system to a macroscopic subsystem was also stated. Some results of mesoscopic, microscopic, and coupled mesoscopic–microscopic calculations were represented. Most of the programs on those calculations were originally coded in our research group. Characteristics of the calculation method adopted here were discussed related to the requirement from the *meso-scale fusion*.

Keywords: Mesoscopic scale; Electronic state; Inhomogeneous materials

1. Introduction: importance of meso-scale inhomogeneity in realistic materials

Many concepts of new types of advanced materials have recently been proposed, i.e., nano-materials, porous/meso-porous materials, multilayered membranes, self-assembled materials, bio-mimetic materials, structure/morphology controlled materials, etc. Most of them are based on a structure describable within the scale of the assembly of molecules. Such a scale is typically 10–100 nm, possibly including 100 or more numbers in terms of small molecules. Because the characteristics of the above advanced materials strongly depend on the structure on such a scale,

E-mail address: e0668@mosk.tytlabs.co.jp (Shi-aki Hyodo).

PII: S0377-0427(02)00523-X

the medium focused on should not be considered a homogeneous one. When the lifetime or the degradation of real materials is considered, the inhomogeneity is an indispensable term also in the usually available materials. Consider destruction by metal fatigue as a typical example. In addition, biological cells should be very important examples for an inhomogeneous structure within the scale noted above. We call such a scale a "meso-scale". We should investigate the meso-scale inhomogeneity as an important factor in advanced and conventional real materials.

Recently, trends in research on theoretical computation at the molecular level seem to be changing. Molecular computation has thus far pursued the expansion of the computation scale by developing the theory and method and improving computer performance. Now there are signs that effective computational forecasting can be done by developing models with an awareness of the hierarchy in the nature. Substances which can be treated homogeneously on a macroscopic scale should be considered as groups of atoms or molecules on the levels of resolution in a high-performance electron microscope. Atoms and/or molecules construct the characteristic structure on the scale of the molecular assembly by themselves. Such a characteristic structure is smeared out to become homogeneous on the macroscopic scale of space. The problem is the composition of a material that appears different depending on the scale of the space under observation. In other words, new trends are aimed at creating an effective method of forecasting the physical properties of a material by combining systematic characteristic methods on several scales of space. Standard computation methods are being developed for several systems of materials [50,21].

In the present paper, a possible procedure to calculate the electronic state at a molecular level under inhomogeneous circumstances will be discussed. In Sections 2–4, the backgrounds of the present problem are introduced and discussed. In Section 2, the necessity for information on the electronic state for practical material design is stated. In Section 3, the basic problem in the transformation of the information between quantum and classical pictures is discussed, and the concept of "*meso-scale fusion*" is introduced. The origin and a possible treatment of a problem with the connection of quantum and classical subsystems are described in Section 4. An example of electronic state calculation under some simple environmental boundary conditions, using a conventional quantum chemical method, is introduced with several related techniques recently proposed. A suitable calculation technique for mesoscopic structure formation and that of electronic state calculation are discussed in Sections 6 and 7, respectively. Examples of the combination of mesoscopic and microscopic calculation, which are progressing in our group, are described in Section 8. In Section 9, the perspective of the present problem and conclusions are given.

2. Necessity for electronic state information for material design

Many simulation techniques have been advanced for macroscopic problems. The dynamics of materials, structural analysis, fluid dynamics, and so on are growing to supply indispensable tools for both academic and industrial fields. The system design based on the information from those numerical calculations is realized commonly now. Introducing the parameterized material constants into a numerical equation, the characteristics of such a system are possibly understood and then the system design can be realized. However, the information, what should be exactly synthesized, cannot be directly introduced from material constants. They can determine only the condition of the material. Synthesis chemists could not choose what kinds of materials are synthesized. Syntheses based on trial and error processes are frequently challenged, and the sufficiency for given conditions is checked in realistic systems, step by step. The name of elements or molecules should be explicitly proposed in this design process. For this purpose, the electronic structure is likely understood to be the most fundamental property of the noticed materials. Reactivity of the states under investigation should also be defined from the electronic state. If the interatomic potential parameters are well established for the noticed material, the thermochemical properties can be deduced by molecular dynamics or molecular mechanics simulations. Such a situation, however, cannot usually be expected. Especially for unknown materials, the parameter-reproducing process should inevitably be analyzed. The information on the electronic state is strongly needed for completeness, general applicability, and the requirement for fundamental knowledge on a realistic material design.

3. Transformation of the information between quantum and classical pictures

On a macroscopic scale that is sufficiently large compared with the characteristic spatial frequency of the inhomogeneity, the inhomogeneous structure is painted over the slowly varying one. A focused material can be considered as nearly homogeneous on such a scale. However, the microscopic condition on the scale of a small molecule is still in the inhomogeneous environment. A part of the material is influenced by the structural inhomogeneity varying with the characteristic spatial frequency. The restriction condition should be defined by the macroscopic boundary condition of the noticed material. The electronic state on a microscopic scale, on the other hand, should be defined within the inhomogeneous environment. Therefore, the restriction condition and the electronic state should be defined individually on each suitable scale, and the information for the definition on each scale should be transformed between them. In order to realistically perform such a process, we define the intermediate scale between macroscopic and microscopic ones. This scale, which we call "meso-scale", is used to characterize the inhomogeneity of the structural environment at the molecular assembly level. We would like to call such a process to define the electronic state in an inhomogeneous environment "*meso-scale fusion*". The process is likely to *fuse* the information in the macroscopic condition and the electronic structure in inhomogeneous materials. Such a process of definition has already been realized by the multigrid technique in numerical calculations for many kinds of differential equations with finite-size grids [6]. Although the numerical calculation technique is already guided by the multigrid scheme, a serious problem still remains with the cooperation of macroscopic and microscopic definitions. This problem usually appears in connecting quantum and classical subsystems. In the next section, the origin and a possible treatment of this problem will be discussed in the relation to the description of quantum composite systems.

4. Problem with connection of quantum and classical subsystems

In order to realize the process of "*meso-scale fusion*", we have to consider the connection of quantum and classical subsystems within a meso-scale region. A quantum system is usually described by wave function, and then it shows the so-called "long-range correlation" originating from the coherent character of wave function. A classical system, on the other hand, can be described without the information of wave function, and it shows no coherency in a quantum mechanics sense. We must connect those two types of subsystems consistently. It is well known that the observation of a quantum mechanical system with a macroscopic device shows the problem of measurement for a quantum mechanical system, which is a long-standing one in quantum theory [40]. This problem introduces some types of a famous paradox in the fundamental theory of quantum mechanics [10,38]. The same problem also appears in the quantum composite system with a macroscopically large scale [26]. The appearance of this serious and complicated problem originates from the connection of quantum and classical systems. Therefore, our "*meso-scale fusion*" may possibly be serious and complicated.

Recently, it was reported that classically definable characteristics of a macroscopic quantum mechanical subsystem can be safely (without any paradox) reproduced by averaging the states on a corresponding subsystem over the generalized density matrix or the statistical operator [26]. The quantum mechanical coherency is removed by this averaging process. In other words, the averaging in a density matrix representation can reproduce a classically definable decoherent state. The condition to satisfy decoherency is that the subsystem contains statistically large degrees of freedom. Although the density matrix representation is familiar in quantum chemistry [8], less consideration has been given in relation to the description of quantum mechanical composite systems.

When we divide a noticed meso-scale system into an objective microscopic subsystem and an environmental subsystem, the same procedure in the problem of a quantum composite system can be applied here. Although explicit formulation for averaged states of a meso-scale subsystem over a density matrix representation has not been derived yet, it can be correctly expected that the subsystem with averaged states is safely reproducible by means of classical expressions. Therefore, it is expected that the structure on a meso-scale can also be defined by classical descriptions, which is suitable for noticed phenomena and is independent of a quantum mechanical description. The electronic structure is transformed into the electron distribution describable by the density matrix. Once the electrostatic and/or the electrodynamic properties on the corresponding subsystem are defined within the mesoscopic environment structure, the electrostatic potential reproduced by them can be added to the Hamiltonian of the focused microscopic subsystem. The long-range property of electrostatic potential demands that we explicitly consider the effects of environmental mesoscopic structure. The "*meso-scale fusion*" can be implemented, therefore, by using averaged properties over a density matrix representation on a meso-scale structural calculation.

In the next section, the effectiveness of averaging the results of quantum mechanical calculation in a density matrix representation is tested as an example. In this case, atomic charges, as a kind of averaged quantity in a density matrix representation, are deduced from quantum chemical calculation, and the electrostatic potential of a mesoscopic structure with those atomic charges is considered. In other parts of the following sections, we presuppose that the electrostatic potential reproduced by a mesoscopic environment structure is added to the Hamiltonian of the focused microscopic subsystem in the forthcoming quantum mechanical calculations.

5. Example of electronic state calculation under environmental boundary condition (with conventional molecular orbital methods)

There are many types of theoretical models for calculating the electronic state in an arbitrary extended medium with large degrees of freedom. The recently developed self-consistent reaction field

(SCRF) [29] approach, the quantum mechanics/molecular mechanics (QM/MM) [45] hybrid method, the reference-interaction-site-model self-consistent field (RISM-SCF) [42] approach, the dipped ad-cluster model (DAM) [31], etc., are notable typical examples. Among those, the image force correction for the adsorption energy of a molecule adsorbed on a metal surface is briefly discussed here.

The electronic state calculation of an adsorbed molecule on a material surface usually uses the cluster model and/or the slab model for the structure of the molecular-adsorbed surface. The cluster model assumes that an isolated assembled system can describe a part of the adsorbed surface structures. The slab model, on the other hand, adopts the two-dimensional periodic boundary condition to the adsorbed surface structures. The amount of surface atoms accounted for in those models is not large because of the limitation of computational resources. The equilibration between the electrons in the bulk metal material and that in the adsorbed molecule cannot be accounted for correctly in those models. The electrostatic boundary conditions, which are macroscopically definable, also cannot be accounted for in those models. Nakatuji has successfully defined a scheme to satisfy the electron equilibration and the electrostatic boundary conditions [31]. Nakatuji and coworkers applied the model, which they called the DAM, based on this definition to many kinds of surface adsorption and successfully described interesting surface phenomena [32]. Hyodo extended the DAM for a shaped metal surface like a spherical conductor [19]. Under the spherical electrostatic boundary condition, inductive image charges reproduce electrostatic potentials as illustrated in Fig. 1. Reproduced electrostatic potentials can be introduced into the Hamiltonian for describing the electronic state of the adsorbed molecule. An example of a CO molecule adsorption on a Pt metal spherical conductor is illustrated in Fig. 2. Based on the resultant molecular orbital of the Pt–C–O molecule, the net atomic charges were calculated from the density matrix on atomic-orbital-like basis-function space of the conventional molecular orbital calculation method. In this procedure, the electron distribution is considered to be averaged into a classically definable point charge at each position of the nucleus. The net atomic charge can be treated as a classical point charge at the position of each nucleus, and the electrostatic potential around a spherical conductor surface can be reproduced to satisfy the present boundary condition. On comparing experimental results [30] for CO adsorbed on a Rh particle, supported on some substrate, the particle radius dependence of the CO stretching vibration frequency is understood to be reasonably predicted. The electronic state calculation procedure adopted here is briefly described in Appendix A.

The electrostatic potential, which satisfies the boundary conditions determined by the macro- or mesoscopic structure in the noticed system, is added to the corresponding Hamiltonian as an external potential term in those calculations. A similar treatment of electrostatic potential is also found in the electronic state calculation of a molecule in solution with dielectric continuum media (SCRF) [29]. In this theoretical model, the dipole moment in a cavity, which is apparently defined in the dielectric continuum medium, is calculated by the conventional molecular orbital method, and the induced electrostatic field in dielectrics and the reactively induced field in the cavity are accounted. This reaction field is added into the Hamiltonian for the conventional molecular orbital calculation as an external potential, and then the molecular structure and the dipole moment are calculated iteratively to find the energy minimum. Those models with the electrostatic potential of the environmental structure have been successfully applied to express the characteristic experimental data for many corresponding systems. Those cases, however, are not as complicated as the meso-scale structural environment. Those have a simple boundary, and most of the region to be considered can be treated as

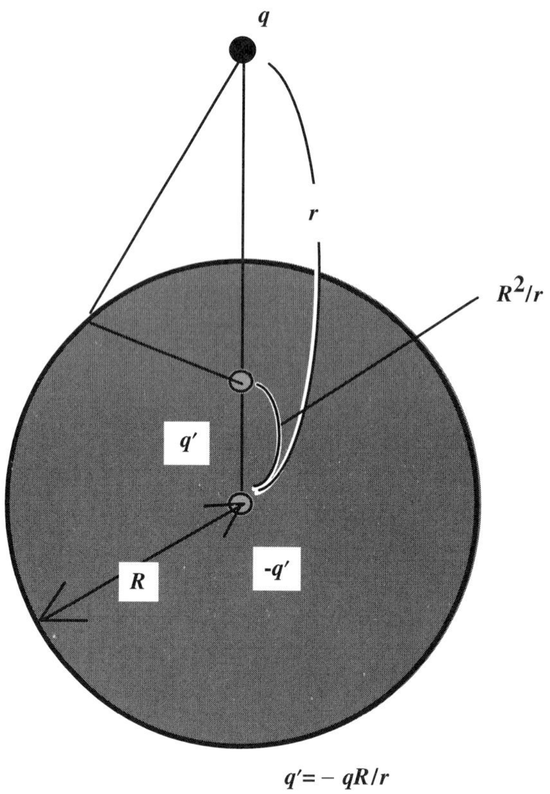

Fig. 1. Schematic illustration of the image charge (q') with its position (R^2/r) originating from a charge q at position r, for which the electrostatic potential is produced outside the spherical conductor with radius R. The system is not grounded.

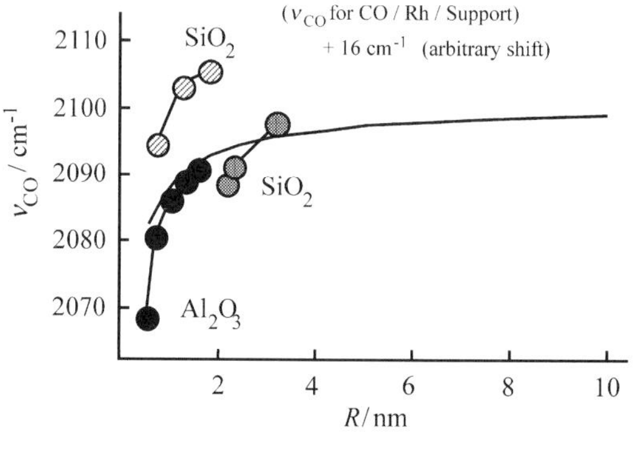

Fig. 2. Particle radius (R) dependence of CO stretching vibration frequency (ν_{CO}) for a CO molecule adsorbed on a spherical metal Pt particle. The circles indicated here are experimental ν_{CO} for a CO molecule adsorbed on a metal Rh particle supported on some various oxides [30] (applied 16 cm^{-1} arbitrary shift).

a homogeneous medium. For general purposes, the process to explicitly define the more complicated mesoscopic structures is needed to be introduced.

6. Simulation of mesoscopic structure formation

In the fluid mechanical analysis of many component systems, for example, the calculation process becomes more complicated as realizing the model for analysis. We need the calculation process that is premised on the analysis of many component systems. One opinion for developing such a process is to employ the time-dependent Ginzburg–Landau theory or the dynamic density functional theory [1,12,44,37], in which the Smoluchowski equation for the density distribution is solved on a lattice as a function of time, by relating the evolution of the local order parameter or densities to variations in the free energy via the proper Onsager kinetic coefficients [22].

As an alternative, to simulate fluid systems in an off-lattice representation, a relatively new method known as dissipative particle dynamics (DPD) [18,14] can be applied here. In this technique, the fluid system is described by particles that act as centers of mass, where each particle represents several numbers of atoms. They repel each particle other with a soft repulsive interaction force, but on top of that, extra noise and friction are added. If this is done in a well-specified way, the time evolution yields the correct equilibrium statistical mechanics in the (N, V, T) ensemble and all hydrodynamic interactions [11,15,27]. The DPD can be naturally applicable to the analysis of many component systems based on its foundation. We applied the DPD to the simulations for the mesoscopic structure formation of polymer-blend and polyelectrolyte aqueous composite systems [47]. The simulation of vesicle formation is also applied as a problem, which has not been simulated explicitly with the technique of molecular dynamics level calculations [49]. Fig. 3 shows the meso-scale structure of a hydrated Nafion membrane reproduced by our DPD simulation code. The interaction parameters of each DPD particle were constructed by interatomic interaction parameters based on the explicit molecular structure of a Nafion oligomer. Reproduced meso-scale structures, like that in Fig. 3, confirmed the results of small angle X-ray diffraction experiments [47]. The DPD is found to be one possible calculation process, which is premised on the analysis of many component systems, like a hydrated Nafion membrane. The procedure of DPD calculation is described in Appendix B.

The phase-field method (PFM) [23] is also applied to the simulation of mesoscopic structure formation. The PFM is a method which solves the time-dependent Ginzburg–Landau equations numerically (see Eqs. (C.1), (C.2), and (C.3) in Appendix C). This method calculates the dynamics of many degrees of order parameters. The structure of a polycrystal and the domain structure of liquid crystals, etc., can be applied by this method, for example. Within the DPD formulation, the internal degree of freedom for each DPD particle is not usually considered. This means that a system having a multicomponent order parameter, such as the orientation of a micro-crystal, etc., is difficult to calculate by DPD, while the DPD makes simple and useful process for calculating the time development of multicomponent density distribution. On the other hand, multicomponents of order parameters can be considered under PFM by assigning them to each lattice point on a numerical spatial grid. We should therefore select a suitable technique for simulation of mesoscopic structure formation for each objective system.

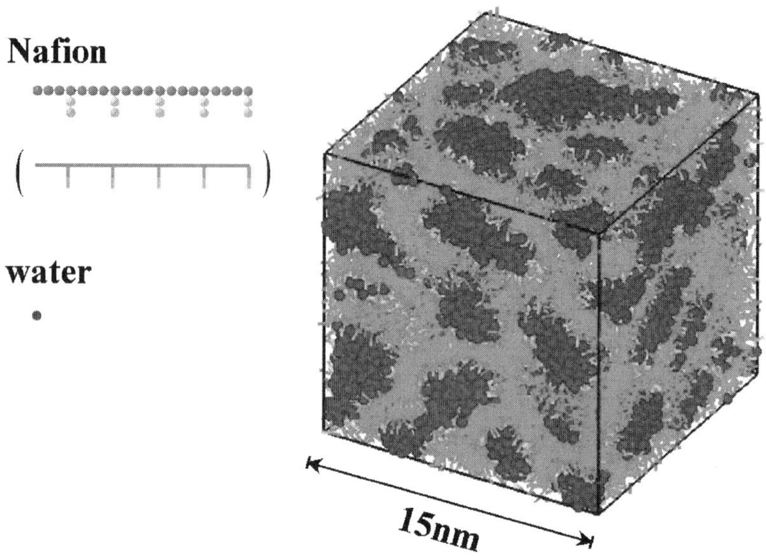

Fig. 3. Meso-scale structure of a hydrated Nafion membrane reproduced by the DPD simulation. The interaction parameters of each DPD particle are defined by molecular dynamics level calculation using the explicit molecular structure of a Nafion oligomer.

7. Finite element method for electronic state calculation

The electrostatic potential is distributed over a microscopic region for electronic state calculation, because it originates from the inhomogeneous structure at the meso-scale level. The conventional molecular orbital methods use analytical basis functions and the analytical formulation of the integrals for solving the eigenvalue problem [41]. In such processes, the external potential should be given analytically for calculation within a realistic computation time. The existence of the boundary in a mesoscopic structure forces the matrix element integration to be developed by complicated calculation routines. In the conventional molecular orbital method [33], the integral routines are the most time-consuming processes in general. Therefore, another procedure should be adopted for our purpose. It can be simply seen that the numerical calculation in the spatial grid technique is applicable for the case with inhomogeneously distributed potentials. As a possible technique, we adopted the finite element method (FEM) [2] as is well known in the structural analysis (i.e., for elastic materials) research field. The eigenvalue problem is easily solved even for the case with the spatially distributed potential.

The FEM has already been applied to the electronic state calculation [39]. For the electronic state calculation of single crystals (so-called "band calculation") [35], the results from this approach have been increasingly appearing. For molecular electronic state calculation, however, the explicit applications have been less publicized [17]. The essential reason for this seems to be in the *chemical accuracy* of the quantum chemical theory. The requirement for highly accurate calculation increases day by day. The improvement of the accuracies of quantum chemical calculation theory accelerates this requirement. The preferable accuracy of this, at the present time, seems to be less than about

0.1 kcal/mol under a total electronic energy typically greater than about 1000 kcal/mol. Such accuracy is called *chemical accuracy*. To achieve *chemical accuracy* by means of FEM, the required number of nodes is extremely large. The very sharp variation in wave function near the nucleus forces this problem to become more serious. Therefore, lower advantage of FEM has been assumed in the electronic state calculation of molecules. Once the applicability of FEM to various boundary conditions is noticed, however, this advantage should be respected rather than its disadvantage in *chemical accuracy*. Avoiding the requirement of *chemical accuracy* level, the number of nodes can be reduced. An example of the means of reduction involves using the effective core potentials [43]. Assuming that the core electrons, which are dominantly distributed near the nucleus, are unchanged in chemical bonding formation, the most serious region in FEM calculation can be omitted.

8. Example progressing in our group

8.1. Cooperation of atomistic and mesoscopic simulations using DPD [46]

We demonstrate DPD simulation coupled with realistic χ-parameters for binary polymer blends of (a) polyethylene (PE)–polypropylene (PP), (b) PE–polystyrene (PS), and (c) PP–PS. The χ-parameter describes the relative strengths of the interaction between different components in a polymer system based on the Flory–Huggins theory [15], which is familiar in polymer science [24,9]. While the parameter χ is positive when two components in a polymer system do not favor contact, χ is negative when they are favored to contact each other. The estimated χ-parameters for these mixtures by atomistic simulations are listed in Table 1. The process for χ-parameter estimation is described by Eqs. (B.2) and (B.3). It is predictable that miscibility is in the (a) $>$ (c) $>$ (b) order. Fig. 4 shows the concentration profiles of each polymer for these mixtures by DPD simulation. It is recognized that the thickness of the interface corresponds well with this miscibility order.

According to the simulated mesoscopic structure by DPD, a molecular structure is generated by Monte Carlo technique. This procedure is as follows:

1. Consider the concentration profile ϕ for the specific composition.
2. Generate an atom of the desired molecule at a random position.
3. Acceptance or rejection for adoption of the generated atom is determined based on the concentration profile of the composition.
4. By repeating 2 and 3 for connecting atoms to the latest accepted one, the molecular structure is generated. In this step, the accepted positions of atoms are not changed.
5. The above operation is continued until the concentration of the specific component reaches the desired level.

Table 1
χ-Parameter of binary blend

Mixture	PE-PP	PE-PS	PP-PS
χ-parameter	0.09	0.71	0.21

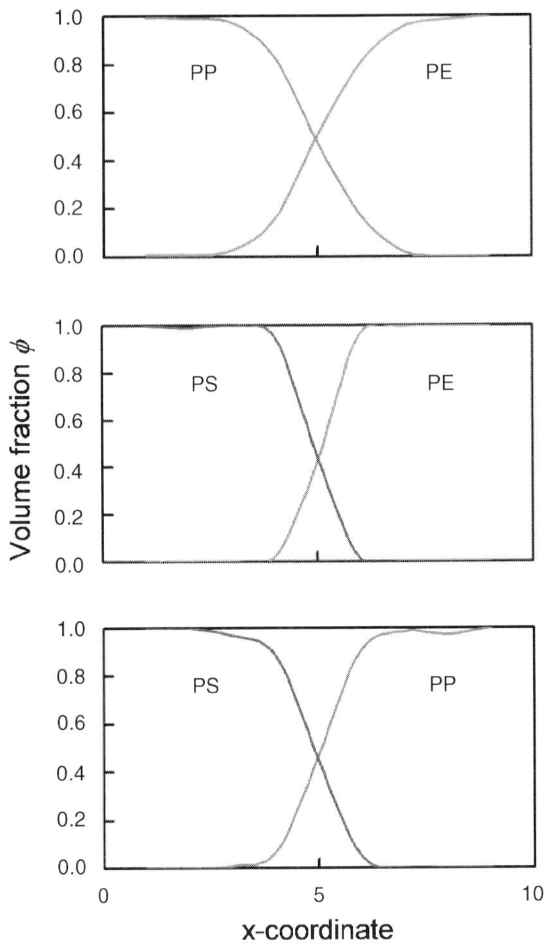

Fig. 4. Concentration profiles at the interface of a binary polymer blend. PP: polypropylene. PE: polyethylene. PS: polystyrene.

6. After the space was filled by atoms following the above procedure, the positions of the atoms are optimized by using the molecular mechanics or the molecular dynamics method. Although the positions of DPD particles are optimized in DPD calculation, the atomic positions are still arbitrary for each DPD particle. Therefore, this step is necessary for finding a stable structure on an atomic scale.

As a simple and typical problem, the structures at the interfaces in binary polymer blend systems were investigated here. Molecular structure is generated by the Monte Carlo technique based on the concentration profile at the interface for binary polymer blends (Fig. 4). Fig. 5 shows the carbon atoms of each polymer backbone and their connecting bonds. The position of each carbon atom is found to be restricted by the mesoscale structure on each corresponding interface. Molecular structures are successfully obtained by this procedure.

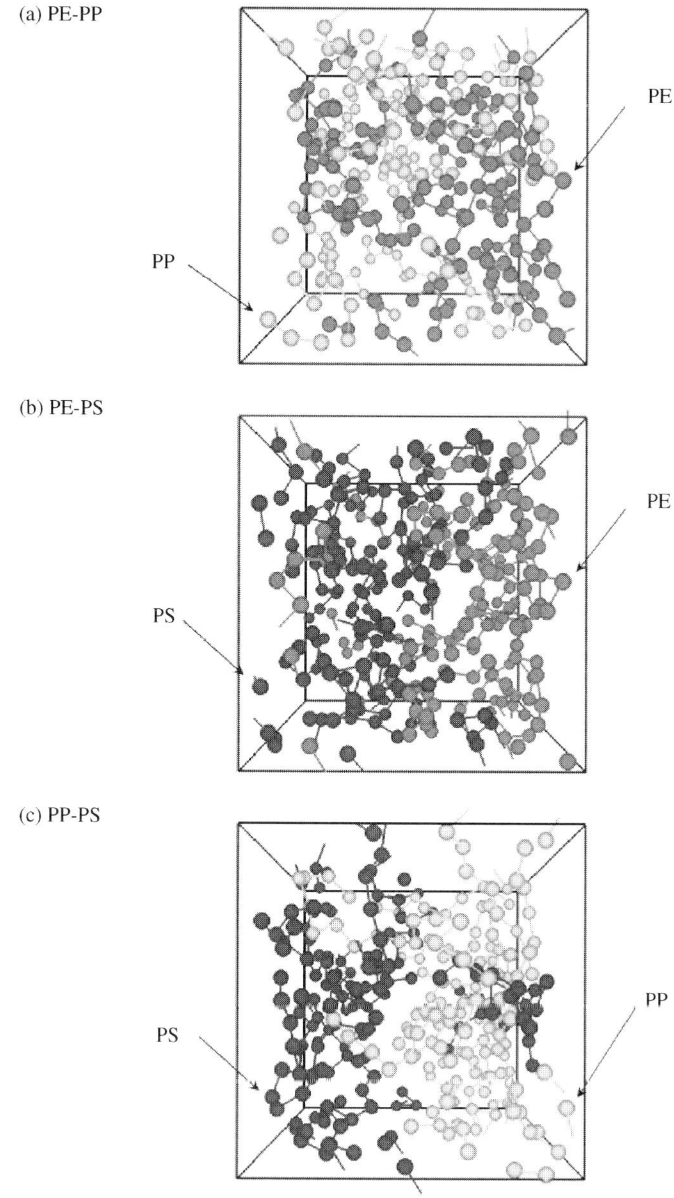

Fig. 5. Molecular structure at the interface of a binary polymer blend based on the mesoscopic structure in Fig. 4. PP: polypropylene. PE: polyethylene. PS: polystyrene.

8.2. Electronic state calculation with an inhomogeneous electrostatic field. Hydrogen atom in metallic polycrystalline-like boundaries as an example [48]

In Section 5, a rather simple example of electronic state calculation under environmental boundary condition is represented. This method is based on the conventional molecular orbital theory. The conventional method may have a heavy computational cost when applied to arbitrary boundary conditions. In order to discuss an electronic state under a more complicated boundary condition, we should introduce another method. The FEM seems to be one possible technique for this purpose as is discussed in Section 7.

Electronic states are affected dominantly by electrostatic potentials. The additional electrostatic potential in the Hamiltonian of an electronic state is sufficient to describe an environmental effect. Therefore, we should define this term based on mesoscopic structural information. Here, an example of electronic wave function of a hydrogen atom in metallic polycrystalline-like boundaries will be depicted. It is well-known that hydrogen transport in a hydrogen-storage alloy is strongly affected by the structure of the polycrystalline boundaries. Also the destruction of metal or ceramic materials is affected by the strength of the polycrystalline boundaries. The example reported here can be considered a preliminary fundamental examination of those problems. The mesoscopic structure was deduced by the phase-field method (PFM) [23], which is based on the numerical evaluation of the time-dependent Ginzburg–Landau equation. The electrostatic potential was obtained by solving the Poisson equation numerically under this mesoscopically defined boundary condition. Sources of the electrostatic field are introduced by the charge distribution deduced from the electronic wave function of a hydrogen atom. The wave function and the electrostatic potential are calculated self-consistently. The electronic wave function was calculated by FEM with this additional electrostatic potential. The procedures of the PFM and FEM are briefly described in Appendices C and D, respectively.

The mesoscopic structure adopted here is shown in Fig. 6. At the present stage of our work, the parameters of PFM for this structure have no physical meaning yet. It is sufficient to deduce only a mesoscopic-scale complicated structure now. Fig. 6 shows a complicated structure such as one found in a polycrystallite. An example of the distribution of the electrostatic potential, which is reproduced by a hydrogen nucleus under the boundary condition shown in Fig. 6, is presented in Fig. 7. Fig. 8 shows the electronic wave function of this hydrogen atom located on the electrostatic potential of the boundary region in the polycrystalline-like structure. Here, the dielectric constant in the boundary region is assumed to be one (the value of air or vacuum). The slight but obvious spreading of the wave function can be seen in Fig. 8. This result is introduced by the distributed electron charge and the nuclear charge of the hydrogen atom and the mesoscopic metallic boundary condition. This is understood to be a first introductory study of the electronic state calculation with the complicated inhomogeneity of a mesoscopic environment.

9. Perspective and conclusions

In the present paper, a possible procedure for calculating an electronic state at a molecular level under inhomogeneous circumstances was discussed. In order to realize material design practically, such a procedure is strongly required. The concept of "*meso-scale fusion*", introduced here, seems to be one type of so-called multiscale simulation techniques. Although the basic problem in transformation of the information between quantum and classical pictures generally still remains, the procedure based on "*meso-scale fusion*" is found to be applicable to practical calculations for materials. Some examples of particulate computation, which were thus far impossible, are introduced by

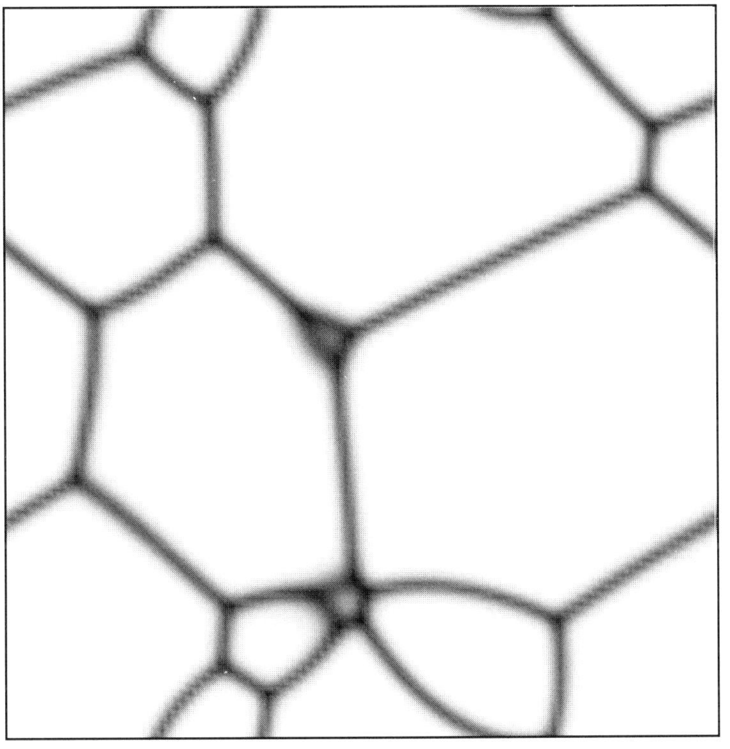

Fig. 6. Mesoscopic polycrystalline-like structure deduced by phase-field method. A two-dimensional section of a three-dimensional calculation is shown. The full length of this picture is about 5 nm. The thickness of the boundaries is about 0.2 nm.

combining the computational methods on different spatial scales. It is thought that such computation procedures may rapidly increase in importance in the future for calculating the physical properties of materials.

The present report describes:

1. the importance of meso-scale inhomogeneity in realistic materials,
2. the necessity for the information on the electronic state for practical material design,
3. the basic problem in transformation of the information between quantum and classical pictures and the interpretation of the problem in the application to practical systems, and
4. the presentation of mesoscopic structure calculations and electronic state calculations in an inhomogeneous environment.

The importance of the electronic state calculation in a mesoscopic inhomogeneous environment is stated in relation to realistic computational material design. The possibility of such calculations is also stated with some fundamental examples. Progress in and extension of the method are, of course, needed.

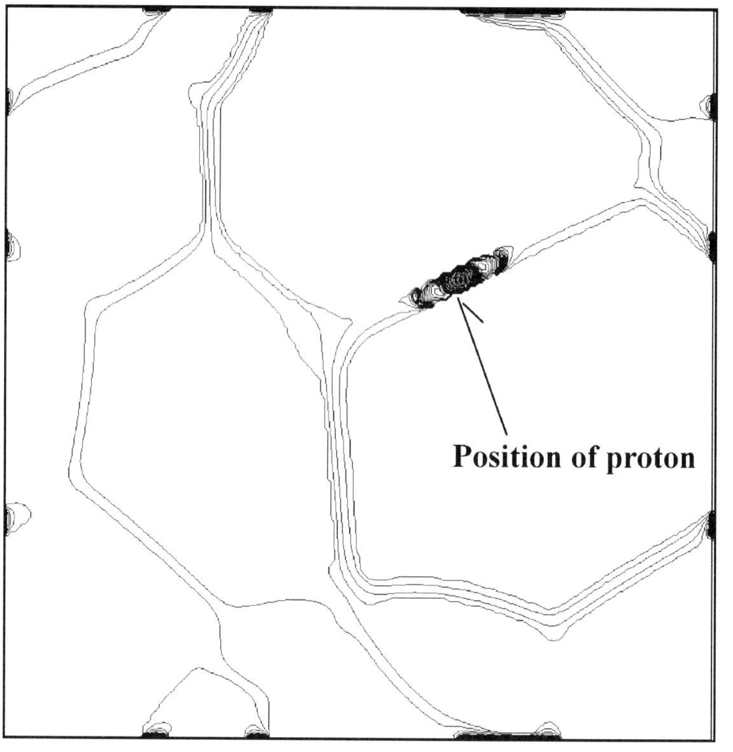

Fig. 7. Electrostatic potential distribution in an arbitrary region in the polycrystalline-like structure of Fig. 6. This figure shows the contribution from proton only. The site of a proton is indicated by an arrow. In the following FEM calculation, the electrostatic potential shown here and that originating from the charge distribution of electron were accounted for as additional potential terms in the Hamiltonian of the electronic state of a hydrogen atom.

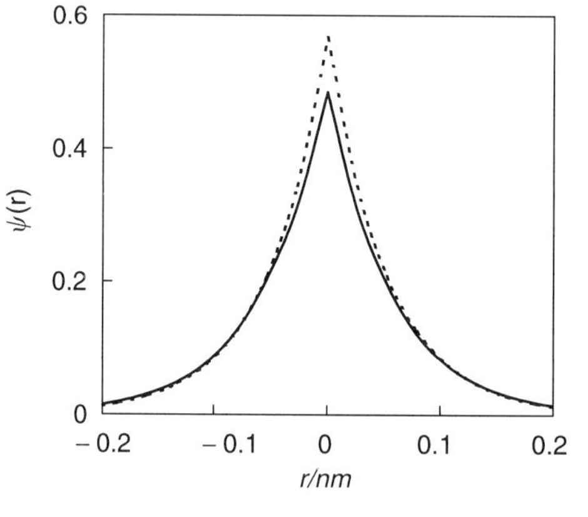

Fig. 8. Electronic wave function, $\psi(r)$, of a hydrogen atom located on the boundary region in the polycrystalline-like structure (Fig. 6). The broken line indicates that of an isolated hydrogen atom. The nucleus is at $r = 0.0$ nm. The direction of the coordinate r is almost perpendicular to the plane of the boundary.

Acknowledgements

The author of the present article is grateful to all the members of Frontier Research Group V of TOYOTA Central R& D Labs., *Inc.*, for their enthusiastic collaboration.

Appendix A. Procedure of the present molecular orbital calculation under the conventional method

The electronic structure calculation was performed within the gradient-corrected local density functional framework. The expressions of the exchange and the correlation potentials are those proposed by Becke [3] and Lee et al. [25], respectively. The program for the calculation is fully coded by the present author. The effective core potentials (ECPs) for Pt by Hay and Wadt [16] and for C and O by Kahn et al. [20] are used with the corresponding basis sets. The basis set for Pt is (3s 3p 3d)/[3s 2p 3d]. The C and O bases are augmented by the d-type polarization functions with the exponents of 0.600 and 1.154, respectively. The ECP of Pt is a relativistic one. The matrix elements of the ECP terms are calculated by essentially the same procedure of Melius [28], but the integration scheme is partly reanalyzed to reduce numerical errors. The matrix element integrals, except for the ECP ones, are reproduced following the method of Huzinaga et al. [34]. The self-consistent field calculation is based on Pulay's method [36]. For solving the eigenvalue problem, the procedure of Beppu and Ninomiya [5] is used. The Euler–Maclaurin and the Gauss–Legendre quadratures [13] are used for the radial and angle grids, respectively, in the numerical integrations. The grid points are 50, 16, and 32 for the radial, polar angle, and azimuthal angle, respectively. Becke's multicenter three-dimensional integration scheme [4] is applied.

Appendix B. Procedure of dissipative particle dynamics method

The DPD is one of the mesoscopic simulation methods and is based on the dynamics of coarse-grained particles interacting by conservative, dissipative, random forces, and a spring force for polymer systems [18]

$$\mathbf{f}_i = \sum_{j \neq i} (\mathbf{F}_{ij}^C + \mathbf{F}_{ij}^D + \mathbf{F}_{ij}^R + \mathbf{F}_{ij}^S). \tag{B.1}$$

The repulsion parameters between particles in the conservative forces are related to the χ-parameters [15], which can be calculated from the energy of mixing ΔE_{mix} by an atomistic simulation [7],

$$\chi = V_{\text{seg}} \left(\frac{\Delta E_{\text{mix}}}{kT} \right), \tag{B.2}$$

where V_{seg} is the volume of one polymer segment corresponding to the particle size in DPD. ΔE_{mix} is defined by the cohesive energy densities E_{coh}/V for pure components and their blends, calculated from the cohesive energy E_{coh} per volume V using molecular mechanics or molecular dynamics techniques

$$\Delta E_{\text{mix}} = \frac{1}{2} \left(\frac{E_{\text{coh}}}{V} \right)_1 + \frac{1}{2} \left(\frac{E_{\text{coh}}}{V} \right)_2 - \left(\frac{E_{\text{coh}}}{V} \right)_{12}. \tag{B.3}$$

Mesoscopic structure is simulated by DPD with realistic χ-parameters in cooperation with an atomistic simulation.

Appendix C. Procedure of phase-field method calculation [23]

The whole system is divided into subsystems, in which the local equilibrium condition holds. Assigning an order parameter (η) to each subsystem, the spatial distribution of η is expressed. The ηs for a polycrystalline material corresponds to the directions of the micro-crystals. The free energy functional of the whole system can be expressed as

$$F = \int \left[f_0(\eta_1(r), \eta_2(r), \ldots, \eta_P(r)) + \sum_{i=1}^{P} \frac{\kappa_i}{2} (\nabla \eta_i(r))^2 \right] \mathrm{d}^3 r \tag{C.1}$$

$$f_0(\eta_1(r), \eta_2(r), \ldots, \eta_p(r)) = \sum_{i=1}^{P} \left(-\frac{\alpha^2}{2}\eta_i^2 + \frac{\beta}{4}\eta_i^4 \right) + \gamma \sum_{i=1}^{P} \sum_{j \neq 1}^{P} \eta_i^2 \eta_j^2 \tag{C.2}$$

The time development of an order parameter η_i is described as

$$\frac{\partial \eta_i(r,t)}{\partial t} = -L_i \frac{\delta F}{\delta \eta_i(r,t)} = -L_i \left[\frac{\partial f_0}{\partial \eta_i} - \kappa_i \nabla^2 \eta_i \right] \tag{C.3}$$

L_i is a changeability parameter for each η_i for time development. The numerical method for solving Eq. (C.3) is based on the finite difference with five-point reference sites. The parameters used here are: $\alpha = \beta = \gamma = L_i = 1.0$ and $\kappa_i = 2.0$. The number of the grid points adopted here is typically $100 \times 100 \times 100$ applied.

Appendix D. Condition of finite element method for electronic state calculation

A three-dimensional mesh with natural boundary condition was adopted to the electronic state FEM calculation on the Cartesian coordinate system. The wave function and the eigenvalue, which are described by the Schrödinger equation, were solved as an eigenvalue problem by FEM. Mesoscopic boundary conditions were introduced as the distribution of an additional electrostatic potential. The total number of nodes was 68 921 of $40\times40\times40$ sections. The shape function of FEM (corresponding to "basis function" in the words of the usual eigenvalue problem) was described as a quadratic function in a local rectangular parallelepiped element with 20 local nodes. The resulting ground state eigenvalue of an isolated hydrogen atom was -0.4899 (the exact value is -0.5000). This result shows sufficient accuracy for the present purpose.

References

[1] M. Bahiana, Y. Oono, Phys. Rev. A 41 (1990) 6763.
[2] K.-J. Bathe, E.L. Wilson, Numerical Methods in Finite Element Analysis, Prentice-Hall, Englewood Cliffs, NJ, 1976.
[3] A.D. Becke, Phys. Rev. A 38 (1988) 3098.
[4] A.D. Becke, J. Chem. Phys. 88 (1988) 2547.
[5] Y. Beppu, I. Ninomiya, Compt. Chem. 6 (1982) 87.

[6] A. Brandt, Math. Comput. 31 (1977) 333;
A. Brandt, S. McCormick, J. Ruge, SIAM J. Sci. Statist. Comput. 4 (1983) 244.
[7] F.H. Case, J.D. Honeycutt, TRIP 2 (1994) 259.
[8] J. Cioslowski (Ed.), Many-Electron Densities and Reduced Density Matrices, Kluwer Academic Publisher, Plenum, New York, 2000.
[9] M. Doi, Introduction to Polymer Physics, Oxford University Press, Oxford, 1995.
[10] A. Einstein, B. Podolsky, N. Rosen, Phys. Rev. 47 (1935) 777.
[11] P. Espanol, P.B. Warren, Europhys. Lett. 30 (1995) 191.
[12] J.G.E.M. Fraaije, J. Chem. Phys. 99 (1993) 9202;
J.G.E.M. Fraaije, B.A.C. van Vlimmeren, N.M. Maurits, M. Postma, O.A. Evers, C. Hoffman, P. Altevogt, G. Goldbeck-Wood, J. Chem. Phys. 106 (1997) 4260.
[13] F.A. Gianturco, N. Sanna, S. Serna, Mol. Phys. 78 (1993) 1015.
[14] R.D. Groot, T.J. Madden, J. Chem. Phys. 108 (1998) 8713.
[15] R.D. Groot, P.B. Warren, J. Chem. Phys. 107 (1997) 4423.
[16] P.J. Hay, W.R. Wadt, J. Chem. Phys. 82 (1985) 270.
[17] D. Heinemann, B. Fricke, D. Kolb, Phys. Rev. A 38 (1988) 4994;
L. Yang, D. Heinemann, D. Kolb, Phys. Rev. A 48 (1993) 2700;
C. Duesterhoeft, D. Heinemann, D. Kolb, Chem. Phys. Lett. 296 (1998) 77.
[18] P.J. Hoogerbrugge, J.M.V.A. Koelman, Europhys. Lett. 19 (1992) 155;
R.D. Groot, P.B. Warren, J. Chem. Phys. 107 (1997) 4423.
[19] S. Hyodo, submitted for publication.
[20] L.R. Kahn, P. Baybutt, D.G. Truhlar, J. Chem. Phys. 65 (1976) 3826.
[21] R.K. Kalia, T.J. Campbell, A. Chatterjee, A. Nakano, P. Vashishta, S. Ogata, Comput. Phys. Commun. 128 (2000) 245.
[22] K. Kawasaki, K. Sekimoto, Macromolecules 22 (1989) 3063.
[23] R. Kobayashi, Physica D 63 (1993) 410;
D. Fan, L.-Q. Chen, Acta Mater. 45 (1997) 611.
[24] D.W. van Krevelen, Properties of Polymers, 3rd Edition, Elsevier, Amsterdam, 1994.
[25] C. Lee, W. Yang, R.G. Parr, Phys. Rev. B 37 (1988) 785.
[26] S. Machida, A. Motoyoshi, Found. Phys. 28 (1998) 45 references therein.
[27] C.A. Marsh, G. Bachx, M.H. Emst, Phys. Rev. E 56 (1997) 1676.
[28] C.F. Melius, Ph.D. Thesis, California Institute of Technology, November 1972.
[29] S. Miertus, E. Scrocco, J. Tomasi, Chem. Phys. 55 (1981) 117;
S. Miertus, J. Tomasi, Chem. Phys. 65 (1982) 239;
M. Cossi, V. Barone, R. Camini, J. Tomasi, Chem. Phys. Lett. 255 (1996) 327.
[30] Y. Mori, T. Mori, A. Miyamoto, N. Takahashi, T. Hattori, Y. Murakami, J. Phys. Chem. 93 (1989) 2039.
[31] H. Nakatsuji, J. Chem. Phys. 87 (1987) 4995.
[32] H. Nakatsuji, H. Nakai, Y. Fukunishi, J. Chem. Phys. 95 (1991) 640;
H. Nakatsuji, H. Morita, H. Nakai, Y. Murata, K. Fukutani, J. Chem. Phys. 104 (1996) 714.
[33] S. Obara, A. Saika, J. Chem. Phys. 89 (1988) 1540.
[34] K. O-ohata, H. Taketa, S. Huzinaga, J. Phys. Soc. Japan 21 (1966) 2306, 2313.
[35] J.E. Pask, B.M. Klein, C.Y. Fong, P.A. Sterne, Phys. Rev. B 59 (1999) 12352;
E.L. Briggs, D.J. Sullivan, J. Bernholc, Phys. Rev. B 54 (1996) 14362.
[36] P. Pulay, J. Comput. Chem. 3 (1982) 556.
[37] S. Qi, Z.G. Wang, Phys. Rev. E 55 (1997) 1682.
[38] E. Schrödinger, Naturwissencshaften 23 (1935) 807, 823, 844.
[39] W. Schulze, D. Kolb, Chem. Phys. Lett. 122 (1985) 271;
F.S. Levin, J. Shertzer, Phys. Rev. A 32 (1985) 3285;
S.R. White, J.W. Wilkins, Phys. Rev. B 39 (1989) 5819;
L.R. Ram-Mohan, S. Saigal, D. Dossa, J. Shertzer, Comput. Phys. (1990) 50.
[40] F. Selleri, Die Debatte um die Quantuntheorie, Frieder, Vieweg & Sohn, Braunschweig, 1983;
J.R. Friedman, V. Patel, W. Chen, S.K. Tolpygo, I.E. Lukens, Nature 406 (2000) 43.

[41] A. Szabo, N. Ostlund, Modern Quantum Chemistry, Macmillan, New York, 1982.
[42] S. Ten-no, F. Hirata, S. Kato, Chem. Phys. Lett. 214 (1993) 391;
S. Ten-no, F. Hirata, S. Kato, J. Chem. Phys. 100 (1994) 7443.
[43] E. Tsuchida, M. Tsukada, J. Phys. Soc. Japan 67 (1998) 3844.
[44] S. Vakulenko, A. Vilesov, B. Stuehn, S. Frenkel, J. Chem. Phys. 106 (1997) 3412.
[45] A. Warshel, M. Karplus, J. Amer. Chem. Soc. 94 (1972) 5612;
M.J. Field, P.A. Bash, M. Karplus, J. Comput. Chem. 11 (1990) 700;
F. Maseras, K. Morokuma, J. Comput. Chem. 16 (1995) 1170.
[46] S. Yamamoto, S. Hyodo, Proceedings of the Seventh Japanese International SAMPE Symposium, November 13–16, 2001.
[47] S. Yamamoto, S. Hyodo, submitted for publication.
[48] S. Yamakawa, S. Hyodo, unpublished.
[49] S. Yamamoto, Y. Maruyama, S. Hyodo, J. Chem. Phys. 116 (2002) 5842.
[50] M. Yoshida, Comput. Phys. Commun. 127 (2000) 100.

ELSEVIER Journal of Computational and Applied Mathematics 149 (2002) 119–129

JOURNAL OF COMPUTATIONAL AND APPLIED MATHEMATICS

www.elsevier.com/locate/cam

Matrix computations for information retrieval and major and outlier cluster detection

M. Kobayashi *, M. Aono, H. Takeuchi, H. Samukawa

Tokyo Research Laboratory, IBM Japan, Ltd., 1623-14 Shimotsuruma, Yamato-shi, Kanagawa-ken 242-8502, Japan

Received 28 October 2001; received in revised form 13 December 2001

Abstract

In this paper we introduce *COV*, a novel information retrieval (IR) algorithm for massive databases based on vector space modeling and spectral analysis of the covariance matrix, for the document vectors, to reduce the scale of the problem. Since the dimension of the covariance matrix depends on the attribute space and is independent of the number of documents, COV can be applied to databases that are too massive for methods based on the singular value decomposition of the document-attribute matrix, such as *latent semantic indexing* (LSI). In addition to improved scalability, theoretical considerations indicate that results from our algorithm tend to be more accurate than those from LSI, particularly in detecting subtle differences in document vectors. We demonstrate the power and accuracy of COV through an important topic in data mining, known as outlier cluster detection. We propose two new algorithms for detecting major and outlier clusters in databases—the first is based on LSI, and the second on COV. Our implementation studies indicate that our cluster detection algorithms outperform the basic LSI and COV algorithm in detecting outlier clusters.

1. Introduction

In recent years, the volume of data stored in electronic databases has become so massive that development of systems to enable fast and accurate retrieval of information that are tailored to the interests of individual users has become imperative. Several mathematical approaches are being taken in the race to build fast, accurate and intelligent knowledge mining and management systems, one of which is vector space modeling [3], introduced by Salton and his colleagues [13] over a quarter century ago. In vector space modeling, each document in a database is modeled by a vector, each coordinate of which represents an attribute. In *Boolean* models, a coordinate of a vector is

* Corresponding author.
E-mail address: mei@jp.ibm.com (M. Kobayashi).

PII: S0377-0427(02)00524-1

naught (when the corresponding attribute is absent) or unity (when the attribute is present). *Term weighting* is a refinement of this model that takes into account the frequency of appearance of words and their location of appearance (e.g., in the title, abstract, or section header). The relevancy ranking of a document with respect to a query is determined by its so-called "*distance*" to the query vector, e.g., the cosine of the angle defined by the query and each document vector. This method for ranking is impractical for very large databases since there are too many computations and subsequent comparisons.

In the late 1980s Deerwester et al. [4] proposed the *latent semantic indexing* (LSI) algorithm as a means of reducing the dimension of the document-attribute matrix to enable real-time information retrieval (IR) from very large databases. The fundamental idea in LSI is to model a database by an M-by-N document-attribute matrix A (the rows of which are vectors that represent the documents in the database) and to reduce the dimension of the IR problem to k, where $k \ll \min(M,N)$, by projecting the problem into the subspace spanned by the rows of the closest rank-k matrix to A in the Frobenius norm. One of the major bottlenecks in applying LSI to massive databases with hundreds of thousands of documents is the need to compute the largest few hundred singular values and corresponding singular vectors of the document-attribute matrix for a database. Even though document-attribute matrices that appear in IR tend to be very sparse (usually 0.2–0.3% nonzero), computation of the top 200–300 singular triplets of the matrix using a powerful desktop PC becomes impossible when the number of documents exceeds several hundred thousand.

In this paper we propose *COV*, an IR algorithm based on spectral analysis of the covariance matrix for the document vectors, that reduces the dimension of IR problems and overcomes the scalability problem associated with LSI [9]. Our algorithm depends on computation of the largest few hundred eigenvalues and corresponding eigenvectors of a square matrix, with dimension less than or equal to that of the attribute space. Since covariance matrices are independent of the number of documents, COV can be used for IR and data mining applications that involve hundreds of thousands of documents as long as the attribute space is relatively small, e.g., up to 20,000 dimensions, which is often the case. The remainder of this paper is organized as follows. In Section 2 we present our COV algorithm, discuss the underlying theoretical concepts and present results from our implementation studies comparing results from COV with those from LSI. Section 3 presents applications of our work to data mining systems. We present two new algorithms (one based on LSI, and a second based on algorithm from Section 2) for detecting major and outlier clusters in massive databases [8].

2. Covariance matrix analysis-based IR

Given a database modeled by an M-by-N document-attribute term matrix A, with M row vectors $\{d_i^{\mathrm{T}} \mid i = 1, 2, \ldots, M\}$ representing documents, each having N dimensions representing attributes, the *covariance matrix* for the set of document vectors is

$$C \equiv \frac{1}{M} \sum_{i=1}^{M} d_i d_i^{\mathrm{T}} - \bar{d}\bar{d}^{\mathrm{T}},$$

where d_i represents the ith document vector and $\bar{d}$ is the component-wise average over the set of all document vectors [10], i.e.,

$$\bar{d} = [\bar{a}_1 \ \bar{a}_2 \ \cdots \ \bar{a}_N]^{\mathrm{T}}, \quad d_i = [a_{i,1} \ a_{i,2} \ \cdots \ a_{i,N}]^{\mathrm{T}} \quad \text{and} \quad \bar{a}_j = \frac{1}{M}\sum_{i=1}^{M} a_{i,j}.$$

Since the covariance matrix is symmetric, positive, semi-definite, it can be decomposed into the product $C = V\Sigma V^{\mathrm{T}}$, where V is an orthogonal matrix that diagonalizes C so that the diagonal entries of Σ are in monotone decreasing order going from top to bottom, i.e., $\mathrm{diag}(\Sigma) = (\lambda_1, \lambda_2, \ldots, \lambda_N)$, where $\lambda_i \geqslant \lambda_{i+1}$ for $i = 1, 2, \ldots, N$. To reduce the dimension of the IR problem to $k \ll M, N$, we project all document vectors and the query vector into the subspace spanned by the k eigenvectors $\{v_1, v_2, \ldots, v_k\}$ corresponding to the largest k eigenvalues $\{\lambda_1, \lambda_2, \ldots, \lambda_k\}$ of the covariance matrix C. Relevancy ranking with respect to the modified query and document vectors is performed in a manner analogous to that before dimensional reduction, e.g., by computing the angle defined by the query and each document vector.

COV is similar to LSI in that, it projects a high-dimensional problem into a subspace to reduce computational requirements for determining relevancy rankings of documents with respect to a query, but large enough to retain enough information about different characteristics of documents to enable accurate retrieval. The algorithms use different criteria to determine a subspace: LSI uses the subspace spanned by the rows of the closest rank-k matrix to A in the Frobenius norm [5], while COV uses the k-dimensional subspace that best represents the full data with respect to the minimum square error [9]. Furthermore, COV shifts the origin of the coordinate system to the "center" of the subspace to spread apart documents as much as possible so that documents can be more easily distinguished from one another (as illustrated in Fig. 1).

Tables 1 and 2 show representative results from one of our numerical IR experiments using LSI and COV with the Reuters-21578 news database [12]. (The documents in the Reuters-21578 collection appeared on the Reuters newswire in 1987 and have been made publically available for academic research purposes.) The format of the query vector is: term 1 followed by its weighting factor, term 2 followed by its weighting factor, term 3 followed by its weighting factor. Results from both algorithms are very close, as expected, since successful algorithms should have similar outputs; for instance, the top 10 relevancy rankings are identical and the relevance scores are within 1–2%. Some rankings are switched from the 11th ranking, but the relevancies are still close, i.e., within 2–3%.

3. Major and outlier cluster detection

Clustering in IR is the automatic creation of groups of documents, known as *clusters*, based on the computed relevancies between documents in a database [11]. One of the goals of clustering is to improve the efficiency and effectiveness of retrieval. Another goal, which relies on successful identification of smaller clusters, is to identify trends and abnormalities in time-series data [2]. Some fairly successful techniques have been developed to identify major clusters (i.e., clusters that are comprised of more than 4% of the documents in a database) [7], however the techniques often fail to identify smaller, so-called *outlier clusters* or *outliers* (i.e., clusters that are comprised of 3–4% of

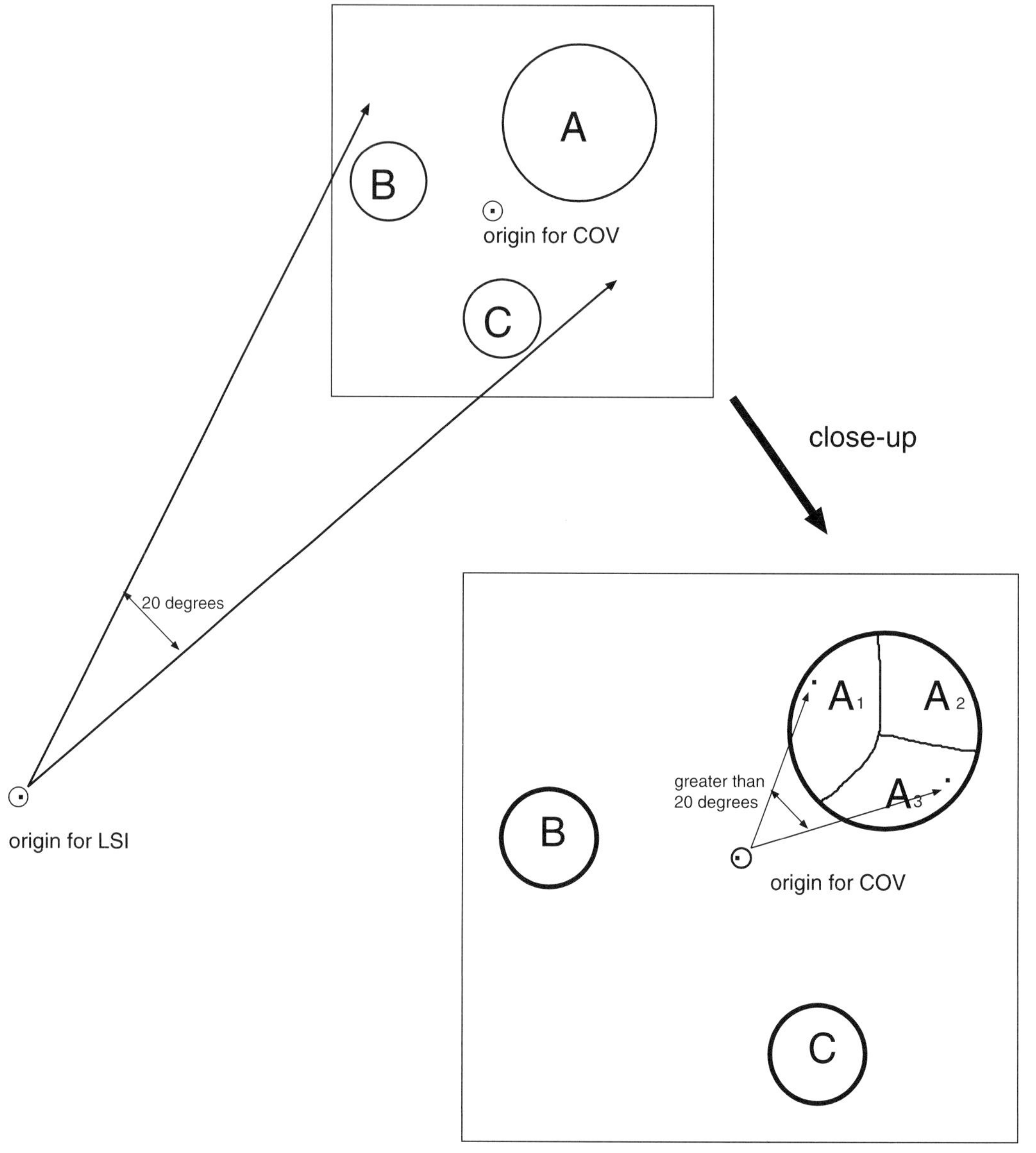

Fig. 1. LSI vs. COV: both algorithms map the relevancy ranking problem into a proper subspace of the document-attribute space. LSI does not move the origin. COV shifts the origin to the "center" of the set of basis vectors in the subspace so that document vectors are more evenly spaced apart, enabling finer detection of differences between document vectors in different clusters and subclusters.

the documents in a database). We present two new algorithms for detecting both major and outlier clusters in the databases. Our algorithms are significant enhancements of LSI and COV which have been used for clustering with limited success.

Table 1
LSI: results for query vector = {Japan 1 car 1 market 1}

Relevancy	Doc #	Title of document
0.8074	208	Japan to try to open market to US car parts
0.7934	8647	Suzuki Motor plans Hungarian joint car venture
0.7239	19637	Japan sets condition for car plant loan to Poland
0.6688	287	Japan distrib may import Mazda US-made cars
0.6610	17195	EC says Japan car export restraint not enough
0.6580	13295	Oki Electric studies import of US car phones
0.6535	10740	Nissan Mexicana to take over car engine exports
0.6434	20022	Nissan starts to markets remodelled 4WD vehicles
0.6338	16790	US urges Japan to open farm market further
0.6338	16763	US urges Japan to open farm market further
0.6282	8090	Foreign share of German car market falls
0.6210	10131	Honda exports first US-made cars
0.6158	17282	EC says Japan car export restraint not enough
0.6020	19021	Japan still asks institutions limit speculative dlrdeals
0.5905	5326	Endotronics heavy losses from withdrawal Japan
0.5870	21276	Takeshita chosen as next Japan Prime Minister
0.5853	4329	Porsche recalls 892 of its 1987 model cars
0.5795	7907	Japan opens home market to US fish
0.5784	19436	Audi of America lists car prices
0.5774	19046	Japan machinery orders fall in April

Table 2
COV: results for query vector = {Japan 1 car 1 market 1}

Relevancy	Doc #	Title of document
0.8074	208	Japan to try to open market to US car parts
0.7932	8647	Suzuki Motor plans Hungarian joint car venture
0.7241	19637	Japan sets condition for car plant loan to Poland
0.6688	287	Japan distrib may import Mazda US-made cars
0.6608	17195	EC says Japan car export restraint not enough
0.6579	13295	Oki Electric studies import of US car phones
0.6532	10740	Nissan Mexicana to take over car engine exports
0.6428	20022	Nissan starts to markets remodelled 4WD vehicles
0.6337	16763	US urges Japan to open farm market further
0.6337	16790	US urges Japan to open farm market further
0.6209	10131	Honda exports first US-made cars
0.6205	8090	Foreign share of German car market falls
0.6155	17282	EC says Japan car export restraint not enough
0.6021	19021	Japan still asks institutions limit speculative dlrdeals
0.5903	5326	Endotronics heavy losses from withdrawal Japan
0.5861	21276	Takeshita chosen as next Japan Prime Minister
0.5846	4329	Porsche recalls 892 of its 1987 model cars
0.5794	7907	Japan opens home market to US fish
0.5784	19436	Audi of America lists car prices
0.5767	13546	Hyundai car exports rise sharply in early 1987

3.1. Prior art

Implementation studies have shown that the basic LSI and COV algorithms are somewhat successful at finding major clusters, however, they usually fail to identify outliers. In fact, both algorithms normally delete information in these outliers, because major clusters and their large sub-clusters dominate the subjects that will be preserved during dimensional reduction. Recently, Ando [1] proposed an algorithm to overcomes this problem in limited contexts. The main intended idea in her algorithm is to prevent major themes from dominating the process of selecting the basis vectors for the reduced dimensional subspace. This is supposed to be carried out during the basis vector selection process by introducing a negative bias to documents that belong to clusters that are well-represented by basis vectors that have already been selected. The negative bias is imparted by computing the magnitude (i.e., the length in the Euclidean norm) of the *residual* of each document vector (i.e., the proportion of the document vector that cannot be represented by the basis vectors that have been selected thus far), then re-scaling the magnitude of each document vector by a power q of the magnitude of its residual.

Ando's Algorithm

$R = A;$

for $(i = 1; i \leqslant k; i++)\{$

$\quad R_s = [|r_1|^q r_1, |r_2|^q r_2, \ldots, |r_M|^q r_M]^{\mathrm{T}};$

$\quad b_i =$ the first eigenvector of $R_s^{\mathrm{T}} R_s;$

$\quad R = R - R b_i b_i^{\mathrm{T}};$ (the residual matrix)

$\}$

The input parameters are the document-term matrix A, the constant scale factor q, and the dimension k to which the IR will be reduced. The *residual matrices* are denoted by R and R_s. We set R to be A initially. After each iterative step, the residual vectors are updated to take into account the new basis vector b_i. After the kth basis vector is computed, each document vector d_j in the original IR problem is mapped to its counterpart $\hat{d}_j$ in the k-dimensional subspace as follows: $\hat{d}_j = [b_1, b_2, \ldots, b_k]^{\mathrm{T}} d_j$. The query vector is mapped to the k-dimensional subspace before relevance ranking is performed. Ando's algorithm is somewhat successful in detecting clusters, however, the following problems can occur: all outlier clusters may not be identified; the procedure for finding eigenvectors may become unstable when the scaling factor q is large; the basis vectors b_i are not always orthogonal; and if the number of documents in the database is very large, the eigenvector cannot be computed on an ordinary PC.

3.2. Our algorithms

We propose two new algorithms for detecting major and outlier clusters that overcome some of the problems associated with Ando's algorithm. The first is a significant modification of Ando's algorithm and the second is based on COV.

Algorithm 1

for $(i = 1; i \leqslant k; i++)\{$
  $t_{\max} = \max(|r_1|, |r_2|, \ldots, |r_M|)$;
  $q = \text{func}(t_{\max})$;
  $R_s = [|r_1|^q r_1, |r_2|^q r_2, \ldots, |r_M|^q r_M]^{\mathrm{T}}$;
  SVD (R_s); (the singular value decomposition)
  $b_i' =$ the first row vector of V^{T};
  $b_i = \text{MGS}(b_i')$; (modified Gram–Schmidt)
  $R = R - Rb_i b_i^{\mathrm{T}}$; (residual matrix)
}

Our first algorithm is based on the observation that re-scaling document vectors after the computation of each basis vector in Ando's algorithm leads to the rapid diminution of documents that have even a moderate-size component in the direction of one of the first few document vectors. To understand how negative biasing can obliterate these vectors, consider the following scenario. Suppose that a document has a residual of 90% after one basis vector is computed, and q is set to be one. Before the next iteration, the vector is re-scaled to length 0.81, after two more iterations $0.81 \times 0.81 < 0.66$, and after n more iterations, less than 0.81 to the nth power. We recognize that biasing can be useful, however the bias factor should dynamically change to take into account the length of the residual vectors after each iterative step to prevent over-biasing. More specifically, in the first step of the iteration, we compute the maximum length of the residual vectors and use it to define the scaling factor q that appears in the second step.

$$q = \begin{cases} t_{\max}^{-1} & \text{if } t_{\max} > 1, \\ 1 + t_{\max} & \text{if } t_{\max} \approx 1, \\ 10^{t_{\max}^{-2}} & \text{if } t_{\max} < 1. \end{cases}$$

As a second modification, we replace the computation of eigenvectors in Ando's algorithm with the computation of the SVD for robustness. Our third modification is the introduction of modified Gram–Schmidt orthogonalization [6] of the basis vectors b_1.

Algorithm 2

for $(i = 1; i \leqslant k; i++)\{$
  $t_{\max} = \max(|r_1|, |r_2|, \ldots, |r_M|)$;
  $q = \text{func}(t_{\max})$;
  $R_s = [|r_1|^q r_1, |r_2|^q r_2, \ldots, |r_M|^q r_M]^{\mathrm{T}}$;
  $\mathrm{C} = \mathrm{COV}(R_s)$; (covariance matrix)
  SVD (C); (the singular value decomposition)
  $b_i' =$ the first row vector of V^{T};
  $b_i = \text{MGS}(b_i')$; (modified Gram–Schmidt)
  $R = R - Rb_i b_i^{\mathrm{T}}$; (residual matrix)
}

Our second algorithm for outlier detection is a modification of COV that is analogous to the modification of LSI to produce Algorithm 1. In this second algorithm, the SVD of the covariance matrix associated with the document vectors is computed (in lieu of the residual matrix in Algorithm 1). Results from our implementation studies (given in the next section indicate that our second algorithm is better than Ando's, LSI, COV, and Algorithm 1 at identifying large and multiple outlier clusters.

3.3. Implementation experiments

To test and compare the quality of results from the algorithms discussed above, we performed two sets of numerical experiments: the first with a small data set to test our concept, and the second with a larger data set to demonstrate the applicability of our method to realistic size data.

For our first set of experiments, we constructed a data set consisting of two large clusters (each of which have three subclusters), four outlier clusters and noise. Each large cluster has two subclusters that are twice as large as the outliers and a subcluster that is the same size as the outliers, as shown below.

Cluster Structure of Data

140 documents, 40 terms
- 25 docs (Clinton cluster) – *major*
  - 10 docs (Clinton + Al Gore only) – *subcluster*
  - 10 docs (Clinton + Hillary only) – *subcluster*
  - 10 docs (Clinton + Al Gore + Hillary) – *subcluster*
- 25 docs (Java cluster) – *major*
  - 10 docs (Java + JSP only) – *subcluster*
  - 5 docs (Java + Applet only) – *subcluster*
  - 10 docs (Java + JSP + Applet) – *subcluster*
- 5 docs (Bluetooth cluster) – *outlier*
- 5 docs (Soccer cluster) – *outlier*
- 5 docs (Matrix cluster) – *outlier*
- 5 docs (DNA cluster) – *outlier*
- 70 docs *noise*

We implemented five algorithms to reduce the dimension of the document-term space: LSI, COV, Ando, and Algorithms 1 and 2. The 40-dimensional term space was reduced to six dimensions, i.e., we set $k = 6$. Table 3 summarizes which clusters were detected as basis vectors were computed (upper case letters are used to label major clusters and lower case for outliers and noise). LSI failed to detect any information in outlier clusters. COV picked up some information in outliers, but failed to detect specific outliers. Ando's algorithm detected two outlier clusters: *Bluetooth* and *Soccer* in b_4 and the two remaining outliers *Matrix* and *DNA* in b_5 and b_6. Our results indicate that after the fourth iteration, the lengths of the residual vectors for documents covering topics other than *Matrix* and *DNA* have been given too much of a negative bias so information in them cannot be recovered. Furthermore, it demonstrates why re-scaling using a constant factor q does not work well when there are multiple outliers. In contrast, Algorithms 1 and 2 successfully detect all major

Table 3
Clustering results from five algorithms

Vector	LSI	COV	Ando's Alg.	Algorithm 1	Algorithm 2
b_1	Clinton	Clinton, Java	Java	Java	Clinton, Java
b_2	Java	Clinton, Java	Clinton	Clinton	*noise*, Out, Clinton, Java
b_3	*noise*	*noise*, Out	*noise*	*noise*	Matrix, DNA
b_4	Clinton	Out, *noise*	Blue, Soccer	Matrix, DNA	Out
b_5	Java	Out, *noise*	Matrix, DNA	Blue, Soccer	Blue, Soccer
b_6	*noise*	Out, *noise*	Matrix, DNA	Out	*noise*, Out

and outlier clusters. Results from using Algorithm 1 are as follows: *Matrix* and *DNA* are detected by b_4; *Bluetooth* and *Soccer* by b_5; and *Out*—all outliers together—by b_6. In short, all outlier clusters are detected evenly. Results for Algorithm 2 are: *Matrix* and *DNA* are detected by b_3; *Bluetooth* and *Soccer* by b_5; and *Out* by b_2, b_4 and b_6, i.e., all outliers are evenly detected, as in Algorithm 1.

In Table 3, the major clusters are: the *Clinton* and *Java* clusters. The outliers are: the *Blue* (for *Bluetooth*), *Soccer*, *Matrix* and *DNA* clusters. *Out* denotes the set of all outlier clusters.

For our second set of experiments, we constructed a dataset consisting of 10,000 documents and 500 attributes (keywords) with 5 major clusters $M1, M2, \ldots, M5$ (each with 500 documents or 5% of the total), 20 outlier clusters $out1, out2, out3, \ldots, out20$ (each with 200 documents or 2% of the total) and 35% noise. We performed numerical experiments to reduce the dimension of the document-attribute space to 25 dimensions and observed how many major and outlier clusters were detected. For the straightforward LSI algorithm, computation of the 25 basis vectors took 0.96 s, for COV 1.05 s, and for COV with iterative re-scaling (Algorithm 2) 263 s. Since the computation time for LSI did not include conversion into Harwell-Boeing format, and the computation times for the COV algorithms include construction of the covariance matrix, the times for LSI and basic COV are comparable. Iterative re-scaling is very expensive for COV because an updated covariance matrix must be constructed after each set of computations to find one eigenvector. However, our numerical experiments show that the extra computation time is worth the effort if the user is interested in finding as many major and outlier clusters as possible. LSI detects only 19 clusters (5 major + 14 outlier), basic COV detects only 21 (5 major + 16 outlier), while COV with re-scaling detects all 25 major and outlier clusters. Our results are summarized in Table 4. Finally, we note that implementation of LSI with re-scaling (Ando's Algorithm and Algorithm 1) would have been too difficult for this dataset because after a fews steps of re-scaling, we would have been required to compute the SVD of a very large, dense matrix.

4. Conclusion

In this paper, we propose three new algorithms: one named "*COV*" for information retrieval (IR) and relevancy ranking based on analysis of the covariance matrix for documents vectors, and two new algorithms for the detection of multiple outlier clusters. COV shifts the origin of the coordinate system to the "center" of the subspace to spread apart documents as much as possible so that

Table 4
Major and outlier clusters identified using basic LSI and COV

Vector	LSI	COV	Algorithm 2
b_1	M1	M3	M1, M3
b_2	M5	M1	M2, M4
b_3	M4	M5	out12, out17
b_4	M2	M4	out14, out17
b_5	M3	M2	out4, out14
b_6	M1	M3	out1, out14
b_7	M5	M1	out11, out18
b_8	M4	M5	out18
b_9	M2	M4	out6
b_{10}	M3	M2	out7, out16
b_{11}	out4	out8, out19	out7, out20
b_{12}	out2	out7, out19	out10, out20
b_{13}	M2	out7	out2, out10
b_{14}	out19	out12, out16	out13
b_{15}	out16	out12, out13	out8, out13
b_{16}	out14	out13, out15	out3, out8
b_{17}	out12	out3, out15	out3, out11
b_{18}	out5	out3, out11	out11, out19
b_{19}	out13	out9, out11	out9, out19
b_{20}	out20	out9, out14	out5
b_{21}	out17	out5, out14	out15
b_{22}	out10	out5, out20	M5
b_{23}	out7	out1, out20	M5
b_{24}	out15	out1, out17	M4
b_{25}	out3	out2, out17	*noise*

subtle differences in documents can be more easily distinguished. Latent semantic indexing (LSI) (an IR algorithm based on vectors space modeling and dimensional reduction using the singular value decomposition [4]) and COV can be used to find major clusters, but they inadvertently discard information in outlier clusters during dimensional reduction. Recently, Ando [1] proposed an algorithm that modifies LSI to retain information about outliers by introducing a negative bias to directions that are already represented during the computation of basis vectors of the subspace to which the IR problem will be mapped. However, the algorithm has some drawbacks. For instance, it can fail to find all clusters because the bias factor is not well-controlled; it may be numerically unstable; the basis vectors it computes may not be orthogonal; and it cannot process massive document-term matrices. The algorithms we propose overcome these problems for moderate-size databases. Our second iterative re-scaling algorithm based on covariance matrix analysis overcomes the scalability problem and can be successfully applied to very large databases as well.

References

[1] R. Ando, Latent Semantic Space, Proceedings of the ACM SIGIR, ACM Press, New York, July 2000, pp. 216–223.
[2] V. Barnett, T. Lewis, Outliers in Statistical Data, 3rd Edition, Wiley, New York, 1994.

[3] M. Berry, S. Dumais, G. O'Brien, Using linear algebra for intelligent information retrieval, SIAM Rev. 37 (1995) 571–595.
[4] S. Deerwester, et al., Indexing by latent semantic analysis, J. Amer. Soc. Inform. Sci. 41 (1990) 391–407.
[5] C. Eckart, G. Young, A principal axis transformation for non-Hermitian matrices, Bull. Amer. Math. Soc. 45 (1939) 118–121.
[6] G. Golub, C. Van Loan, Matrix Computations, 3rd Edition, John Hopkins University Press, Baltimore, MD, 1996.
[7] A. Jain, R. Dubes, Algorithms for Clustering Data, Prentice-Hall, Englewood Cliffs, NJ, 1988.
[8] M. Kobayashi, M. Aono, H. Samukawa, H. Takeuchi, Information retrieval apparatus for accurately detecting multiple outlier clusters, Japanese Patent (July 5, 2001) filed.
[9] M. Kobayashi, L. Malassis, H. Samukawa, Retrieval and ranking of documents from a database, Japanese Patent (June 12, 2000) filed.
[10] K. Mardia, J. Kent, J. Bibby, Multivariate Analysis, Academic Press, New York, 1979.
[11] E. Rasmussen, Clustering algorithms, in: W. Frakes, R. Baeza-Yates (Eds.), Information Retrieval, Prentice-Hall, Englewood Cliffs, NJ, 1992. (Chapter 16)
[12] Reuters-21578 news data base: http://www.research.att.com/~lewis.
[13] G. Salton, The SMART Retrieval System, Prentice-Hall, Englewood Cliffs, NJ, 1971.

ELSEVIER Journal of Computational and Applied Mathematics 149 (2002) 131–145

JOURNAL OF COMPUTATIONAL AND APPLIED MATHEMATICS

www.elsevier.com/locate/cam

GRAPE project

Junichiro Makino

Department of Astronomy, School of Science, University of Tokyo, 7-3-1 Hongo, Bunkyo-ku, Tokyo 113-0033, Japan

Received 6 November 2001; received in revised form 22 January 2002

Abstract

We overview our GRAvity PipE (GRAPE) project to develop special-purpose computers for astrophysical N-body simulations. The basic idea of GRAPE is to attach a custom-build computer dedicated to the calculation of gravitational interaction between particles to a general-purpose programmable computer. By this hybrid architecture, we can achieve both a wide range of applications and very high peak performance. Our newest machine, GRAPE-6, achieved the peak speed of 32 Tflops, and sustained performance of 11.55 Tflops, for the total budget of about 4 million USD.

We also discuss relative advantages of special-purpose and general-purpose computers and the future of high-performance computing for science and technology.

Keywords: Computational science; Special-purpose computer; Numerical algorithms

1. Introduction

In this paper, we discuss a relatively unexploited approach to the computational science, namely to design and build the computer hardwares specialized and optimized for relatively narrow range of problems.

The development of the computers itself has been regarded as not really being a part of the computational science. It has been considered as what industries and/or computer scientists would do for us. This view was okay when relatively simple computers were still very expensive and had to be shared by a number of researchers from a wide variety of fields. For example, Cray-1, which was completed in 1976, was really a simple and small computer by today's standard. It has just 8 MB of SRAM memory in 16 banks, one multiplier unit and one adder unit. The estimated total

E-mail address: makino@astron.s.u-tokyo.ac.jp (J. Makino).

0377-0427/02/$ - see front matter
PII: S0377-0427(02)00525-3

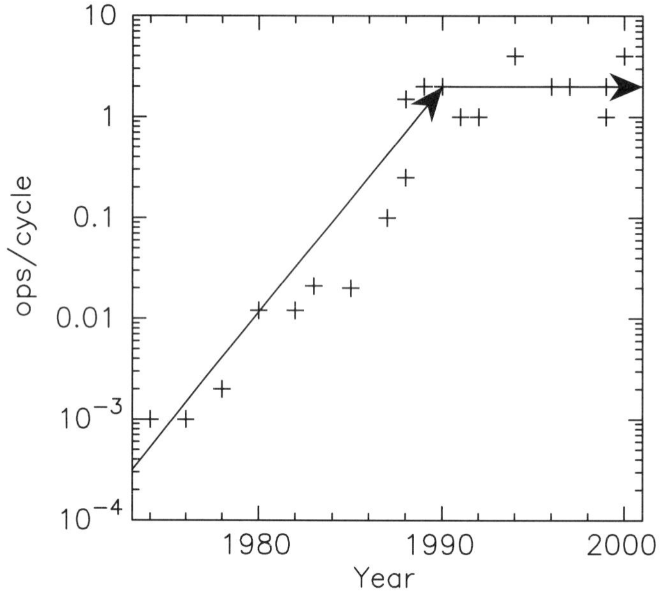

Fig. 1. Number of floating-point operations performed per single machine cycle of representative high-performance microprocessors.

gate count was just 300K. In comparison, today's single microprocessor integrates 40M transistors, which is equivalent to around 10M gates. Thus, the world's largest supercomputer 25 years ago is less than one-tenth of a today's desktop (or even notebook) PC in gate count.

With such a small gate count, it was unpractical to design a machine specialized to one application. Since it was difficult to make just one floating-point calculation unit, the natural way to make a computer is to add control logics to that floating-point unit so that it can perform operations specified by the software.

The available number of transistors, however, has been increasing exponentially in the last half century, in the rate of a factor of 100 per every decade. Thus, present-day microprocessors integrate the number of transistors sufficient to implement several hundreds of floating-point units.

Fig. 1 shows this situation rather clearly for the case of single-chip microprocessors. When it was impossible to implement full-decode multiplier unit into a single chip, we had to use smaller multipliers which needed multiple cycles to perform one operation (before 1989). During this period, we could use the increase in the available number of transistors just to make bigger and bigger multipliers. Thus, in 1980s a factor of 100 improvement in the cycle count, which is consistent with the factor of 100 increase in the number of transistors, was achieved.

In 1990s, however, the decrease in the operation count has practically halted, as clearly seen from Fig. 1. In 1988, Intel shipped the first single-chip microprocessor which can perform more than one floating-point operations per clock cycle, the i860. If the exponential increase in 1980s had continued in 1990s, we would see microprocessors with several hundreds of floating-point units in 2001. Instead, processors like Intel Pentium 4 still perform only a few operations per cycle, even though the number of transistors has increased by a factor of 50 or so.

In other words, practically *all* the increase in the available number of transistors in 1990s was spent for things other than the arithmetic units. These "other" things include large L1 and sometimes L2 caches, multiple execution units for integer operations, datapath and control units for out-of-order execution of multiple instructions, additional logics to implement deeper pipelines and so on.

If we write the same figure for high-end supercomputers, we see essentially the same trend, but shifted by 15 years. With Cray-1 (1976), supercomputers had reached the regime where they can perform one floating-point operations per cycle. Until early 1990s, the evolution of the number of floating-point units on a single supercomputer had been very slow. Machines of early 1990s such as Hitachi S-3800, NEC SX-4 and Cray T-90 had only a few tens of arithmetic units, instead of a few thousands which was theoretically possible.

The most obvious way to use multiple floating-point units in a single computer is to develop a parallel computer. Thus, in 1980s, parallel computers based on one-chip microprocessors and those based on supercomputers were built. Parallel computers based on microprocessors were mostly distributed-memory processors, which were essentially a cluster of simple computers connected by a relatively slow and inexpensive network. Thus, large machines with more than 1000 processors were possible. Parallel computers based on supercomputers were more complicated shared-memory machines. The shared-memory architecture limited the number of processors to 32 or less.

Although microprocessor-based machines used much higher number of processors, they were still slow in absolute performance simply because each microprocessor was slow. Parallel supercomputers offered higher peak performance.

However, in 1980s, the performance of microprocessors had improved drastically, because they could fully utilize the increase in the available number of transistors. Thus, by early 1990s, microprocessor-based machines started to offer better price-performance ratio than parallel supercomputers.

In 1993, Fujitsu announced VPP-500, the first parallel supercomputer with distributed-memory architecture. With this architecture, available number of processors was boosted to more than 100, resulting in a very large improvement in the price performance of supercomputers.

However, this move was counteracted by the birth of "Beowulf-class" machines [2,10], which are really simple cluster of commodity PCs with Intel x86 CPUs connected by standard Ethernet. This architecture offers typically one order of magnitude better price performance than other computer architecture, simply because the production cost is low due to mass production. "High-end" Intel x86 CPUs are made and sold in units of hundreds of millions per year, while supercomputers were sold in units of hundreds. Thus, the sheer number of chips reduces the production cost per chip, and yet huge development cost can be spent for Intel x86 chip since the total revenue would be tens of billion USDs.

Thus, at present the computer architecture which offers both the best price performance and best absolute throughput is a large cluster of PCs with microprocessors of high-volume production. However, as we have already seen at the beginning, this architecture, at present, does not make good use of the transistors available on one chip. To make matters worse, the fraction of transistors used for floating-point operations will decrease in future. In a few years, the architecture called on-chip multiprocessor will be common. This evolutionally path is again following the path of supercomputers, but with the delay of 20 years. Thus, we can "predict" the future of on-chip multiprocessor architecture in the near future rather accurately by just looking at the evolution of supercomputers in 1980s: The evolution will be slow.

So our question here is: Is there any better way to design a computer for scientific simulation? We believe the answer is "yes", and in the rest of this paper we describe why we believe so.

In Section 2, we discuss the potential advantages and drawbacks of designing and developing special-purpose systems. The potential gain is very large, since, at least in some cases, we might be able to use a fair fraction of available transistors to perform useful arithmetic operation. On the other hand, there are many practical difficulties which would offset the potential gain. In Section 3, we discuss our GRAPE project [11,8] as an example of, well, reasonably successful projects to develop special-purpose computers. In Section 4, we speculate on the future of the large-scale scientific computing.

2. Special-purpose computing

2.1. Astrophysical N-body problem

The idea of building special-purpose computers for specific numerical problems is certainly not new. In fact, the very first digital electronic computers (ABC and ENIAC) were designed to solve specific problems. However, the designers of early computers found that their machines could be used for a much wider range of problems, and thus the evolution of the programmable general-purpose computer started.

At the time when even the largest supercomputers did not have a fully parallel multiplier, it did not make sense to design a special-purpose computer. Any computer, either special- or general-purpose, consists of arithmetic units (mainly multipliers), control logic and memory. A special-purpose computer might have simpler control logic or smaller memory than a general-purpose computer. However, if the cost of the arithmetic unit itself is a fair fraction of the total cost of the machine, we cannot gain much by trying to reduce the cost of the rest of the system.

This situation, however, has changed completely, as we have seen in the previous section. To give a concrete example, let us discuss the astrophysical N-body problem and our GRAvity PipE (GRAPE) hardware specialized for that problem. The master equation for the gravitational N-body problem is given by

$$\frac{d^2\mathbf{x}_i}{dt^2} = -\sum_{j\neq i} Gm_j \frac{\mathbf{x}_i - \mathbf{x}_j}{|\mathbf{x}_i - \mathbf{x}_j|^3}, \tag{1}$$

where $\mathbf{x}_i$ and m_i are the position and mass of the particle with index i and G is the gravitational constant. The summation is taken over all particles (typically stars) in the system under study.

The number of particles N varies widely depending on what kind of systems is studied. A small star cluster like Hyades might consist of several thousand stars. Globular clusters typically contain one million stars, and galaxies hundreds of billions of stars. In the case of galaxies, it is impossible to model them on a star-by-star basis, and we model the distribution of stars in the six-dimensional phase space with much smaller number of particles.

If N is larger than several tens, the calculation of the right-hand side of Eq. (1) dominates the total calculation cost. If N is very large, we could use sophisticated algorithms such as the Barnes–Hut treecode [1], FMM [3], or the particle-mesh or particle–particle particle-mesh method [4]. However, with treecode or FMM, the direct evaluation of gravitational interaction between near-neighbor particles still dominates the total cost. Even in the case of the particle-mesh method, if we want high accuracy, near-neighbor interaction must be calculated directly.

The calculation itself is rather simple. All we have to do is to calculate and accumulate the gravitational force between two particles. Also, there is very large degree of potential parallelism, since in principle we can evaluate all N^2 interactions in parallel. Thus, N-body simulations are quite well suited for execution on massively parallel processors or PC clusters.

As discussed earlier, however, our aim here is to achieve a better use of transistors than what is realized with present microprocessors. Since N-body simulations have a large degree of parallelism, we first concentrate on how to make use of large number of transistors available in one chip. We return to the problem of parallelization later.

The problem with present microprocessors is that very small fraction of the available transistors is used to implement floating-point arithmetic operations. The number of floating-point units in a single one-chip microprocessor has been almost constant for the last 10 years, though the number of transistors has increased by almost two orders of magnitude.

Why the number of floating-point units has been just a few, even though there are enough transistors to integrate hundreds of them? There are essentially two reasons. The first reason is that we do not know how to use such a large number of arithmetic units. To be precise, it is of course possible to use multiple arithmetic unit in parallel for applications with high degree of parallelism such as N-body simulations. Most of scientific applications do have high degree of parallelism, and would run with high efficiency on processors with large number of arithmetic units. However, at present, microprocessors are designed to achieve the following two goals. The first is to run the most widely used applications as fast as possible. That is, to run Microsoft Windows operating system and Microsoft Office application package. The other is to run standard benchmarks such as SPEC CPU2000 suites as fast as possible. Of course, these are most important applications and benchmarks from the viewpoint of commercial success. However, it goes without saying that the performance of a microprocessor for scientific applications has rather little to do with its performance in processing MS-Word documents. Clearly, multiple floating-point units would not help much improving the performance of MS-Word.

The second reason is that it is difficult to provide sufficient memory bandwidth to keep processors busy. It is not impossible to integrate 100 floating-point arithmetic unit on a chip. However, if they operates independently, or even if they operate in the SIMD fashion, each unit needs to get two words of data from memory and to store one word of data, to perform one floating-point operation. If these 100 arithmetic units operate on 500 MHz clock, which is rather slow by today's standard, the necessary memory bandwidth would be 150 Gwords/s or 1.2 Tbytes/s, which is about 500 times higher than the speed of the off-chip transfer bandwidth of today's microprocessors. A 2 GHz Intel Pentium 4 processor has the theoretical peak transfer bandwidth of just 3.2 GB/s.

Of course, the necessary bandwidth is smaller for operations like summation. The difference, however, is small compared to the factor of 500 discrepancy shown above. This huge discrepancy means it is not easy to keep just one processor busy. With present clock speed of around 2 GHz, one processor needs the bandwidth of 48 GB/s, or 15 times more than what is actually provided. It is quite understandable that architects of microprocessors are not too interested in adding more floating-point units. External memory cannot feed a single unit. Unless there is a way to vastly increase the bandwidth, little or no gain is achieved by increasing the number of floating-point units, even for scientific applications. It makes a lot of sense that they used a large fraction of the

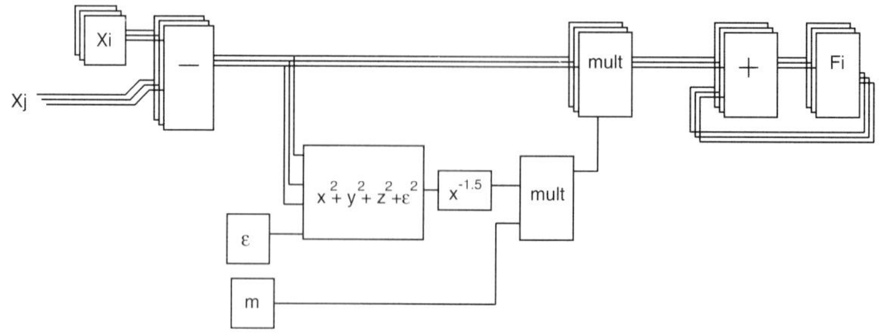

Fig. 2. The force calculation pipeline.

transistors on chip to cache memory, which can somewhat reduce the necessary off-chip memory bandwidth.

To summarize, though the current chip design with just one or two arithmetic units looks like a waste of the transistors, processor architects have good reasons to choose such designs.

2.2. Special-purpose computer

If we are to design a computer specialized for N-body simulation, or rather, the calculation of the gravitational interaction between particles, we can work around the limitations discussed above.

Fig. 2 shows the basic "processor" for our N-body computer. Calculation of the gravitational force from one particle to another requires some 40 floating-point operations, depending on how you count division and square-root operations. If we assign 10 operations to each of them, we end up with about 40 operations per interaction. Instead of using one arithmetic unit to perform these 40 operations, we connect 40 floating-point arithmetic units in the form of a pipeline, which calculate and accumulate the force from one particle at each clock cycle.

With this approach, we can work around the two limitations discussed above. The first one was the efficiency of the system on commercial applications. This we simply ignore. Clearly not everybody who buys a PC wants a fast computer for astrophysical N-body simulation.

The second limitation is the memory bandwidth. With the simple pipeline of Fig. 2, 40 floating-point operations are performed for the input of four words. Thus, there is a saving of a factor of 30, compared to a single arithmetic unit or multiple units working in an SIMD fashion.

When we integrate multiple pipelines, we can let these pipelines to calculate the forces on different particles from the same particle. Thus, we can integrate multiple pipelines into a chip without increasing the necessary memory bandwidth.

It is also possible to reduce the required bandwidth for one pipeline, by letting a single pipeline to calculate forces on multiple particles. This can be done by adding the register files for the input particles and calculated forces, and by switching them at each clock cycle. We call this the virtual multiple pipeline (VMP) architecture [9]. The idea of VMP is quite similar to what is now called as multithredding, but the difference is that VMP reduces the required memory bandwidth, while usual multithredding only relaxes the requirement for the memory latency.

Note that many of the above ideas can be applied to the calculation of particle–particle interaction on general-purpose programmable computer. Thus, with some careful programming (sometimes in

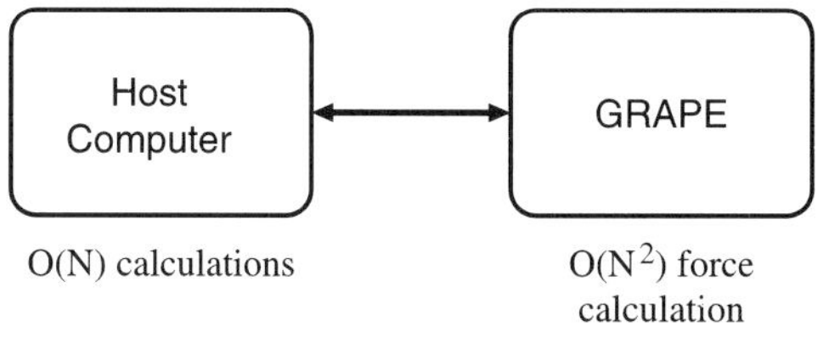

Fig. 3. The basic architecture of GRAPE.

assembly languages), it is possible to achieve the performance close to the theoretical peak performance of the processor. Here, the simple fact that there are only a few arithmetic units on a chip limits the ultimate performance.

Our GRAPE project is based on this idea of integrating multiple (both physical and virtual) pipelines specialized for gravitational force calculation into one chip (or one board, when the number of available transistors was not as large as what is now). To give an idea, the GRAPE-6 system which was completed in July 2001 integrates 1024 pipeline chips, each with 6 pipeline processors which calculate the gravitational force and its first time derivative. As a result, single GRAPE-6 processor chip integrates about 350 floating-point arithmetic units.

The clock speed of GRAPE-6 chips is only 100 MHz, though it uses reasonably advanced 0.25 μm technology. The reason for this rather low clock speed is simply that we did not have sufficient resource to fine tune the design. Even so, with 350 arithmetic units a single chip offers peak speed exceeding 30 Gflops, which is still more than a factor of 10 better than that of the fastest microprocessors. In addition, because of this rather low clock speed the power consumption and therefore heat dissipation are small (around 15 W). This low power consumption is quite important in reducing the overall cost of the system and the running cost (electricity is expensive in Japan).

Since the GRAPE hardware performs only the force calculation, all other operations, such as the time integration of the orbits of stars and analysis of the calculated results, must be done on somewhere else. Therefore, we connect the GRAPE hardware and general-purpose computer by some communication link. This is the basic structure of our GRAPE systems, as shown in Fig. 3.

3. GRAPE project

We started GRAPE project in 1988. The first machine we completed, the GRAPE-1 [5] was a single-board unit on which around 100 IC and LSI chips were mounted and wire-wrapped. We used commercially available IC and LSI chips to implement force calculation pipeline. This choice was a natural consequence of the fact that we lacked both money and experience to design custom LSI chips. In fact, none of the original design and development team of GRAPE-1 had the knowledge of electronic circuit more than what was learned in basic undergraduate course for physics students.

GRAPE-2 is similar to GRAPE-1 since it is also based on commercial LSI chips. The difference is in the numerical accuracy. For GRAPE-1, we used an unusually short word format, to make the hardware as simple as possible. The input coordinates are expressed in 15-bit fixed point format. After subtraction, the result is converted to 8-bit logarithmic format, in which we use just 3-bits for "fractional" part. This format is used until we obtain $1/r^3$. The final accumulation was done in

48-bit fixed point to avoid overflow and underflow. The advantage of the short format like 8-bit logarithmic format is that we could use ROM chips to implement complex functions that require two inputs. Any function of two 8-bit words can be implemented by one ROM chip with 16-bit address input. Thus, all operations other than the initial subtraction of the coordinates and final accumulation of the force were implemented by ROM chips.

The drawback of GRAPE-1 was its limited accuracy, which was insufficient for fairly wide range of astrophysical simulations, though for many other applications accuracy of GRAPE-1 turned out to be sufficient. With GRAPE-2, we used standard IEEE-754 format (64 bit for initial subtraction and accumulation, and 32 bit for all other operations).

GRAPE-3 was our first machine with custom LSI chip. The number format was again the combination of the fixed point and logarithmic format similar to what were used in GRAPE-1, but implementation of the arithmetic operations was quite different since we could not integrate large tables to a custom LSI chip. Conversions between fixed and logarithmic formats were implemented by shifters and small lookup table, and addition in logarithmic format is implemented by two adders and one small lookup table. Chip design was done as a joint research project between University of Tokyo and Fuji Xerox Corp. The chip was fabricated using 1 μm design rule by National Semiconductor. The number of transistors on chip was 110K. Single chip operates at 20 MHz clock speed, offering the speed of about 0.8 Gflops. We designed a printed-circuit board with eight chips, for the speed of 6.4 Gflops per board. Thus, GRAPE-3 is also our first trial to integrate multiple pipelines into a system.

Also, GRAPE-3 was the first GRAPE machine to be manufactured and sold by a commercial company. Nearly 100 copies of GRAPE-3 have been sold to more than 30 institutes (more than 20 outside Japan).

With GRAPE-4, we finally integrated a high-accuracy pipeline into one chip. Also, with this chip we added the additional pipeline to calculate the first time derivative of the force, so that we can implement high-order time integration algorithms in a simple and efficient way [7]. We could not fit a full single pipeline to a chip with the technology available at that time. So we designed a "1/3" version of the pipeline which processes x, y and z components of coordinates serially in three consecutive clock cycles. The chip was fabricated using 1 μm design rule by LSI logic. Total transistor count was about 400K.

For GRAPE-4, we got a rather large grant of about 2 million USD. So we were able to build a large system consisting of 1728 pipeline chips (36 PCB boards each with 48 pipeline chips). GRAPE-4 system operates on 32 MHz clock, delivering the speed of 1.1 Tflops. Completed in 1995, GRAPE-4 was the first computer for scientific calculation to achieve the peak speed higher than 1 Tflops. Also, for 1995 and 1996 it was awarded the Gordon Bell Prize for peak performance, which is given for a real scientific calculation on parallel computer with the highest performance.

Technical details of machines from GRAPE-1 to GRAPE-4 can be found in our book [8] and references therein.

GRAPE-5 [6] was an improvement over GRAPE-3, with a similar, but more accurate number format than that was used for GRAPE-3. Also, it integrated two full pipelines which operate on 80 MHz clock. Thus, single GRAPE-5 chip offered the speed 8 times more than that of the GRAPE-3 chip, or the same speed as that of a 8-chip GRAPE-3 board. GRAPE-5 was awarded the 1999 Gordon Bell Prize for price performance. The GRAPE-5 chip was fabricated with 0.35 μm design rule by NEC.

Table 1
History of GRAPE project

GRAPE-1	(89/4 — 89/10)	120 Mflops, low accuracy
GRAPE-2	(89/8 — 90/5)	40 Mflops, high accuracy (32/64 bit)
GRAPE-1A	(90/4 — 90/10)	240 Mflops, low accuracy
GRAPE-3	(90/9 — 91/9)	14 Gflops, high accuracy
GRAPE-2A	(91/7 — 92/5)	180 Mflops, high accuracy
HARP-1	(92/7 — 93/3)	180 Mflops, high accuracy Hermite scheme
GRAPE-3A	(92/1 — 93/7)	6 Gflops/board some 80 copies are used all over the world
GRAPE-4	(92/7 — 95/7)	1 Tflops, high accuracy Some 10 copies of small machines
MD-GRAPE	(94/7 — 95/4)	1 Gflops/chip, high accuracy programmable interaction
GRAPE-5	(96/4 — 99/8)	5 Gflops/chip, low accuracy
GRAPE-6	(97/8 — 01/7)	32 Tflops, high accuracy

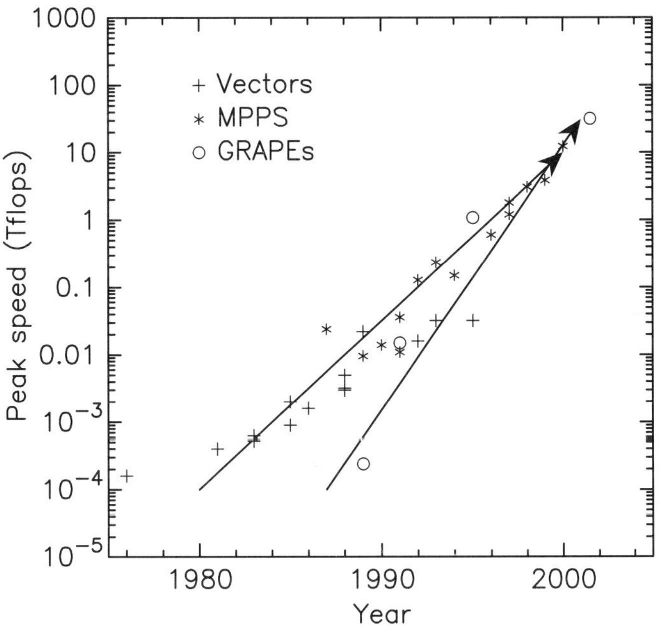

Fig. 4. The evolution of GRAPE and general-purpose parallel computers. The peak speed is plotted against the year of delivery. Open circles, crosses and stars denote GRAPEs, vector processors, and parallel processors, respectively.

GRAPE-6 is similarly the improvement over GRAPE-4. Since it is our newest hardware, we will give a close inspection of its architecture in the next section.

Table 1 summarizes the history of GRAPE project. Fig. 4 shows the evolution of GRAPE systems and general-purpose parallel computers. One can see that evolution of GRAPE is faster than that of general-purpose computers.

These GRAPE hardwares, including GRAPE-6, have been applied to a number of astrophysical problems both by our group and by other researchers worldwide. Since there are too many interesting results, we do not try to list them. Some of the recent results are summarized in the Proceedings of IAU Symposium 208 "Astrophysical Supercomputing Using Particles", held in July 2001. Some more informations can be found at `http://www.astrogrape.org`.

3.1. Machines for molecular dynamics (MD)

Classical MD calculation is quite similar to astrophysical N-body simulations since in both cases we integrate the orbit of particles (atoms or stars) which interact with other particles with simple pairwise force. In the case of Coulomb force, the force law itself is the same as that of the gravitational force, and the calculation of Coulomb force can be accelerated by GRAPE hardware.

However, in MD calculations the calculation cost of van der Waals force is not negligible, though van der Waals force decays much faster than the Coulomb force (r^{-7} compared to r^{-2}).

It is fairly straightforward to design a hardware which can handle particle–particle force, which is some arbitrary function of the distance between particles. We approximate the given function by a table of polynomials. In fact, we use this combination of lookup table and polynomial approximation for the calculation of $1/r^3$ from r^2 in GRAPE hardwares. So the actual change in the design of the hardware is rather minor.

We have designed two machines, GRAPE-2A and MD-GRAPE, following these lines of idea. GRAPE-2 was built using commercial chips and MD-GRAPE used a custom-designed pipeline chip.

Another difference between astrophysical simulations and MD calculations is that in MD calculations usually the periodic boundary condition is applied. Thus, we need some way to calculate Coulomb forces from image particles. The direct Ewald method is rather well suited for implementation in hardware. In 1991 we developed WINE-1, a pipeline to calculate the wavespace part of the direct Ewald method. The real-space part can be handled by GRAPE-2A or MD-GRAPE hardware.

In 1995, a group led by Ebisuzaki in RIKEN started to develop MDM, a massively parallel machine for large-scale MD simulations. Their primary goal is the simulation of protein molecules.

MDM consists of two special-purpose hardwares, massively parallel version of MD-GRAPE (MDGRAPE-2) and that of WINE (WINE-2). MDGRAPE-2 part consisted of 1536 custom chips with four pipelines for the theoretical peak speed of 25 Tflops. WINE-2 part consists of 2304 custom pipeline chips, for the peak speed of 46 Tflops.

MDM shared the 2000 Gordon-Bell performance Prize with GRAPE-6. It also was selected as the finalist for the 2001 Gordon-Bell performance Prize, again along with GRAPE-6.

4. GRAPE-6

In 1997, we started the GRAPE-6 project. It is a 5-year project funded by Japan Society for the Promotion of Science (JSPS), and the planned total budget is about 500M JYE.

The GRAPE-6 is essentially a scaled-up version of GRAPE-4 [9], with the peak speed of around 100 Tflops. As of the time of writing, a 32 Tflops system with 1024 chips is in operation. The peak speed of a single pipeline chip is 31 Gflops. In comparison, GRAPE-4 consists of 1728 pipeline chips, each with 600 Mflops. The increase of a factor of 50 in speed is achieved by integrating

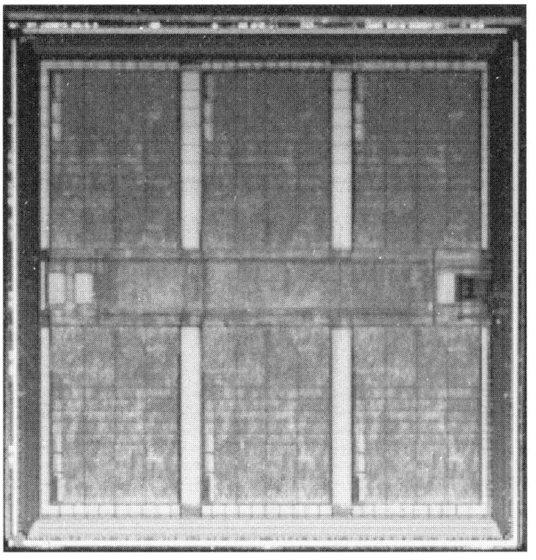

Fig. 5. The GRAPE-6 processor chip.

Fig. 6. The processor board of the GRAPE-6 with 32 processor chips. Four processor chips are mounted on modules, on which eight memory chips are also mounted on the bottom side. One board houses eight modules.

six pipelines into one chip (GRAPE-4 chip has one pipeline which needs three cycles to calculate the force from one particle) and using 3 times higher clock frequency. The advance of the device technology (from 1 to 0.25 μm) made these improvements possible. Fig. 5 shows the processor chip delivered in early 1999. The six pipeline units are visible.

Fig. 7. The 32-board, 32-Tflops GRAPE-6 system with its host computer. The host is a cluster of four PCs with 1.7 GHz Intel P4 processors connected by 100 Mb Ethernet.

Figs. 6 and 7 show the processor board with 32 processor chips and the 32-board system. This 32-board system has the theoretical peak speed of 32 Tflops, and has achieved the sustained speed of 11.5 Tflops for the simulation of 1.4-million-body system.

We plan to extend this system to 80-board, 80-Tflops system by the end of FY 2001. Single-board systems (4–32 chips) are commercially available.

5. Discussion

5.1. Are special-purpose computers difficult to build?

We have seen that it is possible to develop special-purpose computers which offer price performance and also absolute performance better than those of general-purpose computers by one or two orders of magnitude.

However, our GRAPE project is not the only project to develop special-purpose computers, and yet it is not too easy to name other projects which have achieved similar level of success. We briefly discuss why.

There are a variety of reasons why a project ends up as a failure. These include:

(1) *Long development time*: Even though there are enormous inefficiency, the performance of general-purpose computers is still improving at the rate of a factor of 10 in every 5 years.

Therefore, if you spend 5 years to develop a machine, you lose a factor of 10 in relative advantage. Of course, in many cases the designers originally underestimated the development time. So this problem was very difficult to avoid.

(2) *Too small gain*: Even if your machine achieved the advantage of a factor of 10 by the time of completion, in 5 years it will lose all advantages. So the lifetime of your machine is rather short. If the advantage is a factor of 10 *at the time of the design*, the project is guaranteed to fail.
(3) *Too wide applications*: This is not necessarily a failure, if high performance is achieved. However, this is the most common failure which leads to both small gain and long development time.
(4) *Too narrow applications or too difficult to use*: If there is not much scientific outcome from your machine, it would not be regarded as a great success even if it achieves very good performance.
(5) *Obsolete technology/design method*: Device technology is moving very fast, and so is the design software/methodology. So, if the technology you used was 3-year old, you have already lost the relative advantage of a factor of 4.
(6) *Untested technology/design method*: Even though the advance of the device technology is pretty well predictable, in most cases what the manufacturers claim that they can deliver would not be delivered on time. So be careful.

Of course, projects to develop general-purpose computers also suffer most of the above problems. The advantage of the special-purpose design is the possibility to make better use of the available transistors. The disadvantage is the limitations in available design resources (design experts, budget, access to the latest technology, etc.).

5.2. Future prospect

We overviewed the technological and economical trends in high-performance scientific computing. At least for certain range of problems, a special-purpose computer, or a combination of special- and general-purpose computers such as our GRAPE systems, offers a real and proven advantage over traditional general-purpose computers. The relative advantage has been increasing, and will keep to do so for the next one or two decades, unless some radical change in the design method for general-purpose computer would take place.

Currently, it seems the direction of the evolution of general-purpose computer is a cluster of commodity PCs. Current microprocessors for commodity PCs offer very high performance in very low cost, partly because of their low production cost due to mass production, and partly because the investment for the design of the chip is actually very high. The large volume of production justifies the high development cost. However, the fact that microprocessors for commodity PCs offer performance better than any other general-purpose computers on the market does not necessarily imply that their design is optimal. As we have seen, only a tiny fraction of the available transistors is used for actual arithmetic operations, and that fraction has been decreasing quite rapidly.

Special-purpose systems, at least in principle, will not suffer these problems. In practice, however, the approaches we have taken so far are becoming more and more difficult, because the initial development cost of the custom LSI goes up as the technology advances. The development cost goes up because of two reasons. The first is that the amount of work to do the logic design, test design, physical layout and design validation increases as the number of transistors in a chip increases. In the case of special-purpose systems, the amount of work for the logic and test design, which we can do in-house, does not increase too rapidly, but physical layout and design validation, which is the work of the semiconductor company, take a long time and therefore lots of money. The second reason is that the investment needed to build the semiconductor plant increases rapidly as the technology advances.

Roughly speaking, for same physical size of the chip, the amount of money we paid was inversely proportional to the design rule: around a quarter million USD for 1 μm (GRAPE-4) and around 1 million USD for 0.25 μm (GRAPE-6). In both cases, the size of the chip is around 110 mm^2. If this trend continues, the design cost of 0.13 μm would be 2 million USD. The total budget must be significantly larger than the development cost of the chip, since otherwise the price of a single chip would be too high. It is not easy to get such a large fund for a project of theoretical study in pure science.

One possible compromise is the use of field programmable gate array (FPGA) chips as the building blocks of the pipelined processors. This choice will reduce the initial cost from 1 million dollars to less than 10,000 dollars (the price of the design software), since the FPGA chip itself is mass-produced and the design is loaded to a FPGA chip by configuring the switches and lookup tables in the chip.

Roughly speaking, because of this programmability, the calculation speed achieved by one FPGA chip is 100 times lower than that can be achieved with a custom LSI of the same size and same technology. However, this difference can be offset by the possibility of using the most advanced technology, the possibility to fine-tune the design to individual problems, and most importantly, much shorter design cycle time.

To give an example, large FPGA chips at the time of writing (Summer 2001) have the nominal gate count of around 1 million, which is sufficient to implement the logic for a GRAPE-4 chip (100K gates), and the clock speed would be around 75 MHz or more. Thus, one FPGA chip can deliver 1.5 Gflops. This might not sound very impressive compared to 31 Gflops of GRAPE-6 chip or around 2 Gflops of present microprocessors. However, compared to microprocessors, building a massively parallel machine out of FPGA is much easier, and we can expect higher execution efficiency for the same reason, as we have achieved higher efficiency on GRAPE hardwares.

Here again, in the long run we might see the same problem as the general-purpose computer. The on-chip wiring would ultimately limit the speed and the circuit density. Because of the requirement of the programmability, as the number of transistors on the FPGA chips increases, more and more fraction of these chips will be used for wiring. However, for the next several years, this limitation would not be too severe, and it is possible that some new design philosophy will allow us to make better use of FPGA chips.

It is at least possible to use FPGAs for "proof of the concept" studies, where we demonstrate that one particular custom design is actually usable and that it achieves better cost-performance than general-purpose solutions. If that demonstration is successful, the grant large enough to make custom LSI might be offered.

To summarize, the initial cost of the large custom LSI might become too high for the level of the amount of grants we can reasonably expect. However, machines based on FPGAs can be used for small projects. The cost advantage of FPGAs will not be as large as that of custom LSI chips, but compared to general-purpose microprocessors, they still offer large advantage. So we expect to see many successful projects to apply FPGAs for large-scale computing in the near future. The largest ones will be done on custom LSIs, but the rest will be done on FPGAs.

Acknowledgements

We would like to thank Daiichiro Sugimoto, Toshikazu Ebisuzaki, Makoto Taiji, Tomoyoshi Ito, Toshiyuki Fukushige and many others who were involved in the development of the six generations of GRAPE hardwares, and Yoko Funato, Simon Portegies Zwart, Steve McMillan, Piet Hut and again many others for discussions and collaborations in software development and scientific works on GRAPE hardwares. This work is supported by the Research for the Future Program of Japan Society for the Promotion of Science (JSPS-RFTP97P01102).

References

[1] J. Barnes, P. Hut, A hierarchical $o(n \log n)$ force calculation algorithm, Nature 324 (1986) 446–449.
[2] D.J. Becker, T. Sterling, D. Savarese, D.J.E., U.A. Ranawake, C.V. Packer, Beowulf: a parallel workstation for scientific computation, in: Proceedings of the 1995 International Conference on Parallel Processing (ICPP), IEEE Computers Society, Los Alamitos, 1995, pp. 11–14.
[3] L. Greengard, V. Rokhlin, A fast algorithm for particle simulations, J. Comput. Phys. 73 (1987) 325–348.
[4] R.W. Hockney, J.W. Eastwood, Computer Simulation Using Particles, IOP Publishing, Ltd., Bristol, 1988.
[5] T. Ito, J. Makino, T. Ebisuzaki, D. Sugimoto, A special-purpose n-body machine grape-1, Comput. Phys. Comm. 60 (1990) 187–194.
[6] A. Kawai, T. Fukushige, J. Makino, M. Taiji, Grape-5: a special-purpose computer for n-body simulations, Publ. Astronom. Soc. Japan 52 (2000) 659–676.
[7] J. Makino, S.J. Aarseth, On a hermite integrator with ahmad-cohen scheme for gravitational many-body problems, Publ. Astronom. Soc. Japan 44 (1992) 141–151.
[8] J. Makino, M. Taiji, Scientific Simulations with Special-Purpose Computers—The GRAPE Systems, Wiley, Chichester, 1998.
[9] J. Makino, M. Taiji, T. Ebisuzaki, D. Sugimoto, Grape-4: a massively parallel special-purpose computer for collisional n-body simulations, Astrophys. J. 480 (1997) 432–446.
[10] T.L. Sterling, J. Salmon, D.J. Becker, D.F. Savarese, How to Build a Beowulf: A Guide to Implementation and Application of PC Clusters, MIT Press, Cambridge, MA, 1999.
[11] D. Sugimoto, Y. Chikada, J. Makino, T. Ito, T. Ebisuzaki, M. Umemura, A special-purpose computer for gravitational many-body problems, Nature 345 (1990) 33–35.

ELSEVIER Journal of Computational and Applied Mathematics 149 (2002) 147–153

JOURNAL OF COMPUTATIONAL AND APPLIED MATHEMATICS

www.elsevier.com/locate/cam

Future of supercomputing

Yoshio Oyanagi *

Department of Computer Science, University of Tokyo, Hongo 7-3-1, Bunkyo-ku, Tokyo, 113-0033 Japan

Received 26 October 2001; received in revised form 11 February 2002

Abstract

Supercomputing or High Performance Computing plays ever more important roles in industrial fields as well as in basic research. Based upon the history of supercomputers in the last few decades, a personal view of the supercomputing in the first decade of the 21st century is presented.

Keywords: Supercomputers; HPC; Simulation; Computational science; Petaflops computer

1. From supercomputing to high performance computing

The first "supercomputer" Cray-1 was built by Seymour Cray in 1976 and people thought only a few of such supercomputers would meet the demand from science and engineering. At present, however, a notebook PC in your bag supersedes the computing power of Cray-1.

Supercomputing has played important roles in industrial fields such as automobile, aeronautics, building and civil engineering, electric and electronic engineering, material and pharmaceutics, as well as in basic research such as elementary particles, chemistry, condensed matter, genome, protein and complex systems. Supercomputers are regarded as the computers which have an order of magnitude higher performance. They are special expensive facilities which only government laboratories or universities or big companies can afford.

Currently, although supercomputing is playing ever more important roles in various fields, the word "supercomputer" is not very popular. It is replaced with "high performance computer". Needless to say, supercomputers are high performance computers, but the word "high performance computer"

* Tel.: +81-033812-2111; fax: +81-3-3818-1073.
E-mail address: oyanagi@is.s.u-tokyo.ac.jp (Y. Oyanagi).

0377-0427/02/$ - see front matter
PII: S0377-0427(02)00526-5

emphasizes the continuity with ordinary computers. High performance computer (HPC) has become a part of a larger set of continua of computing powers starting with PCs.

2. History of HPC

We would like to review briefly the history of supercomputers or high performance computers. The present author proposed [1] a division of the history into periods each with about five years.

2.1. *Primordial ages*

After an experimental parallel computer Illiac IV (1972), the first successful commercial vector computer Cray-1 was shipped in 1976 by CRI (Cray Research Inc.). It is to be noted that just after a couple of years a number of computers with vector facilities came out from various Japanese vendors. The first vector processor in Japan was Fujitsu's FACOM 230-75 APU (Array Processing Unit) in 1977. Hitachi produced HITAC M-180 IAP (Integrated Array Processor) in 1978, M-200H IAP (1979) and M-280H IAP (1982). NEC also produced ACOS-1000 IAP (1982). In these Japanese machines vector processors were integrated in mainframe computers.

2.2. *The first period*

In the first-half of the 1980s, in the US, CRI shipped Cray XMP (1983) and CDC (Control Data Corporation) shipped Cyber 205 (1981). In this period, Japanese vendors started to produce full-fledged vector computers: Hitachi's S810 (1983), Fujitsu's VP200 (1984) and NEC's SX-2 (1985). There was a striking contrast between American and Japanese supercomputers. American ones had relatively small number of arithmetic units (2 or 3) and adopted parallel architecture, while Japanese ones had as many as 12 arithmetic units and had only one CPU.

2.3. *The second period*

In the second-half of the 1980s, the players in the first period upgraded their machines: CRI's Cray-2 (1985) and Cray YMP (1988), ETA (CDC's subordinate)'s ETA10 (1987), Hitachi's S820 (1987), Fujitsu's VP2600 (1989) and NEC's SX-3 (1991). Although VP2600 and SX-3 adopted parallel architecture, their multiplicity (at most 4) was smaller than that of American machines (8).

In this period, a number of venture companies started in US to develop massively parallel machines like Thinking Machine Corporation's (TMC) CM-2 (1987), Intel's iPSC (1985) or MasPar's MP-1 (1990). On the other hand in Japan, no commercial parallel machines were developed although there were quite a few research parallel machines in universities and laboratories.

2.4. *The third period*

In the first-half of the 1990s, there was a big change in the supercomputer business in US as well as in Japan. After CDC closed ETA in 1989, CRI and CCC (Cray Computer Corporation, founded by Seymour Cray in 1989) were the only two vendors of vector computers in US. CRI shipped new vector computer C90 (1991) and a new series of low-cost, compact CMOS supercomputer J90 (1994). The CCC placed Cray-3 in NCAR (1993) but unfortunately CCC went bankrupt in 1995.

In Japan, vector vendors took different strategies. Hitachi adopted a parallel (shared memory) vector machine S3800 (1993), while Fujitsu produced distributed memory vector processors VPP500 (1995). NEC shipped a cluster of shared memory vector processors SX-4 (1997).

MPPs became successful in US due to their better price/performance ratios. To name a few, Intel's Paragon (1991), TMC's CM-5 (1991) and MasPar's MP-2 (1992). The biggest event was that IBM entered in this field by announcing SP-1 (1993) and then SP-2 (1994). IBM's SP series soon became the best seller of the MPPs. In Japan Fujitsu and NEC began to sell AP1000 (1994) and Cenju-3 (1993), respectively, as a testbed of parallel research.

2.5. The fourth period

The second-half of the 1990s was also an era of big changes. CRI, after shipping T90 (1995) and T3E (1996), was merged by SGI (1996). Cray Unit in SGI started to sell a CMOS vector machine SV1 (1996). Finally Tera Company took Cray Unit and became new Cray Inc. (2000).

In Japan, Fujitsu continued its distributed memory vector processor series with VPP300 (1995), VPP700 (1996) and VPP5000 (1999). Along with vector computers, Fujitsu started to ship AP3000 series (1996), an UltraSPARC parallel server. NEC announced SX-5 (1998) and SX-6 (2001) as successors to SX-4. NEC also launched a parallel computer Cenju-4 (1997) based on MIPS R10000. Hitachi started SR2201 (1996) and SR8000 (1998) RISC parallel machines based on pseudo-vector processing.

3. Trends in the last 10 years

3.1. Top 500

We would like to see the history of the development of HPC. The recent progress can be seen in the so-called Top 500 list [3], which lists the 500 fastest computers according to the performance of the LINPACK benchmark since 1993. This benchmark measures the performance of a computer in terms of the speed to solve a dense linear equation by the direct method. The size of the equation can be as large as each computer can handle. Although such a simple problem cannot represent the real world computing, it is considered a reasonable measure of the capability of each computer.

3.2. Performance trend

In June 1993 version of the list, the highest performance was 59.7 GFlops attained by Intel Paragon XP/S140 at Sandia Lab. and the 500th one was 0.4 GFlops. After eight years, the highest performance in June 2001 was 7.23 TFlops of ASCI White at Lawrence Livermore Lab. The average annual increase is about 182%. The 500th one in June 2001 was 67.8 GFlops, which could stand at the top if it were in 1993.

3.3. Technology trend

The drastic increase in the performance is based on the computer technology. As shown in the previous section, the first-half of the 1990s is characterized by the shift from vector computers to parallel computers based on COTS (Commodity-Off-The-Shelf) processors. In June 2001, more than 90% of the listed computers were based on RISC chips.

Not only the processors but also the architectures changed. In 1993, half of the machines were shared memory vector processors and one quarter were MPPs. The rest of the machines on the list consisted of single-vector processors and SIMD processors. In 2001, however, more than half were MPPs and 30% were clusters of WS/PC. The manufacturers also changed.

4. HPC in the 10 years to come

4.1. HPC in 2005

Prediction of the future in information technology is very dangerous. Who expected the popularization of web technology ten years ago? If we extrapolate this trend, however, the highest performance in 2005 will be 80 TFlops and the 500th one might exceed 1 TFlops.

There are several projects to construct very high-performance machines. The Earth Simulator of Japan [5], which is being constructed in Yokohama near Tokyo, will be completed in February 2002 and will have the peak performance of 40 TFlops. In US, the final system in the ASCI (Advanced Strategic Computing Initiative) program, which is scheduled to be completed in 2005, will have the peak performance of 100 TFlops.

Scalability is now a very important factor in the HPC, which means high performance microprocessors cover from the low-end PCs to the high-end machines. By that the large expenditure to develop microprocessors is amortized over millions of PC or WSs. For this reason it would be economically difficult to build a machine of 100 TFlops or more using conventional vector technology, although vector machines are best in wide range of scientific and engineering computations.

Special purpose computers dedicated to some special mission will reach the region of Petaflops in terms of equivalent performance. IBM announced they would build Blue Gene computer dedicated to protein folding simulation by 2005. Although the memory is relatively small, its peak performance will be one Petaflops.[1] The GRAPE computer, on which Prof. Makino talked in this conference, will reach sub-Petaflops performance soon. Hybrid system of dedicated and multipurpose computers connected via high-speed network would be a hot topic in various fields.

4.2. Petaflops in 2010

If we further extrapolate the trend to 2010, the top performance then is expected to reach 1 PFlops (Peta Flops = 1000 TFlops). What would the Petaflops machine be like? In US, a research

[1] On November 9, 2001, IBM announced a revised version of the Blue Gene with 200 TFlops performance. It is named "Blue Gene/L" and will be designed with LLNL.

project has been running to find possible designs of Petaflops computer [2]. Several ideas have been proposed.

(1) *COTS plan*: Since commercial processors are expected to have a very high performance in coming years, one might build a Petaflops machine using COTS. The SIA (Semiconductor Industry Association of the US) prediction says extreme minute semiconductor technology (50–70 nm) will be available in 2009. If a chip containing 500 million transistors runs at 6 GHz, it would comprise four CPUs with 30 GFlops peak speed. Petaflops will be available using 8192 of such chips. A number of break throughs will be necessary to attain this goal, i.e. quantum limit, heat dissipation, fabrication limit, etc.
(2) *Optical connection plan*: One of the biggest barriers might be the pin neck. In order to attain a sufficient bandwidth between the chip and the memory, huge number of pins are necessary. The pin size, however, does not decrease parallely with the transistor size. Steve Wallach proposed to adopt optical connection between chips [4]. If the WDM (wavelength division multiplexing) technology can be incorporated in the chip, the communication bandwidth for a single fiber will be more than 1 Tb/s. One big problem will be the routing, which will be solved by the Internet technology.
(3) *PIM plan*: Another plan to cope with the pin neck is to let the processor and the memory cohabit in a single chip. This technology is called PIM (Processor in Memory). Then the latency will become one tenth and the bandwidth hundred times wider. The PIM technology is already popular in embedded systems or game machines. A new architecture is needed in HPC PIM.
(4) *HTMT* (*Hybrid technology multi-threaded*) *plan*: A hybrid combination of superconductive devices, PIM, optical connection and hologram memory has been proposed for the Petaflops machine. Using superconductive devices, one can make a 200 GHz processor with 0.25 μm technology. An assembly of 4196 of such processors would enable Petaflops performance. The memory system will consist of ten levels and that will make programming very difficult. Technological plausibility is questionable and it seems this plan no longer enjoys vigorous support, which it once had.

4.3. Application of Petaflops

Computational science and technology have always demanded thousand times faster computer than the current ones. Petaflops computer would open variety of new research and development. It is needless to say that Petaflops machine should not be alone but it lies at the top of the continuum of high performance computers, i.e. although Petaflops machine may be the only one in the world, there should be a number of 100 Teraflops computers all around the world.

4.3.1. Simulation from the first principles

The most important application will be the simulation from the first principles, which predicts the macroscopic behaviour of the system in terms of microscopic fundamental laws of nature. Since our computing capacity is not enough for computing everything from the quantum mechanics, various modeling and approximation is very important.

Biological simulation will be an important target of Petaflops computing. Although human genome is decoded, there is a long way to the functions of human body. One big challenge is to predict the three-dimensional structure and the chemical reaction on it in terms of the sequence of amino acids. Currently we can simulate the dynamics in a nano second (10^{-9} s) using classical forces. Protein folding takes milli second so that we need million times more computing power. If quantum mechanics is used in calculating forces, we would need ever more high performance. Drug design would require a quantum simulation of chemically active regions.

4.3.2. Coupled simulation

Real world is complex and is governed by various laws of nature. Conventional simulation is applied to single phase stationary systems based on a single law such as fluid dynamics or structural mechanics. Real airplane wing, for example, is subject not only to the aerodynamics but also to the structural mechanics. It is not easy to incorporate both laws, since huge degrees of freedom should be handled and they have different time and space scales.

More complex is the simulation of internal combustion engine. Fluid dynamics, structural dynamics, chemical reaction and heat transfer are coupled in the system. Simulation of crack in materials would need quantum, molecular and structural analysis coupled with each other. Petaflops level computer will enable us to simulate such complex systems and change not only basic research but also industry.

4.4. Qualitative change in research and development

In the early stage of simulation technology, due to the limitation of the computing speed, simulation could only reproduce the data which were already known by experiments. At present, however, new inovative data are available by simulation. Moreover phenomena in extreme conditions (high pressure, high temperature, etc.) can only be "observed" by computational methodology.

Computational methods have a number of advantages. In physical experiments, one can measure only a limited amount of physical observables. In computer experiments, although they are limited by the precision of assumed models and numerical accuracy, one can sometimes have more precise data than the physical experiments. Computational methods have a big impact not only on scientific research, but on industrial design.

5. HPC in 20 years from now

If we further extrapolate to 2020, what might be expected? Does Moore's law about the exponential expansion of semiconductor technology still continue up to this point? Will there be no flattening? If the exponential growth continues, we will have an Exa Flops (10^{18} floating operations per second) computing in 2020!

What would be the technology for the Exaflops computing? Quantum limit is not very far and we might have to use a single electron to compute. Heat dissipation is a very big difficulty. Molecular and/or quantum computing will be a hot topic in this area. Biocomputing may be another candidate. We have to be reminded that in the history of semiconductor technology many no-go predictions have been overthrown by new technologies.

Who will use Exaflops machines? There will be no shortage of users in Exaflops machines. In three-dimensional models, 1000 times computing power would enable computation of only five or six times larger systems. In quantum mechanical models the improvement in the number of atoms would be similar.

6. Conclusion

As was shown in many talks in this conference, simulation technology gives us the problem solving technique for complex and nonlinear systems with huge degrees of freedom, which appear in natural science, engineering, socio-economic problems and so on. Large-scale high-speed simulation technology and high performance computing as its basis are the key to the scientific and technological development in the 21st century. Japan will have a 40 TFlops machine (Earth Simulator in Yokohama) next February but it has no definite plan thereafter. The author believes Japan should start a project aiming at Petaflops as soon as possible.

References

[1] Y. Oyanagi, Development of supercomputers in Japan: hardware and software Parallel Comput. 25 (1999) 1547–1567 further references therein.
[2] Proceedings of PETAFLOPS II, Second Conference on Enabling Technologies for Peta(fl)ops Computing, Santa Barbara, February 15–19, 1999.
[3] E. Strohmaier, J.J. Dongarra, H.W. Meuer, H.D. Simon, The marketplace of high-performance computing, Parallel Comput. 25 (1999) 1517–1544 see also http://www.top500.org.
[4] S. Wallach, Petaflops in the Year 2009, Keynote Talk at the SC2000 Conference, Dallas, November 4–10, 2000.
[5] M. Yokokawa, S. Habata, S. Kawai, H. Ito, K. Tani, H. Miyoshi, Basic design of the earth simulator, in: C. Polychronopoulos, K. Joe, et al., (Eds.), High Performance Computing, Lecture Notes in Computer Science, Vol. 1615, Springer, Berlin, 1999, pp. 269–280.

ELSEVIER

Journal of Computational and Applied Mathematics 149 (2002) 155–169

JOURNAL OF COMPUTATIONAL AND APPLIED MATHEMATICS

www.elsevier.com/locate/cam

Parallel evolutionary algorithms for optimization problems in aerospace engineering

J.F. Wang[a,b], J. Periaux[b,*], M. Sefrioui[b]

[a] *College of Aerospace Engineering, NUAA, Nanjing, People's Republic of China*
[b] *Dassault-Aviation, Direction de la Prospective, 78 Quai Marcel Dassault, F-92214 Saint-Cloud, France*

Received 14 February 2002; received in revised form 28 March 2002

Abstract

This paper presents the recent developments in hierarchical genetic algorithms (HGAs) to speed up the optimization of aerodynamic shapes. It first introduces HGAs, a particular instance of parallel GAs based on the notion of interconnected sub-populations evolving independently. Previous studies have shown the advantages of introducing a multi-layered hierarchical topology in parallel GAs. Such a topology allows the use of multiple models for optimization problems, and shows that it is possible to mix fast low-fidelity models for exploration and expensive high-fidelity models for exploitation. Finally, a new class of multi-objective optimizers mixing HGAs and Nash Game Theory is defined. These methods are tested for solving design optimization problems in aerodynamics. A parallel version of this approach running a cluster of PCs demonstrate the convergence speed up on an inverse nozzle problem and a high-lift problem for a multiple element airfoil.

Keywords: Evolutionary algorithms; Hierarchical genetic algorithms; Variable fidelity models; Nash game; High-lift; Distributed parallelization; Computational fluid dynamics; Inverse problems

1. Introduction and motivation

As part of evolutionary computing, genetic algorithms (GAs) have now established themselves as a powerful tool for design optimization in various fields [6,5]. However, they have been applied to aerodynamics only quite recently [9]. Some of the major advantages of GAs are their robustness and their flexibility to handle multi-objective criteria problems. Realizing this we have been active in developing evolutionary design strategies for optimizing aerodynamic shapes in different environments [8].

* Corresponding author.
E-mail address: periaux@rascasse.inria.fr (J. Periaux).

PII: S0377-0427(02)00527-7

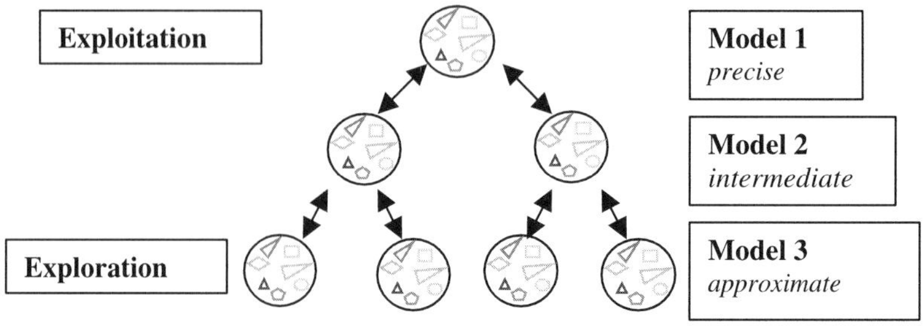

Fig. 1. Schematic of hierarchical genetic algorithm.

However, one of the main drawbacks of GA is the fact that they can require a rather large number of evaluations. The problem is particularly crucial in aerodynamics, since CFD codes are very time consuming. For instance, codes that solve nonlinear Navier–Stokes equations take a considerable time to perform a single simulation of the flow past an airfoil. A typical optimization using GAs may involve several hundreds or even a few thousand computations, thus limiting their application.

The main focus of our research has been to identify strategies to overcome this drawback. One such strategy is hierarchical genetic algorithms (HGAs), who offer the possibility of using alternatively low-cost models for exploration and high-fidelity (HF) models for refinement.

The other focus of this article is the introduction of multiple-objective optimization via Nash equilibrium and Game Theory. This technique is used not as a stand alone, but as a part of a larger hierarchical structure.

Some results on case studies in CFD optimization are presented, namely an inverse transonic nozzle problem and the maximization of the lift for multi-element airfoils for landing and take off conditions. These examples illustrate the potentiality and robustness of the various methodologies we introduce.

2. Hierarchical genetic algorithms

HGAs are inspired from parallel genetic algorithms, which are in turn based on a network of several GAs running in parallel [8,10]. Parallel GAs generally have small sub-populations and evolve independently most of the time, except that they exchange solutions now and then—the migration phase. Hierarchy is introduced via an architecture with a binary tree spanning three *layers*. The advantage is that during the migration phase, the sub-populations do not swap their solutions more or less at random with their neighbors. Instead, the solutions go up and down the layers and the best ones keep going up till they are refined (see Fig. 1). The bottom layer carries out most of the exploration, by ranging widely in the search space via a large mutation span. The intermediate layer is a compromise between exploration and exploitation. The top layer takes small steps in mutation and refines the solution.

Going a step further, we realized that the exploration layer does not need to use a very precise model (hence time consuming), as the idea is only to get a good guess of the lay of the search space. Accordingly, at the bottom layer we use a very approximate model for fitness evaluation. Then in the intermediate layer, a slightly better model is employed. As the role of the top layer is to refine the solution, we use the most accurate model. The following algorithm describes more in details the workings of HGAs:

```
for all sub-populations {Initialization}
  Randomly generate solutions
  Layer 1 uses Model 1 to evaluate fitness
  Layer 2 uses Model 2 to evaluate fitness
  Layer 3 uses Model 3 to evaluate fitness
end for
repeat
      for i = 1 to Epoch do {Isolation phase}
        for all sub-populations
          Evolve 1 generation
        end for
      end for
Start migration {Migration phase}
Layer 1: gets best solutions from Layer 2
         reevaluates them using Model 1.
Layer 2: gets random solutions from Layer 1
         gets best solutions from Layer 3
         reevaluates them using Model 2.
Layer 3: gets random solutions from Layer 2
         reevaluates them using Model 3.
until Stopping criterion is met
```

For example, one could use panel methods, Euler equations and Navier–Stokes equations in the bottom, intermediate and top layers, respectively. Alternately, one could have a coarse, a medium and a fine mesh for each layer using the same model. Possibilities abound.

2.1. Test-case definition

In this section, we present results obtained on a test case consisting in the reconstruction of a transonic nozzle with a shock at the downstream of the throat (Fig. 2). The target function is the Mach number distribution along the length of the nozzle obtained by computing the flow under one-dimensional approximation using the equation

$$\frac{\partial W}{\partial t} + \frac{\partial F}{\partial x} + H = 0 \tag{1}$$

with standard boundary conditions. The computations were performed with a code developed using CUSP formulation [11]. An iterative technique is used to solve the equation.

2.2. HGAs with multiple grid sizes

The reconstruction problem was carried out for the following cases:

(a) Conventional GAs using a single population.

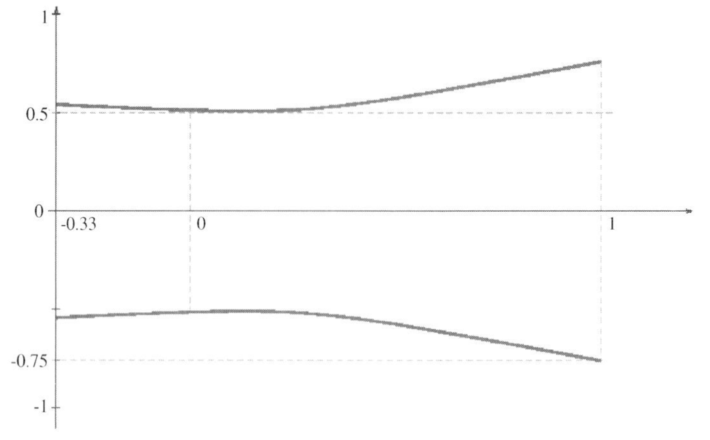

Fig. 2. Converging–diverging nozzle.

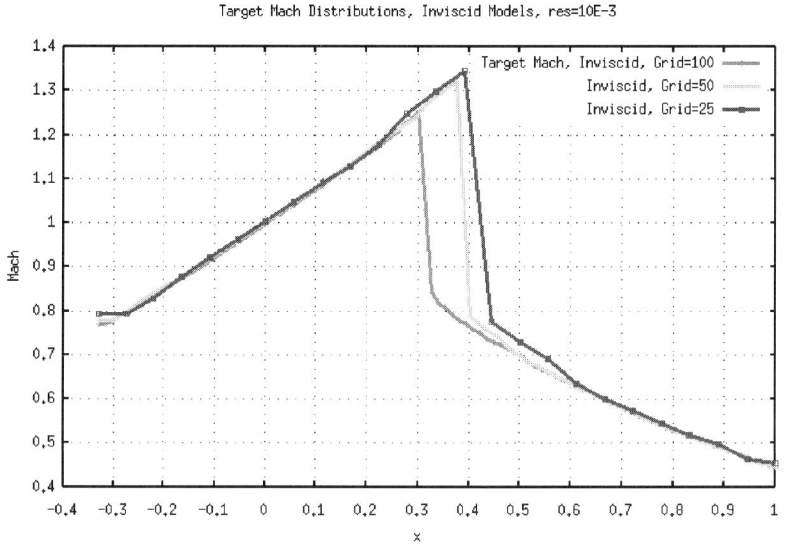

Fig. 3. Target Mach for three different grid sizes.

(b) HGAs with the same grid size for all the layers (HGAs, single model)
(c) HGAs with different grid sizes in different layers (HGAs, multiple models)

The runs are based on real coded GAs, and the shape of the nozzle is described by a Bézier curve of order 5. Conventional GAs uses a population of size 20, whereas the HGA uses seven sub-populations of size ten each, with a migration phase every 20 generations.

For cases (a) and (b), the computational grid has 100 points equally spaced along the nozzle. For case (c), the grid has 25, 50 and 100 points in the bottom, intermediate and top layers, respectively. Since the position of the shock in computations varied appreciably depending on the grid, each layer had its own target Mach number distribution (Fig. 3).

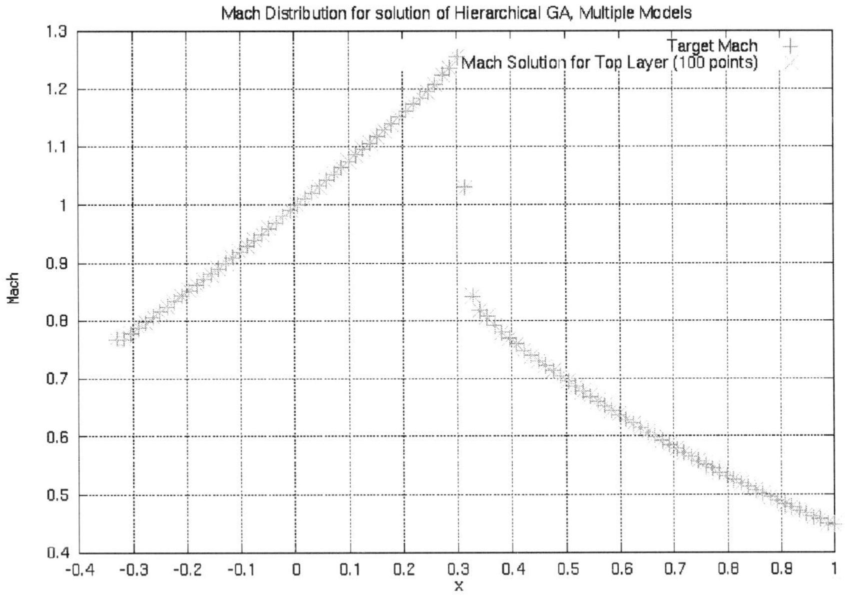

Fig. 4. Target and converged Mach distribution.

Table 1
Reconstruction using different grids in layers

Model	Number of evaluations	CPU (s)
a. Single population	1080	6460
b. HGAs, single model	1485	5200
c. HGAs, multiple model	1128	1850

Fig. 4 shows the converged results obtained. The resulting shape of the nozzle coincides exactly with that of the target. The CPU times required for the single and the multiple models are given in Table 1.

It is clear that the time taken for reconstruction decreases appreciably for HGAs with a single model. For HGAs with multiple model the decrease in time is very substantial, a factor of 2.8 compared to HGAs single model. It may be noted that the converged solution is identical to the target Mach number distribution for the finest grid (100 points) in each of the three cases.

2.3. HGA with multiple solvers

This example introduces a further level of complexity by reconstructing the nozzle for a viscous flow. Each layer may have a different flow model—inviscid or viscous. Since the pressure distribution can vary with the model, each layer has its own target Mach number distribution. The viscous solution was obtained by modifying Eq. (1) to account for viscous effects.

Table 2 gives the CPU times required for this example.

Table 2
Convergence times for viscous reconstruction

Single GAs with viscous model	6420 s
Viscous model in all three layers	2880 s
Inviscid in bottom and intermediate layers and viscous in top layer	1920 s

While HGAs with a single model give a speed up of more than two compared to conventional GAs, HGAs with multiple models give an additional increase in speed (30%) in the present example.

With mixed inviscid/viscous layers, all the exploration is done using inviscid layers. But when the best solutions are sent to the top layer, a viscous model is used to reevaluate them. Hence, we achieve the quality of a viscous modeling, but as a much lower cost. With this approach, we end up converging towards a fine mesh grid with the viscous model in approximately the same time it took for the inviscid only model.

At present tests are in progress with two-dimensional flows past a bump and through a converging–diverging nozzle.

3. Hierarchical Nash GAs for multi-objective optimization

3.1. Description of a Nash/GAs game

Due to the large dimension of global optimization problems and access to low-cost distributed parallel environments—such as a cluster of PCs—it is quite natural to try replacing global optimization by decentralized local sub-optimizations. One way to do that is to turn optimization problems into Nash games, and use tools from Game Theory [13].

Nash equilibrium is the solution of a noncooperative strategy of multi-objective optimization first introduced by J.F. Nash in 1951 [7]. Since it originated in Game Theory, the notion of player is often used in the sequel. For a multi-criteria optimization with N objectives, a Nash game consists in having N players, each in charge of one objective and able to modify their sub-set of variables. During the game, each player looks for the best strategy (i.e. the optimal values for the variables he is in charge of) in his search space in order to improve his own criterion while the other criteria of others players are fixed. When no player can further improve his criterion, the system has reached a state of equilibrium named Nash equilibrium.

GAs can be used to build the Nash equilibrium, thus making a Nash/GAs (N-GAs) approach able to deal with multiple objective problems without having to compute directly the Nash equilibrium. During a Nash game, each player uses GAs to improve his own criterion along generations constrained by the strategies of the other players. In applications, the design variables are geometrically split between players who exchange their best strategies (best solutions) at each generation.

Fig. 5 explains more in detail the working of the N-GAs algorithm.

Let $S=XY$ be the string representing the potential solution for a two objective optimization, where X corresponds to the subset of variables optimized by Player 1 and Y to the subset of variables optimized by Player 2. Player 1 optimizes X according to criterion f_1 (with Y fixed by Player 2)

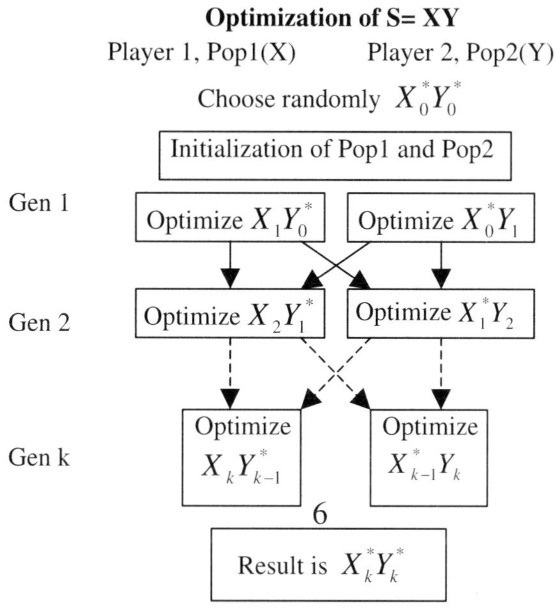

Fig. 5. Nash/GAs strategy.

and Player 2 optimizes Y according to criterion f_2 (with X fixed by Player 1). Each player has his own GAs. The process is iterated until no player can further improve his criterion (see [14] for more details). At this stage the system has reached the Nash equilibrium. One of the evident properties of NGAs is their inherent parallel structure and symmetry during evolution.

3.2. Nash hierarchical GAs

In hierarchical algorithms, each model (sub-process) could have an algorithm with different control parameters, such as population size, probability of crossover and mutation, etc. For the sake of simplicity, we consider in the sequel a hierarchical structure with only two layers: a bottom layer with a low-fidelity (LF) model which corresponds to a cheap and approximate description of the environment and a top layer with a high-fidelity model (HF, which corresponds to a fine and expensive description of the environment as Model 1 in Fig. 1) [1]. For LF, a relatively large population is used for the sake of exploring the search space, but the population size of the HF layer is kept small it is expensive in CPU time.

Then the next step is to mix this two-layered hierarchical structure to the strategy already outlined to implement Nash gas strategies. Fig. 6 presents the way these approaches are linked.

The optimization begins with the LF model. At the end of a sub-optimization process, the LF model sends his best individuals to the HF model (the number of individuals being denoted as the migration size). These new individuals participate in a *survival competition* with those of the HF model, and only the winners become members of the new population of HF. Then the new sub-optimization process starts using the HF model. If the optimization converges with the HF model of the top layer, this process corresponds to a single-direction evolution; otherwise, the HF

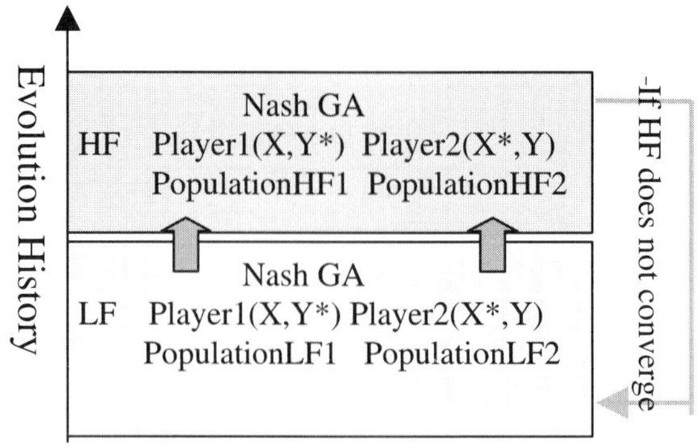

Fig. 6. Nash hierarchical GA for two players.

model sends his best individuals to the bottom layer using the LF model the population of which in turn restarts a new sub-optimization. The whole process is iteratively repeated until the HF top layer converges. In this case, an evolution loop is simulated. Fig. 6 describes the two-layered optimization process for a hierarchical NGAs with two players and a LF–HF model.

3.3. High lift multi-airfoil problems

Increased performance requirements for high-lift aircraft system and innovative evolutionary tools have renewed recently a strong interest in the aerodynamic optimization of multi-element configurations [15]. High-lift systems consist generally of leading and trailing edge devices (denoted slat and flap, respectively). Leading edge devices increase the maximum lift of an airfoil by delaying its stall angle. Trailing edge devices produce also a lift increment. There is a commercial benefit to increase the maximum lift at prescribed landing and take off speeds since it increases the available payload of the aircraft. Trailing edge devices are often designed to produce a lift increment while maintaining a high L/D ratio during the take off of the aircraft.

The global lift coefficient of such a system is the combination of the lift coefficients of each element, taking into account their interaction. The lift characteristics of the flow around each element strongly depend on their relative position. Thus, optimizing a configuration with respect to the fitness—or objective—lift function $C_{\mathrm{L}}(X)$, where X denotes the feasible design variables set can be seen as a combinatorial problem. Furthermore, the variation of $C_{L\,\max}$ can be strongly nonlinear, for instance the separation point on profiles can move rapidly from one position to another due to the wake/boundary layer interaction. The fitness function may therefore have several local optima and classical deterministic optimization algorithms can be trapped locally and fail capturing the global solution. This situation explains is the choice of GAs-based tool to solve this nonconvex optimization problem.

3.4. The DAMIEN flow analysis solver

The real cost of an evolutionary optimizer for CFD problems depends on the number of fitness function evaluations using the flow analysis solver, which is determined by the population size and the number of generations.

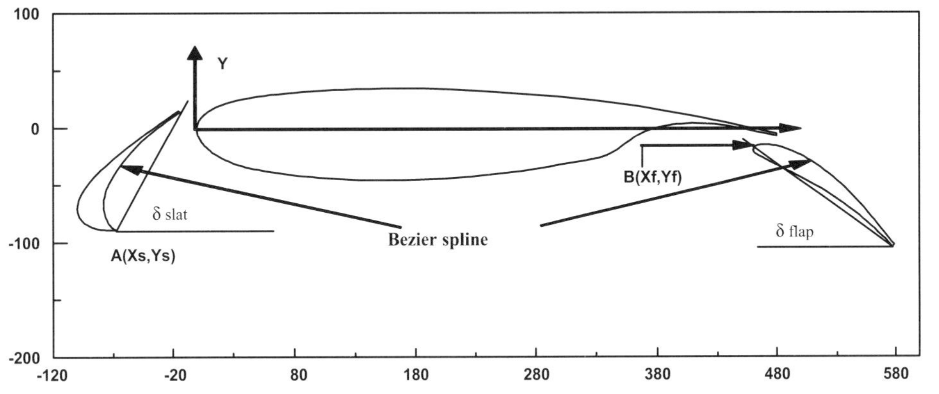

Fig. 7. Position definition and shape representation.

The DAMIEN flow analysis solver considered in this section is an in-house two-dimensional solver for multi-element airfoil systems developed by Dassault Aviation from existing theories in the late 1970s. The code uses a panel method for perfect flow with viscous laminar/turbulent boundary layers and wake effects using integral methods. These models have been implemented numerically by Dassault aviation, and some specific developments have been made concerning the introduction of transition criteria, the treatment of stagnation line for boundary layer initialization, the treatment of separated zones and the wake/boundary layer interaction. Details can be found in [3,4].

The global validation of the code has been calibrated and assessed through numerous numerical comparisons with ONERA wind tunnel results. Specific treatment of the separation phenomenon based on a large amount of available results is implemented as well as a comprehensive model of merged layers to treat the impingement of a separated layer with a boundary layer developing on the adjacent downstream element. A coupling between perfect flow and boundary layer/wake provides a low-cost simulation of viscous effects.

In case of multi-element flow analysis, a wake-boundary layer interaction can take into account separation of the flow induced by the wake. It is obvious that the multi-flow model DAMIEN solver is highly nondifferentiable and therefore considered as a 'black box' in the sequel during the evolutionary optimization process using GAs.

3.5. Problem definition for multi-airfoil optimization

The deployed configuration of a multi-element airfoil is shown in Fig. 7 [15]. The design variables for the problems considered in this paper are twofold:

- shape design variables representing the intrados of a slat and the extrados of a flap are modeled by Bezier splines of 11th order, each spline with ten variables;
- position design variables representing the positions of a slat and flap with respect to the primary airfoil. The parameterization of the positions with three real variables are described in Fig. 7.

The role of players using evolutionary algorithms combined with Nash games depends on the physical problem to be solved. In the sequel, a rational set of tasks is based on the geometrical and

Table 3
Definition of players and their variables

	Player J1	Player J2
Shape	V_{Bs} (slat, ten variables)	V_{Bf} (flap, ten variables)
Position	V_{Ps} (slat, three variables)	V_{Pf} (flap, three variables)
Nb. Var.	13	13

kinematics definition of multi-element configuration. Two players are defined in Table 3, where B and P denotes, respectively, Bezier and Position and s, f, respectively, the slat and the flap.

3.6. Numerical methods and parallelization analysis

The evolutionary methods used in this section for the speed-up comparison consist of the three following algorithms:

- Genetic algorithms using a single HF model (GAs);
- Nash-GAs using a one-layer HF model (N-GAs);
- Hierarchical N-GAs using two-layer LF model/HF model (HN-GAs) and its parallel analogue denoted (PHN-GAs).

Three different parallelized modules associated to parallel hierarchical N-GAs (PHN-GAs), from bottom to top layer are considered in the sequel and denoted VFM (variable fidelity model), PN-GAs and EM (evaluation module):

- the VFM module contains the parallelization of different models of VFM;
- the PN-GAs or parallel Nash gas module, included in the VFM module; and
- the EM or evaluation module in charge of the parallelization of the fitness used by players.

In each module, the association of algorithms, either in sequential or parallel computations depends on the resource of the parallel computing environment, the cost of communications between inter-processors and also load balancing [14].

It should be mentioned that for complex optimization with large dimension, the parallelization of the fitness of individuals cannot always provide a significant speed up due to the large number of inter-processors communications. According to the previous remark, an efficient strategy is to parallelize the fitness evaluation only on sub groups. Several groups (or subsets of individuals) are distributed to different processors. This strategy allows to improve the parallelization efficiency via computations in parallel and also the reduction of inter-processors communications.

3.7. Optimization problem settings

The optimization problem consists in maximizing the lift of multi-element airfoil configuration operating under the following viscous flow conditions:

$$M_\infty = 0.12, \quad \alpha_\infty = 17.18^\circ, \quad \mathrm{Re} = 4 \times 10^6,$$

where M, α and Re denotes Mach number, angle of attack and Reynolds number, respectively.

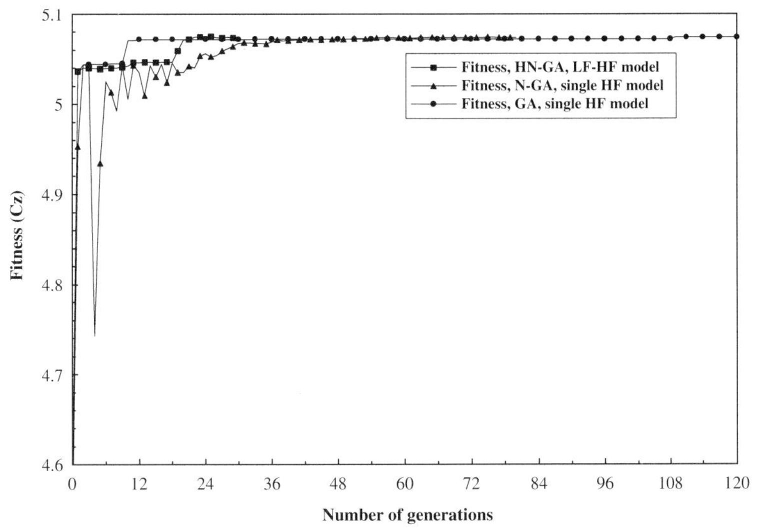

Fig. 8. Comparison of convergence.

The lift coefficient C_L for a given multi-element airfoil is obtained using the non-differentiable DAMIEN flow analysis solver described above. The cost functions for different models (represented by different CFD solvers) are defined below:

- Low-fidelity model (LF): $f_{LF}(\gamma) = C_{L,LF}(V_{Ps} + V_{Bs}, V_{Pf} + V_{Bf})$;
- High-fidelity model (HF): $f_{HF}(\gamma) = C_{L,HF}(V_{Ps} + V_{Bs}, V_{Pf} + V_{Bf})$.

In the case of a Nash game, players optimize their own design variable set (denoted as V spline for shape and V position for the relative slat/primary airfoil or flap/primary airfoil position) to maximize the global cost function.

3.8. Speed-up comparison for sequential algorithms

Fig. 8 compares the convergence speed achieved with GAs, N-GAs and HN-GAs software in the case of a sequential computation in order to obtain similar results. The CPU time needed for each algorithm is described in Table 4. Figs. 9 and 10 show, respectively, a comparison of the optimized airfoil configuration with enlargement in the vicinity of the slat and the flap and associated pressure distribution. A significant speed up is obtained by N-GAs and HN-GAs as clearly indicated in Table 4.

Table 4
Speed-up comparison for sequential algorithm (GA vs. N-GA vs. HN-GA)

	GA	N-GA	HN-GA
CPU	550 mn	330 mn	93 mn
Speed up	1	1.7	5.9

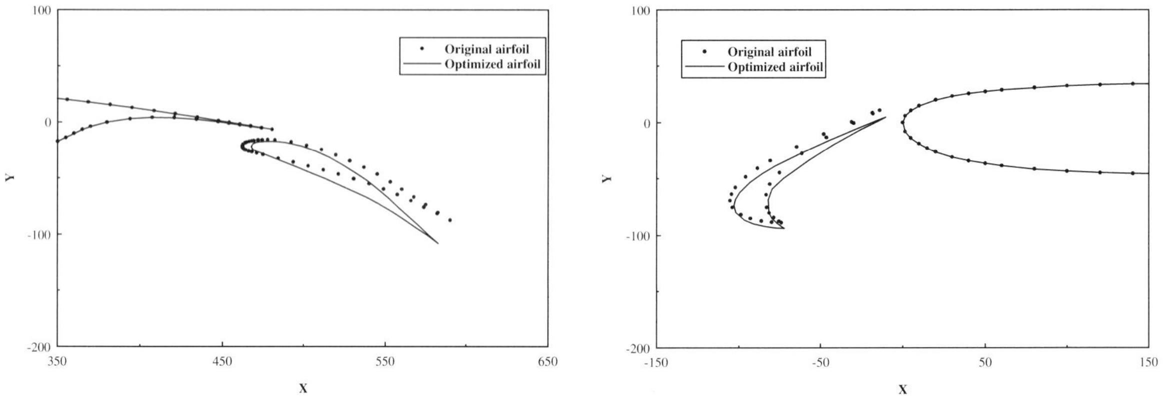

Fig. 9. (a) Zoom for slat; (b) zoom for flap.

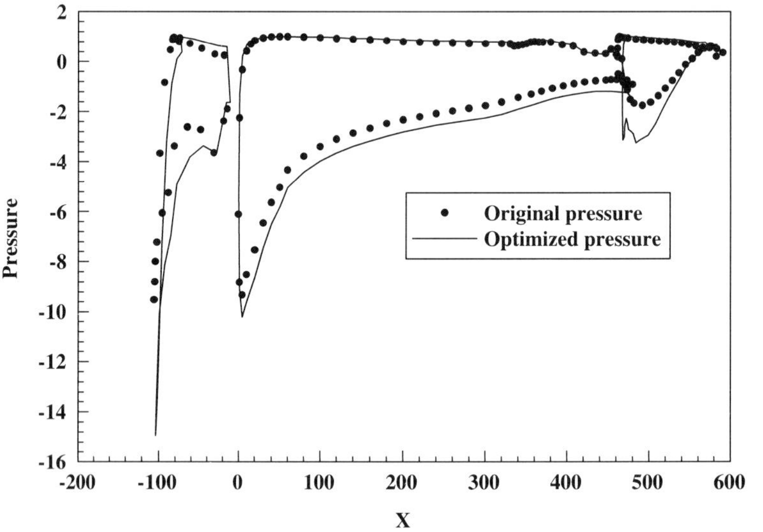

Fig. 10. Comparison of pressure.

3.9. Speed-up comparison for parallel algorithm

A parallel version of HN-GAs (denoted PHN-GAs) has been implemented on a cluster with four PCs (PentiumII 512 Mb/440 MHz, bi-processors) using a three-level parallelization. The Damien solver can be accessed in each PC for fitness evaluations using the PVM library (Parallel Virtual Machine). The speed-up ratio achieved with respect to the number of processors is shown in Fig. 11. CPU time comparisons and speed up obtained with different number of processors are presented in Table 5.

The results obtained in Sections 3.8 and 3.9 using hierarchical models coupled with Game Theory clearly demonstrate that the new approach combining parallel hierarchical algorithms and game

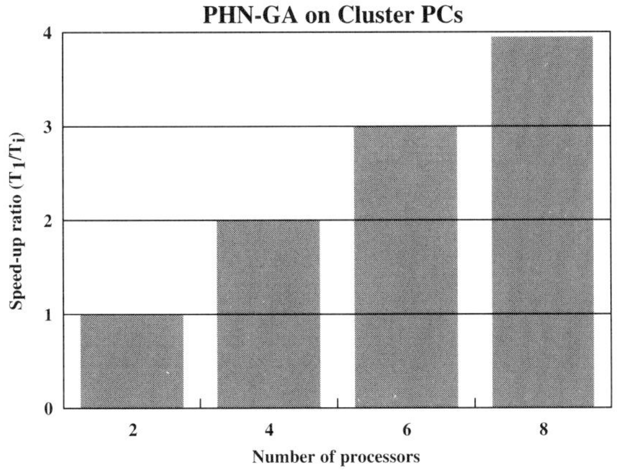

Fig. 11. Speed-up comparison.

Table 5
CPU time and speed-up ratio

N_{proc}	CPU (mn)	Speed up T_l/T_i
2 (1 PC)	133	1
4 (2 PCs)	67	1.98
6 (3 PCs)	45	2.96
8 (4 PCs)	34	3.91

strategies can provide much higher efficiency that classical sequential GAs to capture the same optimized solution of complex problems. Other realistic aerodynamic problems to be solved in an industrial environment and using the same methodology are presently under investigation.

4. Conclusion

The hierarchical GAs developed as a part of this study are shown to be very promising for complex optimization problems. They are directly applicable to design problems (and particularly multi-criteria/multi-disciplinary problems), which can be formulated with varying degrees of fidelity (different models, different meshes, different optimizers, etc.). Modern advanced multi-disciplinary design optimization falls into this category.

Numerical experiments are presented in this paper using parallel hierarchical algorithms. Comparisons of numerical results obtained by different algorithms demonstrate that parallel hierarchical algorithms could get the same results as sequential computation, but with higher efficiency in convergence and CPU time. This is in fact the main motivation for introducing hierarchical algorithms with an open structure.

If using a distributed parallel environment, the choice of the topology connecting the processors could improve significantly the calculation efficiency. For parallel computation, three major parameters such as system resources, communication of inter-processors and load balancing should be considered. Combining hierarchy and competitive Nash games provides definitely another significant speed up with good parallel properties and outlines the road map to distributed optimization under conflict.

Another area we are investigating is linked to asynchronicity. It uses an evolution strategy [2] and a hierarchical structure, but the difference is that there is no longer a need to synchronize the exchanges between sub-populations [12]. This approach exploits distributed parallel architectures, like meta or grid computing in order to carry out a number of trials as a part of program execution.

Acknowledgements

The authors would like to thank J.A. Desideri for fruitful comments and suggestions. Numerical experiments have been obtained with the DAMIEN code developed at Dassault Aviation. We are grateful to J.C. Courty and C. Tran, for providing the code and also proposing realistic aerodynamic applications along this research, and also to P. Thomas for installing the DAMIEN in the PC cluster environment. We would also like to thank K. Srinivas and E. Whitney from University of Sydney for their contribution on the Nozzle problem.

The first author thanks Dassault Aviation for supporting his research during his Ph.D. at Pole Scientifique Dassault-Aviation/UPMC.

References

[1] N.M. Alexandrov, R.M. Lewis, C.R. Cumbert, L.L. Green, P.A. Newman, Optimization with variable-fidelity models applied to wing design, in: 38th Aerospace Sciences Meeting and Exhibit, AIAA 2000-0841, Reno, 2000.

[2] T. Baeck, H.P. Schweffel, Evolution strategies I: variants and their computational implementation, in: Genetic Algorithms in Engineering and Computer Science—EUROGEN 95, Wiley, Las Palmas, 1995, pp. 111–126.

[3] P. Caupenne, Calcul des Décollements par Méthodes de Singularités, in: 13eme colloque d'Aérodynamique Appliquée, Ecole Centrale de Lyon, 1976.

[4] J.C. Courty, Problèmes de confluence sillage-couche limite à l'aide d'un modèle élaboré de turbulence, in: 13eme Colloque d'Aérodynamique Appliquée, Ecole Centrale de Lyon, 1976.

[5] D.E. Goldberg, Genetic Algorithms in Search, Optimization, and Machine Learning, Addison-Wesley, Reading, MA, 1989.

[6] J.H. Holland, Adaptation in Natural and Artificial System, University of Michigan Press, Ann Arbor, MI, 1975.

[7] J.F. Nash, Noncooperative games, Ann. Math. 154 (1951) 289.

[8] J. Periaux, B. Mantel, M. Sefrioui, B. Stoufflet, J.-A. Desideri, S. Lanteri, N. Marco, Evolutionary computational methods for complex design in aerodynamics, in: 96th American Institute for Aeronautics and Astronautics Conference, AIAA-98-0222, Reno, 1998.

[9] C. Poloni, Hybrid GAs for multi-objective aerodynamic optimization shape, in: Genetic Algorithms in Engineering and Computer Science—EUROGEN 95, Wiley, Las Palmas, 1995, pp. 397–415.

[10] M. Sefrioui, J. Periaux, A hierarchical genetic algorithm using multiple models for optimization, in: Parallel Problem Solving from Nature—PPSN 6, Springer, Paris, 2000.

[11] K. Srinivas, Computation of cascade flows by a modified CUSP scheme, Comput. Fluid Dynamics J. 8 (2) (1999) 285–295.

[12] K. Srinivas, E. Whitney, J. Periaux, M. Sefrioui, Advances in hierarchical, parallel genetic/evolutionary algorithms for aerodynamic shape optimization, Proceeding of the Advanced Fluid Information Conference, Sendai, Japan, 2001.
[13] J. Von Neumann, Zur theorie der Gesellschaftsspiele, Math. Ann. 100 (1928) 295–320.
[14] J.F. Wang, Distributed Multi criteria optimization using GAs and game theory and related applications to high lift CFD problems in Aerodynamics, Thèse de doctorat, University of Paris, 2001, p. 6.
[15] J.F. Wang, J. Periaux, GAs and game theory for high lift multi airfoil problems in aerodynamics, in: M. Hafez, K. Morinishi, J. Periaux (Eds.), Proceedings of CFD21 Conference, Vieweg Verlag, Kyoto, 2001.

Journal of Computational and Applied Mathematics 149 (2002) 171–191

JOURNAL OF COMPUTATIONAL AND APPLIED MATHEMATICS

www.elsevier.com/locate/cam

Adjoint sensitivity analysis for differential-algebraic equations: algorithms and software ☆

Yang Cao, Shengtai Li, Linda Petzold *

Department of Computer Science, University of California, Santa Barbara, CA, Engr III, Room 124, 93106-5070, USA

Received 6 May 2001; received in revised form 10 February 2002

Abstract

An efficient numerical method for sensitivity computation of large-scale differential-algebraic systems is developed based on the adjoint method. Issues that are critical for the implementation are addressed. Complexity analysis and numerical results demonstrate that the adjoint sensitivity method is advantageous over the forward sensitivity method for applications with a large number of sensitivity parameters and few objective functions.

Keywords: Sensitivity analysis; Adjoint method; Differential-algebraic equations

1. Introduction

Recent work on methods and software for sensitivity analysis of differential-algebraic equation DAE systems [7,9,13–16] has demonstrated that forward sensitivities can be computed reliably and efficiently via automatic differentiation [2] in combination with DAE solution techniques designed to exploit the structure of the sensitivity system. The DASPK3.0 [15,13] software package was developed for forward sensitivity analysis of DAE systems with index up to two, and has been used in sensitivity analysis and design optimization of several large-scale engineering problems [11,18]. DASPK3.0 is an extension of the DASPK software [1,3,4] developed by Brown et al. for the solution

☆ This work was supported by grants: EPRI WO-8333-06, NSF CCR 98-96198, NSF/ARPA PC 239415, NSF/KDI ATM-9873133 and LLNL ISCR 00-15.

* Corresponding author. Tel.: +1-805-893-5362; fax: +1-805-893-5435.

E-mail address: petzold@engineering.ucsb.edu (L. Petzold).

0377-0427/02/$ - see front matter © 2002 Elsevier Science B.V. All rights reserved.
PII: S0377-0427(02)00528-9

of large-scale DAE systems. For a DAE depending on parameters,

$$F(x, \dot{x}, t, p) = 0,$$

$$x(0) = x_0(p), \tag{1}$$

these problems take the form: find dx/dp_j at time T, for $j = 1, \ldots, n_p$. Their solution requires the simultaneous solution of the original DAE system with the n_p sensitivity systems obtained by differentiating the original DAE with respect to each parameter in turn. For large systems this may look like a lot of work but it can be done efficiently, if n_p is relatively small, by exploiting the fact that the sensitivity systems are linear and all share the same Jacobian matrices with the original system.

Some problems require the sensitivities with respect to a large number of parameters. For these problems, particularly if the number of state variables is also large, the forward sensitivity approach is intractable. These problems can often be handled more efficiently by the adjoint method [8]. In this approach, we are interested in calculating the sensitivity of an objective function

$$G(x, p) = \int_0^T g(x, t, p)\,dt \tag{2}$$

or alternatively the sensitivity of a function $g(x, T, p)$ defined only at time T. The function g must be smooth enough that g_p and g_x exist and are bounded. While forward sensitivity analysis is best suited to the situation of finding the sensitivities of a potentially large number of solution variables with respect to a small number of parameters, reverse (adjoint) sensitivity analysis is best suited to the complementary situation of finding the sensitivity of a scalar (or small-dimensional) function of the solution with respect to a large number of parameters.

In [6] we derived the adjoint sensitivity system for DAEs of index up to two (Hessenberg) and investigated some of its fundamental properties. In this paper, we address some of the issues for the numerical solution and demonstrate the effectiveness of the adjoint method via our implementation of the adjoint DAE solver DASPKADJOINT.

The outline of this paper is as follows. In Section 2 we outline the adjoint sensitivity method for DAEs and summarize some of the relevant results from [6]. In Section 3 we describe how the adjoint DAE may be evaluated accurately and efficiently by an automatic differentiation tool. In Section 4 we study the consistent initialization of the adjoint DAE. Section 5 outlines some important considerations for implementation of the adjoint sensitivity method for DAEs and how they are addressed in our software, DASPKADJOINT. Finally, the algorithms and software are tested for several examples in Section 6. The numerical results show that the adjoint sensitivity method is advantageous over the forward sensitivity method for applications with a large number of sensitivity parameters and few derived functions.

2. The adjoint DAE system and sensitivity calculation

In this section we define the adjoint sensitivity system for DAEs and summarize some of the relevant results concerning initial values, stability and numerical stability from [6].

The adjoint system for the DAE

$$F(t, x, \dot{x}, p) = 0$$

with respect to the derived function $G(x,p)$ (2) is given by

$$(\lambda^* F_{\dot{x}})' - \lambda^* F_x = -g_x, \tag{3}$$

where * denotes the transpose operator and prime denotes the total derivative with respect to t.

The adjoint system is solved backwards in time. For index-0 and index-1 DAE systems, the initial conditions for (3) are taken to be $\lambda^* F_{\dot{x}}|_{t=T} = 0$, and the sensitivities of $G(x,p)$ with respect to the parameters p are given by

$$\frac{\mathrm{d}G}{\mathrm{d}p} = \int_0^T (g_p - \lambda^* F_p)\,\mathrm{d}t + (\lambda^* F_{\dot{x}})|_{t=0}(x_0)_p. \tag{4}$$

For Hessenberg index-2 DAE systems, the initial conditions are more complicated, and will be described in detail along with an algorithm for their computation in Section 4. For index-2 DAE systems, if the index-2 constraints depend on p explicitly, an additional term must be added to the sensitivity (4) [6].

For a scalar derived function $g(x,T,p)$, the corresponding adjoint DAE system is given by

$$(\lambda_T^* F_{\dot{x}})' - \lambda_T^* F_x = 0, \tag{5}$$

where λ_T denotes $\partial\lambda/\partial T$. For index-0 and index-1 DAE systems, the initial conditions $\lambda_T(T)$ for (5) satisfy $(\lambda_T^* F_{\dot{x}})|_{t=T} = [g_x - \lambda^* F_x]|_{t=T}$. We note that the initial condition $\lambda_T(T)$ is derived in such a way that the computation of $\lambda(t)$ can be avoided. This is the case also for index-2 DAE systems. The full algorithm for consistent initialization of the adjoint DAE system will be described in Section 4. The sensitivities of $g(x,T,p)$ with respect to the parameters p are given for index-0 and index-1 DAE systems by

$$\frac{\mathrm{d}g}{\mathrm{d}p} = (g_p - \lambda^* F_p)|_{t=T} - \int_0^T (\lambda_T^* F_p) + (\lambda_T^* F_{\dot{x}})|_{t=0}(x_0)_p. \tag{6}$$

Note that the values of both λ at $t=T$ and λ_T at $t=0$ are required in (6). If $F_p \neq 0$, the transient value of λ_T is also needed. For an index-2 system, if the index-2 constraints depend on p explicitly, an additional term must be added to the sensitivity (6).

We focus on the adjoint system with respect to the scalar objective function $g(x,T,p)$ throughout this paper. If the objective function is of the integral form $G(x,p)$ (2), it can be computed easily by adding a *quadrature variable*, which is equal to the value of the objective function, to the original DAE. For example, if the number of variables in the original DAEs is N, we append a variable x_{N+1} and equation

$$\dot{x}_{N+1} = g(x,t,p).$$

Then $G = x_{N+1}(x,T,p)$. In this way, we can transform any objective function in the integral form (2) into the scalar form $g(x,T,p)$. The quadrature variables can be calculated very efficiently [15] by a staggered method in DASPK3.0; they do not enter into the Jacobian matrix.

From [6] we know that for DAE systems of index up to two (Hessenberg), asymptotic numerical stability in solving the forward problem is preserved by the backward Euler method, but only (for fully implicit DAE systems) if the discretization of the time derivative is performed 'conservatively', which corresponds to solving an *augmented adjoint* DAE system

$$\dot{\bar{\lambda}} - F_x^* \lambda = 0,$$
$$\bar{\lambda} - F_{\dot{x}}^* \lambda = 0. \tag{7}$$

It was shown in [6] that the system (7) with respect to $\bar{\lambda}$ preserves the stability of the original system. Note that the augmented system (7) is of (one) higher index than the original adjoint system (5). This is not a problem in the implementation since the newly high-index variables do not enter into the error estimate (see Section 5.4) and it can be shown that basic DAE structures such as combinations of semi-explicit index-1 and Hessenberg index-2 are preserved under the augmentation. Also, the linear algebra is accomplished in such a way (see Section 5.1) that the matrix needed is the transpose of that required for the original systems. Thus, there are no additional conditioning problems for the linear algebra due to the use of the augmented adjoint system.

3. Evaluation of the adjoint DAE

As we have seen, the adjoint DAE must be solved backward for its solution at $t = 0$. Since not every DAE solver can take backward steps during the integration, we apply a time reversing transformation $\tau = T - t$ to the adjoint system (5). This yields

$$\dot{\lambda}_T^* F_{\dot{x}}(T-\tau, x) + \lambda_T^* \left(F_x(T-\tau, x) - \frac{\mathrm{d}F_{\dot{x}}}{\mathrm{d}t}(T-\tau, x) \right) = 0, \tag{8}$$

where $\dot{\lambda}_T^* = \mathrm{d}\lambda_T^*/\mathrm{d}\tau$, $\dot{x} = \mathrm{d}x/\mathrm{d}t$, and $\mathrm{d}F_{\dot{x}}/\mathrm{d}t$ is the total derivative of $F_{\dot{x}}$ with respect to t. If we assume that the DAE system is linear with respect to $\dot{x}$, then for any constant vector v, this term may be evaluated by

$$v \frac{\mathrm{d}F_{\dot{x}}}{\mathrm{d}t} = vF_{\dot{x}t} + (vF_{\dot{x}})_x \dot{x}. \tag{9}$$

For many applications, $F_{\dot{x}}$ is constant and Eq. (8) becomes

$$\dot{\lambda}_T^* F_{\dot{x}} + \lambda_T^* F_x(T-\tau, x) = 0. \tag{10}$$

In these cases, we do not need to evaluate the term $\mathrm{d}F_{\dot{x}}/\mathrm{d}t$. If $F_{\dot{x}}$ is time-varying, the augmented adjoint system (7), which becomes

$$\dot{\bar{\lambda}} + F_x^* \lambda = 0,$$
$$\bar{\lambda} - F_{\dot{x}}^* \lambda = 0 \tag{11}$$

under the time-reversing transformation, is used to preserve the stability.

Eqs. (10) and (11) involve matrix–vector products from the left side (referred to as vector–matrix products in the following). Although a matrix–vector product $F_x v$ can be approximated via a directional derivative finite difference method, it is difficult to evaluate the vector–matrix product vF_x directly via a finite difference method. The vector–matrix product vF_x can be written as a gradient of the function $vF(x)$ with respect to x. However, N evaluations of $vF(x)$ are required to calculate the gradient by a finite-difference method if we do not assume any sparsity in the Jacobian. Therefore, automatic differentiation (AD) is necessary to improve the computational efficiency. A forward mode AD tool cannot compute the vector–matrix products without evaluation of the full Jacobian. It has been shown [10] that an AD tool with reverse mode can evaluate the vector-Jacobian product as efficiently as a forward mode AD tool can evaluate the Jacobian-vector product. In our implementation with DASPK3.0, we use the AD tool TAMC [10] to calculate the vector–matrix products.

ADIFOR [2] is another AD tool. Until ADIFOR3.0 is released, ADIFOR includes only the forward mode. If the SparseLinC option is used, ADIFOR can evaluate the non-zero elements of the Jacobian efficiently. After the Jacobian is computed, the vector-Jacobian product can be easily obtained. This method still requires the evaluation of a full Jacobian, which is far more expensive than a vector–matrix product.

4. Initialization of the adjoint DAE

In this section, we consider the initialization of the adjoint DAE for the scalar objective function $g(x,T,p)$. All of the adjoint DAE systems in this section are of the form (8), which means that the time reversing transformation has already been applied.

4.1. ODE systems

From Section 2, we know that the adjoint variables must satisfy initial conditions at time $t=T$. For index-0 or index-1 DAEs, $\lambda^*(T)$ satisfies $\lambda^* F_{\dot{x}}|_{t=T}=0$. If the original DAE is of ODE standard form or regular implicit form (non-singular mass matrix $F_{\dot{x}}$), we can take $\lambda(T)=0$. Then $\lambda_T(T)$ is given by

$$\lambda_T(T) = -\frac{\mathrm{d}\lambda}{\mathrm{d}t}|_{t=T} = \mathrm{d}\lambda/\mathrm{d}\tau|_{t=T} = (F_{\dot{x}}^*)^{-1} g_x|_{t=T}.$$

An important point to note is that computation of $\lambda(t)$ is not required for the initialization of λ_T. This will be the case also for higher-index DAEs, although there it may be less obvious.

4.2. Semi-explicit index-one DAE systems

For index-1 DAE systems of the form

$$\begin{aligned} M(t,x^d)\dot{x}^d &= f(t,x^d,x^a,p), \\ 0 &= h(t,x^d,x^a,p), \end{aligned}$$

where x^d and x^a denote the differential and algebraic solution variables, respectively, $\partial h/\partial x^a$ is non-singular, and the mass matrix $M(t,x^a)$ is a non-singular square matrix, a two-step process can be used. First we initialize the adjoint DAE for the objective function of integral form $G=\int_0^T g(x,t,p)\,\mathrm{d}t$

$$\begin{aligned} M(t,x^d)^*\dot{\lambda}^d &= (f_{x^d} + \mathrm{d}M/\mathrm{d}t - \mathrm{d}(M\dot{x}^d)/\mathrm{d}x^d)^*\lambda^d + h_{x^d}^*\lambda^a + g_{x^d}, \\ 0 &= f_{x^a}^*\lambda^d + h_{x^a}^*\lambda^a + g_{x^a} \end{aligned}$$

with $\lambda^d(T)=0$ to obtain consistent initial values for $\dot{\lambda}^d$ and λ^a. $\dot{\lambda}^d$ is the derivative of λ^d with respect to τ. During the initialization, the values of the differential variables $\lambda^d(T)$ are fixed. Second, we set $\lambda_T^d(T)=\dot{\lambda}^d(T)$ and initialize the adjoint DAE

$$M(t,x^d)^*\dot{\lambda}_T^d=(f_{x^d}+\mathrm{d}M/\mathrm{d}t-\mathrm{d}(M\dot{x}^d)/\mathrm{d}x^d)^*\lambda_T^d+h_{x^d}^*\lambda_T^a,$$
$$0=f_{x^a}^*\lambda_T^d+h_{x^a}^*\lambda_T^a$$

for the objective function $g(x,T,p)$. The differential variables $\lambda_T^d(T)$ are fixed during the second step. Each initialization can be done easily in DASPK3.0 [13].

4.3. Fully implicit index-one DAE systems

For fully implicit index-1 DAE systems

$$F(x,\dot{x},t,p)=0,$$

the initialization can be posed as a least-squares problem for

$$A^*\dot{\lambda}+B^*\lambda=g_x^*,$$
$$A^*\lambda=0,$$

where $A=\partial F/\partial\dot{x}|_{t=T}$ and $B=\partial F/\partial x|_{t=T}$. The solution of this problem is not currently implemented in DASPKADJOINT.

4.4. Hessenberg index-two DAE systems

For Hessenberg index-2 DAE systems

$$\dot{x}^d=f(t,x^d,x^a,p),$$
$$0=h(t,x^d,p), \tag{12}$$

the adjoint DAE system for the objective function $g(x,T,p)$ is given by

$$\dot{\lambda}_T^{d*}=-\lambda_T^{d*}f_{x^d}-\lambda_T^{a*}h_{x^d},$$
$$0=\lambda_T^{d*}f_{x^a}.$$

The initial values for the adjoint variable λ_T^d satisfy [6]

$$\lambda_T^d(T)=P^*(g_{x^d}^*+f_{x^d}^*\lambda^d-\dot{h}_{x^d}^*(f_{x^a}^*h_{x^d}^*)^{-1}g_{x^a}^*)|_{t=T} \tag{13}$$

or

$$\lambda_T^{d*}(T)=(g_{x^d}+\lambda^{d*}f_{x^d}-g_{x^a}(h_{x^d}f_{x^a})^{-1}\dot{h}_{x^d})P|_{t=T}, \tag{14}$$

where $\lambda^{d*}=-g_{x^a}(h_{x^d}f_{x^a})^{-1}h_{x^d}$ is the adjoint variable for the objective function $G=\int_0^T g(t,x,p)\,\mathrm{d}t$, $P=I-f_y(h_xf_y)^{-1}h_x$ is a projection matrix for the original index-2 system, and $\dot{h}_{x^d}$ is the total derivative of h_{x^d} with respect to t, $\dot{h}_{x^d} = h_{x^d_t} + (h_{x^d}\dot{x}^d)_{x^d}$. Note that f_{x^d}, f_{x^a}, h_{x^d} and $\dot{h}_{x^d}$ are matrices, and the initialization in this case is much more complicated. In the following, with the help of an AD tool and the options of DASPK3.0, we give a matrix-free implementation.

The matrix-free initialization procedure can be split into four steps. First we compute $\lambda^{d*}(T) = -g_{x^a}(h_{x^d}f_{x^a})^{-1}h_{x^d}|_{t=T}$ by solving the following initialization problem:

$$\dot{\lambda}_1^{d*} = \lambda_1^{d*} f_{x^d} + \lambda_1^{a*} h_{x^d},$$
$$0 = \dot{\lambda}_1^{d*} f_{x^a} - \lambda_1^{d*} \dot{f}_{x^a} - g_{x^a} \tag{15}$$

for $\dot{\lambda}_1^d(T)$ and $\lambda_1^a(T)$ with $\lambda_1^d(T) = 0$ fixed. In (15), $\dot{f}_{x^a}$ represents the total derivative of f_{x^a} with respect to t. Note that $\lambda_1^{a*}(T) = g_{x^a}(h_{x^d}f_{x^a})^{-1}|_{T=T}$, $\lambda^{d*}(T) = -\dot{\lambda}_1^{d*}(T) = -g_{x^a}(h_{x^d}f_{x^a})^{-1}h_{x^d}|_{t=T}$. If $g_{x^a} = 0$, then $\lambda^{d*}(T) = 0$.

In step 2, we calculate

$$v_1 = g_{x^d} + \lambda^{d*} f_{x^d} - g_{x^a}(h_{x^d}f_{x^a})^{-1}\dot{h}_{x^d}|_{t=T} = g_{x^d} + \lambda^{d*} f_{x^d} - \lambda_1^{a*}\dot{h}_{x^d}|_{t=T}. \tag{16}$$

If $g_{x^a} = 0$, then $v_1 = g_{x^d}|_{t=T}$ and we can go directly to step 3. $\lambda^{d*} f_{x^d}$ can be calculated easily by an AD tool with reverse mode. The calculation of $\lambda_1^{a*}\dot{h}_{x^d}$ is more troublesome. If h_{x^d} is a constant matrix then $\lambda_1^{a*}\dot{h}_{x^d} = 0$. Otherwise, an AD tool with a combination of reverse and forward modes is used to evaluate $\lambda_1^{a*}\dot{h}_{x^d}$. $\lambda_1^{a*}h_{x^d}$ is first calculated by a reverse mode AD tool. Then the vector–matrix product is differentiated explicitly by a forward mode AD-tool with respect to t and x^d.

In step 3, we calculate the initial values of λ_T^d

$$\lambda_T^{d*}(T) = v_1 P|_{t=T} = v_1(I - f_{x^a}(h_{x^d}f_{x^a})^{-1}h_{x^d})|_{t=T} \tag{17}$$

by initializing the following system:

$$\dot{\lambda}_2^{d*} = \lambda_2^{d*} f_{x^d} + \lambda_2^{a*} h_{x^d} + v_1,$$
$$0 = \dot{\lambda}_2^{d*} f_{x^a} - \lambda_2^{d*} \dot{f}_{x^a} \tag{18}$$

with $\lambda_2^{d*}(T) = 0$ fixed. Noting that $\lambda_2^{a*}(T) = -v_1 f_{x^a}(h_{x^d}f_{x^a})^{-1}|_{t=T}$ and $\dot{\lambda}_2^{d*} = v_1(I - f_{x^a}(h_{x^d}f_{x^a})^{-1}h_{x^d})|_{t=T}$, we set $\lambda_T^d(T) = \dot{\lambda}_2^{d*}(T)$.

In step 4, we initialize the adjoint system for the objective function $g(x,T,p)$ by fixing the value of $\lambda_T^{d*}(T)$. The index-2 constraints must be differentiated during the initialization, which results in

$$\dot{\lambda}_T^{d*} = \lambda_T^{d*} f_{x^d} + \lambda_T^{a*} h_{x^d},$$
$$0 = \dot{\lambda}_T^{d*} f_{x^a} - \lambda_T^{d*} \dot{f}_{x^a}, \tag{19}$$

where $\lambda_T^d(T)$ is fixed and $\dot{\lambda}_T^d(T)$ and $\lambda_T^a(T)$ are computed.

The DAE systems (15) and (18) have the same format except for the forcing terms. Therefore, we can solve them efficiently with the same algorithm. The second equation of the adjoint DAE system (19) is actually the derivative of the index-2 constraint $\lambda^d f_{x^a}$ with respect to reversed time τ. DASPKADJOINT has a mechanism to differentiate the index-2 constraints and then to perform the initialization for index-2 DAE systems in DASPK3.0 [15].

For index-2 DAE systems which are not explicitly in Hessenberg form, but which have a non-singular square mass matrix, i.e.,

$$\begin{aligned} M(t,x^d)\dot{x}^d &= f(t,x^d,x^a,p), \\ 0 &= h(t,x^d,p), \end{aligned} \tag{20}$$

where $M(t,x^d)$ is a non-singular square matrix, the adjoint system is

$$\begin{aligned} M(t,x^d)^*\dot{\lambda}_T^d &= (f_{x^d} + \dot{M} - (M\dot{x}^d)_{x^d})^*\lambda_T^d + h_{x^d}^*\lambda_T^a, \\ 0 &= f_{x^a}^*\lambda_T^d. \end{aligned} \tag{21}$$

The initial values for the adjoint variables λ_T^d satisfy

$$\begin{aligned} \lambda_T^{d*} &= \left(\dot{\lambda}^d - g_{x^a}(h_{x^d}M^{-1}f_{x^a})^{-1}\frac{\mathrm{d}(h_{x^d}M^{-1})}{\mathrm{d}t}\right)P \\ &= ((g_{x^d} + \lambda^{d*}(f_{x^d} + \dot{M} - (M\dot{x}^d)_{x^d}))M^{-1} - g_{x^a}(h_{x^d}M^{-1}f_{x^a})^{-1}(\dot{h}_{x^d}M^{-1} - h_{x^d}M^{-1}\dot{M}M^{-1}))P \\ &= (g_{x^d} + \lambda^{d*}(f_{x^d} - (M\dot{x}^d)_{x^d}) - g_{x^a}(h_{x^d}M^{-1}f_{x^a})^{-1}\dot{h}_{x^d})M^{-1}P \\ &= (g_{x^d} - \lambda^{d*}F_{x^d} - \lambda_1^{a*}\dot{h}_{x^d})M^{-1}P \\ &= v_1M^{-1}P \end{aligned}$$

at $t = T$, where $\lambda^{d*} = -g_{x^a}(h_{x^d}M^{-1}f_{x^a})^{-1}h_{x^d}M^{-1}$, $P = I - f_{x^a}(h_{x^d}M^{-1}f_{x^a})^{-1}h_{x^d}M^{-1}$, $F = M\dot{x}^d - f(x^d,x^a,p)$, $\lambda_1^{a*} = g_{x^a}(h_{x^d}M^{-1}f_{x^a})^{-1}$, and $v_1 = g_{x^d} - \lambda^{d*}F_{x^d} - \lambda_1^{a*}\dot{h}_{x^d}$. The above initialization procedures for Hessenberg form can still be used by replacing the adjoint DAE system with (21).

5. Implementation

In this section we outline some of the issues for implementation of the adjoint method for DAE sensitivity analysis. Although any DAE solver could in principle be used for the forward and backward integration of the adjoint DAE system, we describe the implementation via our software DASPK3.0 [15].

In the adjoint system (5) and the sensitivity calculation (6), the derivatives F_x, $F_{\dot{x}}$ and F_p may depend on the state variables x, which are the solutions of the original DAEs. Ideally, the adjoint DAE (5) should be coupled with the original DAE and solved together as we did in the forward sensitivity method. However, in general it is not feasible to solve them together because the original DAE may be unstable when solved backward. Alternatively, it would be extremely inefficient to solve the original DAE forward any time we need the values of the state variables.

The implementation of the adjoint sensitivity method consists of three major steps. First, we must solve the original ODE/DAE forward to a specific output time T. Second, at time T, we compute the consistent initial conditions for the adjoint system. The consistent initial conditions must satisfy the boundary conditions of (3). Finally, we solve the adjoint system backward to the start point, and calculate the sensitivities.

With enough memory, we can store all of the necessary information about the state variables at each time step during the forward integration and then use it to obtain the values of the state variables by interpolation during the backward integration of the adjoint DAEs. For example, we can store x and $\dot{x}$ at each time step during the forward integration and reconstruct the solution at any time by cubic Hermite interpolation[1] during the backward integration. The memory requirements for this approach are proportional to the number of time steps and the dimension of the state variables x, and are unpredictable because the number of time steps varies with different options and error tolerances of the ODE/DAE solver.

To reduce the memory requirements and also make them predictable, we use a two-level checkpointing technique. First we set up a checkpoint after every fixed number of time steps during the forward integration of the original DAE. Then we recompute the forward information between two consecutive checkpoints during the backward integration by starting the forward integration from the checkpoint. This approach needs to store only the forward information at the checkpoints and at fixed number of times between two checkpoints.

In the implementation we allocated a special buffer to communicate between the forward and backward integration. The buffer is used for two purposes: to store the necessary information to restart the forward integration at the checkpoints, and to store the state variables and derivatives at each time step between two checkpoints for reconstruction of the state variable solutions during the backward integration.

In order to obtain the fixed number of time steps between two consecutive checkpoints, the second forward integration should make exactly the same adaptive decisions as the first pass if it restarts from the checkpoint. Therefore, the information saved at each checkpoint should be enough that the integration can repeat itself. In the case of DASPK3.0, the necessary information includes the order and stepsize for the next time step, the coefficients of the BDF formula, the history information array of the previous k time steps, the Jacobian information at the current time, etc. To avoid storing Jacobian data (which is much larger than other information) in the buffer, we enforce a reevaluation of the iteration matrix at each checkpoint during the first forward integration.

If the size of the buffer is specified, the maximum number of time steps allowed between two consecutive checkpoints and the maximum number of checkpoints allowed in the buffer can be easily determined. However, the total number of checkpoints is problem-dependent and unpredictable. It is possible that the number of checkpoints is also too large for some applications to be held in the buffer. We then write the data of the checkpoints from the buffer to a disk file and reuse the buffer again. Whenever we need the information on the disk file, we can access it from the disk. We assume that the disk is always large enough to hold the required information.

[1] We could of course consider basing the interpolant on the interpolating polynomial underlying the BDF formula, but this is more complicated, requires more storage, and it not as smooth.

5.1. Jacobian or preconditioner evaluation and Newton solver

For an implicit ODE/DAE solver (e.g., BDF, SIRK), a Jacobian (or preconditioner) is required for solving the non-linear system at each time step. The Jacobian for the adjoint DAE is exactly the transpose of the Jacobian for the original DAE when the mass matrix $F_{\dot{x}}$ is constant. Therefore, we can evaluate the Jacobian of the original DAE and transpose it to obtain the Jacobian for the adjoint DAE.

If $F_{\dot{x}}$ depends on either t or x, we can use the adjoint DAE (7) or (11) for discretization. The Jacobian for system (11) is

$$\begin{pmatrix} \alpha I & F_x^* \\ I & -F_{\dot{x}}^* \end{pmatrix},$$

where α is a scalar coefficient in DASPK3.0. If we solve only for λ, then the second column is multiplied by $-\alpha$ and added to the first column, which yields the Jacobian for λ only, $\alpha F_{\dot{x}}^* + F_x^*$, which remains the transpose of the original Jacobian for the forward integration.

The adjoint system is a linear time-varying system. The Newton solver can converge in one iteration if the Jacobian is up to date. In DASPK3.0, the convergence test of the Newton solver is

$$\frac{\rho}{1-\rho}\|y^{(m+1)} - y^{(m)}\| < 0.33, \tag{22}$$

where the rate of convergence ρ is given by

$$\rho = \left(\frac{\|y^{(m+1)} - y^{(m)}\|}{\|y^{(1)} - y^{(0)}\|}\right)^{1/m}.$$

If the Newton solver converges in one iteration, ρ can be very small and is passed on to future time steps. The near-zero ρ can make (22) satisfied even if $\|y^{(m+1)} - y^{(m)}\|$ is large. The large value of $\|y^{(m+1)} - y^{(m)}\|$ can yield bad results if the Jacobian is not current. This is a deficiency in the convergence test of DASPK3.0 (and previous versions), but to our knowledge it has never before been the source of difficulty because unlike the adjoint DAE, the vast majority of DAEs solved in practice are non-linear.

There are two options to fix the convergence test for the adjoint system: force a Jacobian evaluation on every step, or force a recalculation of the convergence rate on every step. Because a Jacobian evaluation takes much more time than a function evaluation, we force a recalculation of the convergence rate on every step for the adjoint method using DASPK3.0. The initial value of ρ is always 0.99.

5.2. Krylov iterative method

For the Krylov iterative method, we need to evaluate the matrix–vector product

$$u = (\alpha F_{\dot{x}}^* + F_x^*)v. \tag{23}$$

Since we already have the mechanism to evaluate $vF_{\dot{x}}$ and vF_x when we evaluate the adjoint DAE, (23) can be evaluated directly.

A preconditioner must be provided by the user for the Krylov iterative method in DASPK3.0. The preconditioner provided by the user usually works well for the forward integration of the state variables and for the forward sensitivity computation. However, it is not always a simple matter to apply this preconditioner for the adjoint system. Because the Jacobian for the adjoint system is the transpose of the forward Jacobian, a transpose of the preconditioner is needed. If a matrix-free method has been used to construct the preconditioner for the forward system, it is difficult to transpose the preconditioner and apply it to the adjoint system. On the other hand, if the Jacobian matrix is constructed during computation of the preconditioner as, for example, with the incomplete LU factorization (ILU) preconditioner, the preconditioner for the adjoint system can be easily obtained by transposing the Jacobian first and then doing the incomplete LU factorization.

5.3. Sensitivity calculation

The integral in Eq. (6) can be computed either outside DASPK after each time step or as a quadrature variable during the solution of the adjoint DAE. If the quadrature variable method is chosen, we append a variable, say λ_{N+1}, and an equation

$$\dot{\lambda}_{N+1} = \lambda_T^* F_p$$

to the original adjoint system. Then

$$\lambda_{N+1}|_{t=0} = \int_0^T \lambda_T^* F_p \, \mathrm{d}t.$$

λ_{N+1} is called a quadrature variable in DASPK3.0 and can be calculated efficiently [15] without participating in Newton iterations and Jacobian evaluations via a staggered method. The other terms in (6) can be easily obtained: $\lambda^* F_p$ and $\lambda_T F_{\dot{x}}$ can be computed by an AD tool with reverse mode.

$(x_0)_p$ is the Jacobian matrix of the initial condition x_0 with respect to the sensitivity parameter p. When p is the initial condition for an ODE (or implicit ODE), $(x_0)_p = I$ is the identity matrix. For an index-1 or index-2 system, $(x_0)_p$ must be consistent with the algebraic constraints and/or any hidden constraints. $(x_0)_p$ can be input either by the user or evaluated by an AD tool. When the number of parameters is large, the matrix $(x_0)_p$ is huge. However, the number of non-zero elements in $(x_0)_p$ may be small. A sparse format to store and compute $(x_0)_p$ is necessary.

For the Hessenberg index-2 DAE system (12), the value of λ^{a*} is required when $h_p \neq 0$, because $\lambda^{a*} h_p$ should be calculated in the sensitivity (6). To solve for λ^{a*} correctly in (16), we need the value of $\mathrm{d}g_{x^a}/\mathrm{d}t$, which is difficult for the user to provide. However, there is an additional term for the sensitivity (6) if $h_p \neq 0$, which is $\mathrm{d}(-g_{x^a}(h_{x^d} f_{x^a})^{-1} h_p)/\mathrm{d}t$. Then the combination of the two terms will be

$$\begin{aligned}
&(\lambda^{d*} f_p + \lambda^{a*} h_p) + \frac{\mathrm{d}}{\mathrm{d}t}(-g_{x^a}(h_{x^d} f_{x^a})^{-1} h_p) \\
&= \lambda^{d*} f_p - \lambda_1^{a*} \dot{h}_p + \left(\lambda^{a*} - \frac{\mathrm{d}g_{x^a}}{\mathrm{d}t}(h_{x^d} f_{x^a})^{-1} + g_{x^a}(h_{x^d} f_{x^a})^{-1} \left(\dot{h}_{x^d} f_{x^a} + h_{x^d} \dot{f}_{x^a} \right) (h_{x^d} f_{x^a})^{-1} \right) h_p \\
&= \lambda^{d*} f_p - \lambda_1^{a*} \dot{h}_p + (\lambda^{a*} h_{x^d} - \dot{\lambda}^{d*} + \lambda_1^{a*} \dot{h}_{x^d}) f_{x^a} (h_{x^d} f_{x^a})^{-1} h_p \\
&= \lambda^{d*} f_p - \lambda_1^{a*} \dot{h}_p + \lambda_2^{a*} h_p.
\end{aligned}$$

where $\lambda^{d*} f_{x^a} + g_{x^a} = 0$ is used and $\lambda_1^{a*} = -g_{x^a}(h_{x^d} f_{x^a})^{-1}$ and $\lambda_2^{a*} = -(g_{x^d} + \lambda^{d*} f_{x^d} - \lambda_1^{a*} \dot{h}_{x^d}) f_{x^a}(h_{x^d} f_{x^a})^{-1}$ can be obtained from Eqs. (15) and (18), respectively. It can be verified that the last equation is also valid for the case of non-Hessenberg form given by (20).

5.4. Error estimation

Two of the most important decisions that an adaptive DAE solver must make on each step are whether to accept the results of the current step and what stepsize should be used on the next step. Both of these decisions are based on the error estimate. DASSL/DASPK estimates the local truncation error, but it is not obvious how this should be implemented for the adjoint solution.

To get a better understanding, consider applying the implicit Euler method to the augmented adjoint system (11),

$$\frac{\bar{\lambda}_{n+1} - \bar{\lambda}_n}{h_{n+1}} + F_x^* \lambda_{n+1} = 0,$$

$$\bar{\lambda}_{n+1} - F_{\dot{x}}^* \lambda_{n+1} = 0. \tag{24}$$

The true solution to (11) satisfies

$$\frac{\bar{\lambda}(t_{n+1}) - \bar{\lambda}(t_n) - (h_{n+1}^2/2)\bar{\lambda}''(\xi)}{h_{n+1}} + F_x^* \lambda(t_{n+1}) = 0,$$

$$\bar{\lambda}(t_{n+1}) - F_{\dot{x}}^* \lambda(t_{n+1}) = 0. \tag{25}$$

Subtracting (25) from (24), we obtain

$$\frac{\bar{e}_{n+1} - \bar{e}_n + (h_{n+1}^2/2)\bar{\lambda}''(\xi)}{h_{n+1}} + F_x^* e_{n+1} = 0,$$

where $\bar{e}_n = \bar{\lambda}_n - \bar{\lambda}(t_n)$ and $e_n = \lambda_n - \lambda(t_n)$. Thus, the local truncation error (the amount by which the true solution fails to satisfy the difference formula) depends on $\bar{\lambda}''$ rather than on λ''. Therefore, the error estimate should be based on $\bar{\lambda}$. Since the errors in the algebraic variables on previous time steps do not directly influence the errors in any of the variables at the current time [17], we can consider deleting the λ from the error estimate in order to promote the smooth operation of a code. However, the value of λ is important and determines the accuracy of the sensitivities. So for the λ variables which are index-1 in the augmented adjoint system, we opted to include them and to exclude those which are index-2.

5.5. Efficiency of the adjoint sensitivity method

In this subsection we compare the adjoint sensitivity method with the forward sensitivity method. Suppose the cost of the forward integration for N state variables is C_f, the cost of the forward integration for N sensitivity variables is C_{fs}, the cost of the backward integration for N adjoint variables is C_{bs}, the number of parameters is n_p, and the number of objective functions is n_f.

Then our implementation of the adjoint sensitivity method includes two forward integrations and one backward integration.[2] The total cost is roughly

$$C_{asm} = 2C_f + C_{bs}^* + (n_f - 1)C_{bs},$$

where C_{bs}^* denotes the cost for the first objective function, which is more costly than the others because it involves the Jacobian evaluation. The total cost for the forward sensitivity method to perform the equivalent computation is roughly

$$C_{fsm} = C_f + n_p C_{fs}.$$

The computational efficiency and memory requirements for the forward sensitivity method are roughly proportional to the number of sensitivity parameters and are insensitive to the number of objective functions. For the adjoint sensitivity method, the computational efficiency and memory requirements are proportional to the number of objective functions and are insensitive to the number of sensitivity parameters. Thus the two methods are complementary. The adjoint sensitivity method is advantageous over the forward sensitivity method when the number of sensitivity parameters is large and the number of objective functions is small.

The adjoint sensitivity method has a disadvantage that it can only compute the sensitivity at a specific output time. Unlike the forward sensitivity method, the intermediate results of the adjoint variables have no physical meaning.

6. Numerical experiments

In this section we give some examples to demonstrate the effectiveness and efficiency of the adjoint sensitivity method as implemented in DASPKADJOINT. In all of our examples, the tolerance for the adjoint variables has been taken to be double the tolerance for the state variables, because we have found that it is helpful in the solution of the adjoint equations to have solved for the state variables with slightly greater accuracy than we request for the adjoint variables. The integration methods used are the direct method (D) and Krylov method (K). The sensitivity methods are the forward sensitivity method (F), using the staggered corrector option, from DASPK3.0 [14] and the adjoint sensitivity method (A). Therefore, we use AD to represent the adjoint direct method, AK to represent the adjoint Krylov method, FD to represent the forward direct method, and FK to represent the forward Krylov method. The computation was performed on a Linux machine with Pentium III 450 MHZ CPU.

6.1. ODE case

6.1.1. 2-D heat equation

We first consider the heat equation

$$u_t = p_1 u_{xx} + p_2 u_{yy} \tag{26}$$

[2] This assumes that checkpointing is used. If checkpointing is not needed, it requires only one forward integration.

Table 1
Results for heat equation example. NP is the number of sensitivity parameters

METH	NP	g_1 w.r.t. p_1	g_2 w.r.t. p_1	RWORK size	IWORK size	CPU
FD	20	−15.21783	−2.72675	669025	10630	44.15
FD	10	−15.21783	−2.72675	457205	10630	24.40
AD	1766	−15.21831	−2.72685	563563	12462	21.11
FK	10	−15.21782	−2.72675	329420	83868	15.39
AK	1766	−15.21794	−2.72667	235388	167751	6.63

posed on the 2-D unit square with zero Dirichlet boundary conditions. An $M+2$ by $M+2$ mesh is used with uniform spacing $1/(M+1)$. The spatial derivatives are represented by standard central finite difference approximations. At each interior point of the mesh, the discretized PDE becomes an ODE for the discrete value of u. At each point on the boundary, we pose the equation $u_t = 0$. The discrete values of u form a vector U, ordered first by x, then by y. The result is an ODE system $G(t,U,U') = 0$ of size $NEQ = (M+2)\times(M+2)$. Initial conditions are posed as

$$u_{t=0} = 16x(1-x)y(1-y).$$

The sensitivity parameters consist of $p_1 = p_2 = 1.0$ and the initial values $u_{t=0}$. The problem was solved previously by DASPK3.0 with $M = 40$ in [15]. To compute the sensitivities by the adjoint method, we used the time interval $[0, 0.16]$, and the error tolerances for DASPK3.0 were taken as RTOL = ATOL = 1.0e − 5. The size of the buffer was set to such a number that it allows nine time steps between two consecutive checkpoints, and the maximum number of checkpoints the buffer allows was set to 3. There are a total of 10 checkpoints during the forward integration of the state variables and the information at the checkpoints has to be written to the disk file three times. For the direct method, we used the ADIFOR option with SparseLinC to generate the Jacobian. For the Krylov method, we used the incomplete LU (ILU) preconditioner, which is part of the DASPK package. The Jacobian for the ILU preconditioner was also evaluated by ADIFOR with the SparseLinC option. We compared the results with that of the forward sensitivity method where the sensitivity residuals are evaluated by ADIFOR with the seed matrix option. Due to the memory restriction, we used only 20 sensitivity parameters (including p_1 and p_2) in the forward sensitivity method. For comparison, we used two objective functions:

$$g_1 = \sum_1^{\mathrm{NEQ}} u_i^2,$$

$$g_2 = \int_0^T \sum_1^{\mathrm{NEQ}} u_i \,\mathrm{d}t,$$

where g_2 is treated as a quadrature variable. Table 1 gives the results of the adjoint and forward methods.

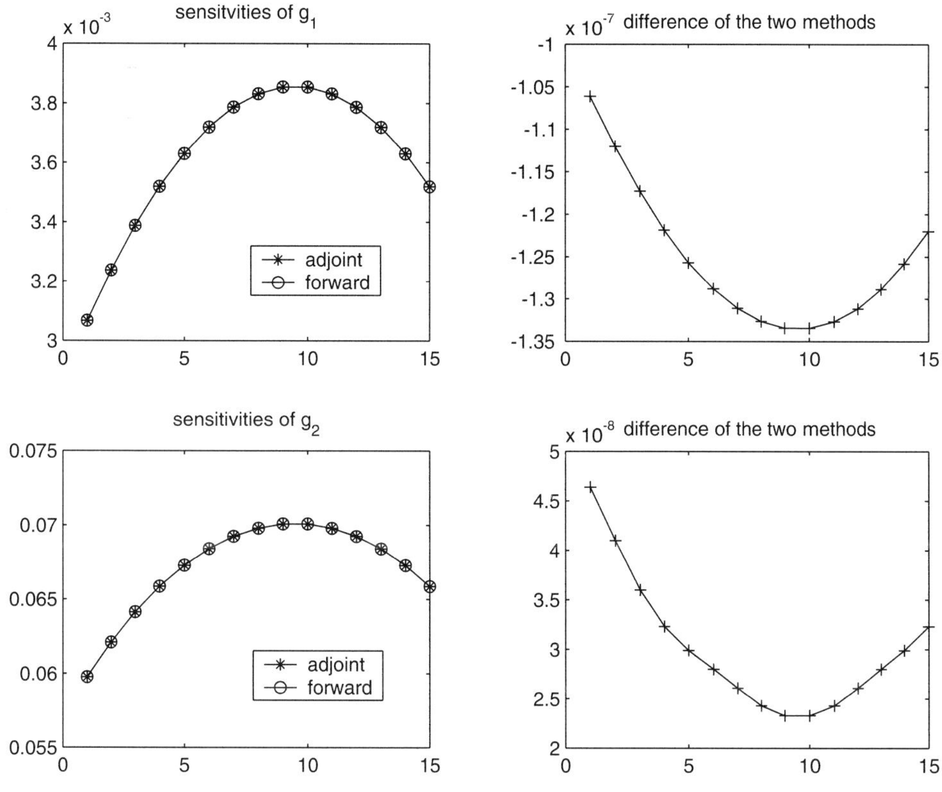

Fig. 1. Sensitivities of the objective functions with respect to the initial conditions for the heat equation example. The x-axis represents the order number of the points.

Fig. 1 shows the sensitivities of the objective functions with respect to the initial conditions. We chose the points between 13 and 27 in the 20th row on the 42×42 mesh as our sample points for the figure.

6.1.2. Index-0 example with state-dependent mass matrix

We also tested the case when the coefficient matrix $F_{\dot{x}}$ is not constant. Consider the system

$$\begin{pmatrix} y_1 & y_2 \\ -y_2 & y_1 \end{pmatrix} \begin{pmatrix} \dot{y}_1 \\ \dot{y}_2 \end{pmatrix} = \begin{pmatrix} 0 \\ -(y_1^2 + y_2^2) \end{pmatrix} \tag{27}$$

with initial conditions $y_1(0) = 0$, $y_2(0) = 1$. We used the time interval [0,1.57] and the objective function $g(y) = y_1 + y_2$ in the test. The sensitivity parameters were taken to be $y_1(0)$ and $y_2(0)$. The augmented adjoint system for g is

$$\bar{\lambda} - \begin{pmatrix} y_1 & -y_2 \\ y_2 & y_1 \end{pmatrix} \lambda = 0,$$

Table 2
Results for example (27). Err_{max} is the maximum error in λ. NSTP is the number of time steps, NRES is the number of function evaluations, NJAC is the number of Jacobian evaluations, NETF is the number of error test failures and NNL is the number of non-linear iterations. There are no convergence test failures in any of the tests

Error control	Recalculate ρ	NSTP	NRES	NJAC	NETF	NNL	Err_{max}
$\bar\lambda$	No	64	245	31	7	121	0.6025
λ, $\bar\lambda$	No	144	383	31	7	259	8.0e-7
λ	No	2662	28486	4174	2660	11790	5.65e-4
$\bar\lambda$	Yes	61	254	28	7	142	7.0e-7
λ, $\bar\lambda$	Yes	86	317	28	7	205	1.4e-7
λ	Yes	2662	28487	4174	2660	11791	5.65e-4

$$\dot{\bar\lambda} + \begin{pmatrix} y_1' & y_2' + 2y_1 \\ y_2' & -y_1' + 2y_2 \end{pmatrix} \lambda = 0.$$

We used the error tolerances $\mathrm{ATOL} = 10^{-9}$, $\mathrm{RTOL} = 10^{-7}$ for DASPK3.0 in the test. The results were

METH	g w.r.t. $y_1(0)$	g w.r.t. $y_2(0)$
FD	−0.999204383	1.00079725
AD	−0.999203795	1.00079653
True solution	−0.999203356	1.00079601

This example illustrates the need for the modifications to the DASPK3.0 error test and Newton convergence test strategies described earlier for solving the augmented adjoint system. It can be shown that when $y_1 = \sin(t)$ and $y_2 = \cos(t)$, $\dot{\lambda} = 0$. If we base the error estimate only on λ, which may at first glance seem to be natural, it results in a large number of error test failures. The reason is that according to the dynamics of λ, DASPK3.0 tries with this error estimate to double the stepsize at almost every time step. However, the local truncation error is determined by $\bar\lambda$ instead of λ, so these large stepsizes fail based on the dynamics of $\bar\lambda$. We also observed that if we did not force a recalculation of the convergence rate ρ during the Newton iteration, a large error can occur in λ if we base the error test only on $\bar\lambda$. The results are good and the code operates efficiently when we base the error test on both λ and $\bar\lambda$, or base the error test only on $\bar\lambda$ and recalculate the convergence rate. Table 2 gives the results of the different options for the error control and convergence tests.

6.2. Index-1 DAE case

6.2.1. 2-D food web problem

We consider a multi-species food web [5], in which mutual competition and/or predator-prey relationships in the spatial domain are simulated. Specifically, the model equations for the concentration vector $c = (c^1, c^2)^{\mathrm{T}}$ are

$$c_t^1 = f_1(x, y, t, c) + d_1(c_{xx}^1 + c_{yy}^1),$$

$$0 = f_2(x, y, t, c) + d_2(c_{xx}^2 + c_{yy}^2)$$

with

$$f_i(x, y, t, c) = c^i \left(b_i + \sum_{j=1}^{2} a_{ij} c^j \right).$$

The coefficients a_{ij}, b_i, d_i are:

$$a_{ii} = -1, \quad i = 1, 2;$$

$$a_{12} = -0.5 \times 10^{-6}, \quad a_{21} = 10^4;$$

$$a_{ij} = e, \quad i > p \text{ and } j < p;$$

$$b_1 = 1 + \alpha xy + \beta \sin(4\pi x)\sin(4\pi y)) = -b_2,$$

$$d_1 = 1, \quad d_2 = 0.05.$$

The domain is the unit square $0 \leqslant x, y \leqslant 1$. The boundary conditions are of Neumann type with normal derivative equal to 0. The PDEs are discretized by central differencing on an M by M mesh. We have taken $M = 20$. Therefore, the resulting DAE system has size $\mathrm{NEQ} = 2M^2 = 800$. The DASPK3.0 tolerances used were $\mathrm{RTOL} = \mathrm{ATOL} = 10^{-5}$.

For sensitivity analysis, α and β were taken as the sensitivity parameters, with nominal values $\alpha = 50$ and $\beta = 100$. The initial conditions were taken as

$$c_0^1 = 10 + (16x(1 - x)y(1 - y))^2,$$

$$c_0^2 = 100,$$

which do not satisfy the constraint equations. Therefore, a consistent initialization is required. For comparison, we also took the initial values c_0^1 as sensitivity parameters. Unlike the ODE case, we cannot take both c_0^1 and c_0^2 as independent sensitivity parameters, because c_0^2 are index-1 variables and they depend on c_0^1. We used the time interval $[0, 5]$, and tolerances for DASPK3.0 of $\mathrm{RTOL} = \mathrm{ATOL} = 10^{-5}$. For the adjoint method, the size of the buffer was set to such a number that it allows 12 time steps between two consecutive checkpoints. The maximum number of checkpoints the buffer allows was set to 4. There are a total of 10 checkpoints during the forward integration of the state variables, and the information at the checkpoints has to be written to the disk file twice. We used two identical objective functions

$$g_1 = \sum_{1}^{\mathrm{NEQ}} u_i^2,$$

$$g_2 = g_2.$$

Table 3 gives the results of the adjoint and forward methods. Note that the storage requirement (RWORK) and the CPU time for the forward sensitivity method are proportional to the number of sensitivity parameters, whereas for the adjoint method they remain the same.

Table 3
Results for food-web problem. NP is the number of sensitivity parameters

METH	NP	g_1 w.r.t. p_1	g_1 w.r.t. p_2	RWORK size	IWORK size	CPU
FD	20	6467.01	3287.73	312890	4840	26.77
FD	10	6467.01	3287.73	208870	4840	14.73
FD	2	6467.01	3287.73	125654	4840	6.09
AD	402	6467.12	3287.79	249350	7313	13.14

6.2.2. Index-1 example with state-dependent mass matrix

We also tested a simple index-1 example with non-constant matrix $F_{\dot{x}}$. The equations are

$$\begin{pmatrix} y_2 & 0 \\ 0 & 0 \end{pmatrix} \begin{pmatrix} \dot{y}_1 \\ \dot{y}_2 \end{pmatrix} = \begin{pmatrix} -y_2(y_2-1) \\ y_2 - y_1 - 1 \end{pmatrix} \tag{28}$$

with initial values $y_1(0)=1$ and $y_2(0)=2$. We used the time interval [0,1] and the objective function $g(y) = y_1 + y_2$. We cannot select both $y_1(0)$ and $y_2(0)$ as sensitivity parameters for a well-posed problem. Thus we chose $y_1(0)$ as the sensitivity parameter. The DASPK3.0 error tolerances are $\text{ATOL} = 10^{-9}$, $\text{RTOL} = 10^{-7}$. The results are

METH	g w.r.t. $y_1(0)$
FD	0.73575887
AD	0.73575898
True solution	0.73575882

6.3. Index-2 DAE case

We consider an index-2 DAE from mechanics. This problem is selected from the set of initial value test problems [12]. It is of the form

$$M\frac{\mathrm{d}y}{\mathrm{d}t} = f(y), \quad y(0) = y_0, \quad y'(0) = y_0'$$

with $y, f \in R^{160}$, $t \in [0, 1000]$. M is a constant mass matrix given by

$$\begin{pmatrix} I_{120} & 0 \\ 0 & 0 \end{pmatrix},$$

where I_{120} is the identity matrix of dimension 120. For the definition of the function f, we refer to [12]. The first 120 components are of index-0, the last 40 of index 2.

The components $y_{0,i}$ of the initial vector y_0 are defined by

$$\begin{pmatrix} y_{0,3(j-1)+1} \\ y_{0,3(j-1)+2} \\ y_{0,3(j-1)+3} \end{pmatrix} = \begin{pmatrix} \cos(\omega_j)\cos(\beta_j) \\ \sin(\omega_j)\cos(\beta_j) \\ \sin(\beta_j) \end{pmatrix} \quad \text{for } j = 1,\dots,20,$$

where

$$\beta_j = \alpha_1\pi \quad \text{and} \quad \omega_j = \frac{2j}{3}\pi + \alpha_2\pi \quad \text{for } j = 1,\dots,3,$$

$$\beta_j = \alpha_3\pi \quad \text{and} \quad \omega_j = \frac{2(j-3)}{7}\pi + \alpha_4\pi \quad \text{for } j = 4,\dots,10,$$

$$\beta_j = \alpha_5\pi \quad \text{and} \quad \omega_j = \frac{2(j-10)}{6}\pi + \alpha_6\pi \quad \text{for } j = 11,\dots,16,$$

$$\beta_j = \alpha_7\pi \quad \text{and} \quad \omega_j = \frac{2(j-17)}{4}\pi + \alpha_8\pi \quad \text{for } j = 11,\dots,16.$$

For the remainder of the initial conditions, the reader is referred to [12]. The vector α_i contains the sensitivity parameters, with nominal values

$$\alpha_1 = \tfrac{3}{8}, \quad \alpha_2 = \tfrac{1}{13}, \quad \alpha_3 = \tfrac{1}{8}, \quad \alpha_4 = \tfrac{1}{29},$$

$$\alpha_5 = -\tfrac{2}{15}, \quad \alpha_6 = \tfrac{1}{7}, \quad \alpha_7 = -\tfrac{3}{10}, \quad \alpha_4 = \tfrac{1}{17}.$$

The origin of the problem is to compute the elliptic Fekete points. For any configuration $x := (x_1, x_2, \dots, x_N)^{\mathrm{T}}$ in a unit sphere in $\Re^3$, the points $\hat{x}_1, \hat{x}_2, \dots, \hat{x}_N$ are called the elliptic Fekete points of order N if the function

$$V(x) := \prod_{i<j} \|x_i - x_j\|_2$$

reaches its global maximum at x. This optimization problem can be formulated as an index-2 DAE [12]. Since it is a global optimization problem, the value of $V(x)$ at the final time should not depend on the initial conditions, i.e., the sensitivity of $V(x)$ with respect to the sensitivity parameters α_i should be zero or close to zero. The buffer size for the adjoint method was set to be large enough to hold the information for 30 time steps and 10 checkpoints. There are a total of eight checkpoints. Thus no temporary disk file was written.

We first computed the sensitivities by the forward sensitivity method (see Table 4). Then the adjoint sensitivity method was used. The tolerances for both methods were RTOL = ATOL = 10^{-4}. Table 4 gives the results of the adjoint and forward methods. Note that the results for the adjoint method are not as good as that of the forward sensitivity method. However, they are within the error tolerances. The CPU time used for the forward method was 11.49, and for the adjoint method 6.78. The CPU time will be almost the same for the adjoint method if more sensitivity parameters are considered, whereas it will increase for the forward method.

Table 4
Results for Fekete problem

Sensitivity	FD	AD
V_{α_1}	−3.7873e-11	6.3065e-4
V_{α_2}	0	4.8950e-5
V_{α_3}	4.4342e-11	2.5240e-4
V_{α_4}	0	2.4969e-5
V_{α_5}	4.4272e-11	−9.3772e-4
V_{α_6}	−3.7543e-11	−1.4526e-4
V_{α_7}	−3.8355e-11	5.4515e-4
V_{α_8}	−3.7363e-11	1.0331e-4

Acknowledgements

The authors would like to thank Muruhan Rathinam and Radu Serban for their interesting discussions and useful comments.

References

[1] U.M. Ascher, L.R. Petzold, Computer Methods for Ordinary Differential Equations and Differential-Algebraic Equations, SIAM, Philadelphia, PA, 1998.
[2] C. Bischof, A. Carle, G. Corliss, A. Griewank, P. Hovland, ADIFOR-Generating derivative codes from Fortran programs, Scientific Programming 1 (1992) 11–29.
[3] K.E. Brenan, S.L. Campbell, L.R. Petzold, Numerical Solution of Initial-Value Problems in Differential-Algebraic Equations, 2nd Edition, SIAM, Philadelphia, PA, 1996.
[4] P.N. Brown, A.C. Hindmarsh, L.R. Petzold, Using Krylov methods in the solution of large-scale differential-algebraic systems, SIAM J. Sci. Comput. (1994) 1467–1488.
[5] P.N. Brown, A.C. Hindmarsh, L.R. Petzold, Consistent initial condition calculation for differential-algebraic systems, SIAM J. Sci. Comput. 19 (5) (1998) 1495–1512.
[6] Y. Cao, S. Li, L. Petzold, R. Serban, Adjoint sensitivity analysis for differential-algebraic equations: the adjoint DAE system and its numerical solution, SIAM J. Sci. Comput., to appear.
[7] M. Caracotsios, W.E. Stewart, Sensitivity analysis of initial value problems with mixed ODEs and algebraic equations, Compu. Chem. Eng. 9 (4) (1985) 359–365.
[8] R.M. Errico, What is an adjoint model? Bulletin of the American Meteorological Society 78 (1997) 2577–2591.
[9] W.F. Feehery, J.E. Tolsma, P.I. Barton, Efficient sensitivity analysis of large-scale differential-algebraic systems, Appl. Numer. Math. 25 (1997) 41–54.
[10] R. Giering, T. Kaminski, Recipes for adjoint code construction, ACM Trans. Math. Software 24 (1998) 437–474.
[11] D. Knapp, V. Barocas, K. Yoo, L. Petzold, R. Tranquillo, Rheology of reconstituted type I collagen gel in confined compression, J. Rheology 41 (1997) 971–993.
[12] W.M. Lioen, J.J.B. de Swart, Test set for initial value problem solvers, Technical Report MAS-R9832, CWI, Amsterdam, 1998.
[13] S. Li, L. Petzold, Design of new DASPK for sensitivity analysis, Technical Report, Department of Computer Science, UCSB, 1999.
[14] S. Li, L. Petzold, W. Zhu, Sensitivity analysis of differential-algebraic equations: a comparison of methods on a special problem, Appl. Numer. Math. 32 (2000) 161–174.

[15] S. Li, L. Petzold, Software and algorithms for sensitivity analysis of large-scale differential-algebraic systems, J. Comput. Appl. Math. 125 (2000) 131–145.
[16] T. Maly, L.R. Petzold, Numerical methods and software for sensitivity analysis of differential-algebraic systems, Appl. Numer. Math. 20 (1997) 57–79.
[17] L. Petzold, P. Lötstedt, Numerical solution of non-linear differential equations with algebraic constraints II, SIAM J. Sci. Statist. Comp. 7 (1986) 720–733.
[18] L.L. Raja, R.J. Kee, R. Serban, L. Petzold, Dynamic optimization of chemically reacting stagnation flows, Proc. Electrochem. Soc. (1998).

JOURNAL OF COMPUTATIONAL AND APPLIED MATHEMATICS

ELSEVIER Journal of Computational and Applied Mathematics 149 (2002) 193–205

www.elsevier.com/locate/cam

Applied optimal shape design

B. Mohammadi[a,1], O. Pironneau[b,*,1]

[a] *Mathematics, University of Montpellier II, France*

[b] *Analyse Numérique, University of Paris VI, 175 rue du Chevaleret, F-75013, Paris, France*

Received 2 November 2001; received in revised form 30 January 2002

Abstract

This paper is a short survey of optimal shape design (OSD) for fluids. OSD is an interesting field both mathematically and for industrial applications. Existence, sensitivity, correct discretization are important theoretical issues. Practical implementation issues for airplane designs are critical too.

The paper is also a summary of the material covered in our recent book, Applied Optimal Shape Design, Oxford University Press, 2001.

Keywords: Optimization; Shape design; Finite elements; Mesh adaptation; Aerodynamics

1. Industrial demand

Since control is a natural desire once the simulation is completed, the applications of OSD are uncountable. For instance the design of a harbor which minimizes the waves coming from far can be done at little cost by standard optimization methods once the numerical simulation of Helmholtz equation is mastered (see Ref. [1]) as shown in Fig. 1. Other applications include

- weight reduction in car engine, aircraft structures, etc,
- electromagnetically optimal shapes, such as in stealth airplanes,
- wave canceling fore bulbe in boat design,
- drag reduction for airplanes, cars and boats.

However, there are no automatic solutions to these problems because most of the time engineering design is made of compromises due to the multi-disciplinary aspects of the problems (see Fig. 2),

* Corresponding author.

E-mail address: pironneau@ann.jussieu.fr (O. Pironneau).

[1] Both authors are member of the Institut Universitaire de France.

PII: S0377-0427(02)00529-0

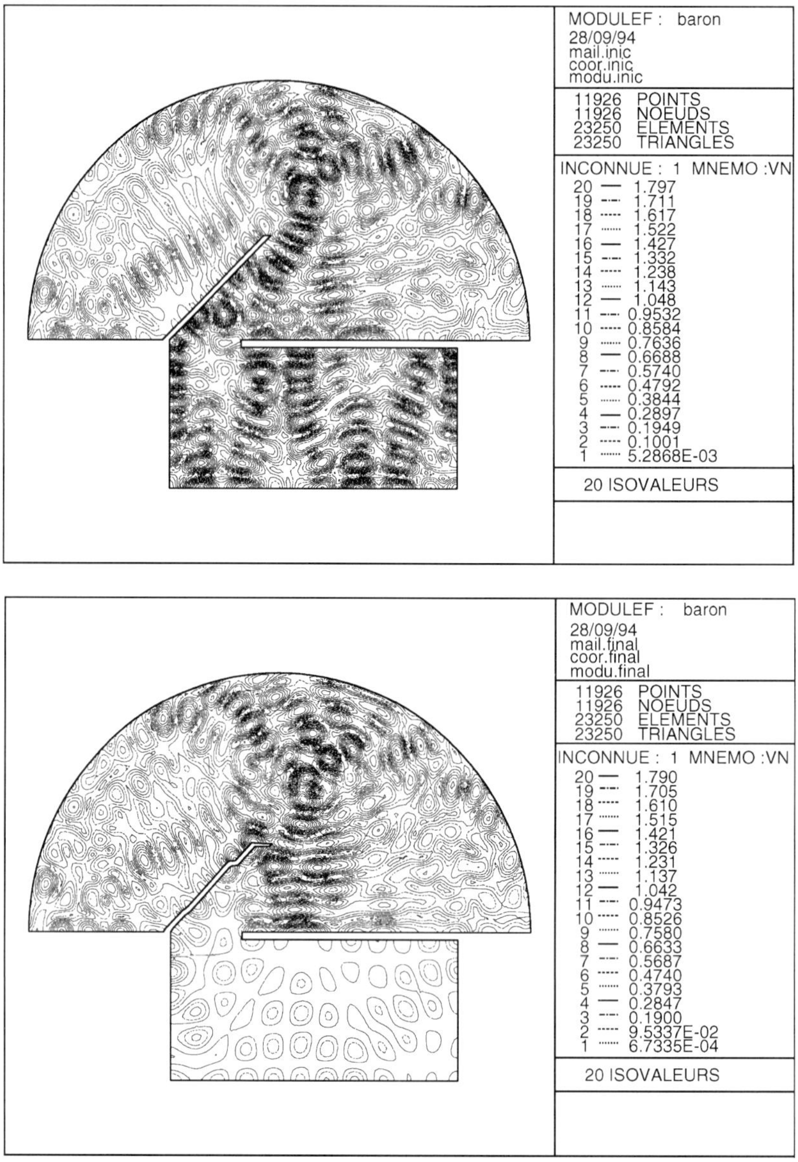

Fig. 1. Optimization of a brake-water (computed by A. Baron).

the necessity of doing *multi-point constrained design* and because the solvers are not always made in house and appear as *black-box* solvers.

1.1. An example in 1D

To understand the difficulties of OSD problem let us consider the problem of design a string of thickness α and length s for a musical instrument which gives a response as close to ψ_d as possible:

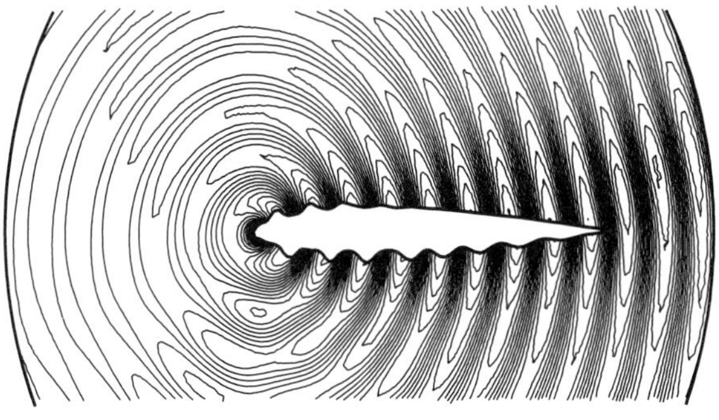

Fig. 2. Optimal design of an airfoil to minimize in a sector, with angle between 180° and 225°, the reflection of a monochromatic incident radar wave. The optimal shape is not admissible from the aerodynamic view point; a multi-disciplinary optimization is necessary (computed by A. Baron).

Find $s > b$ and $\alpha > 0$ such that

$$\min_{\alpha,\mathbf{s}\in\mathcal{S}_d}\left\{\int_a^b |\psi-\psi_d|^2 : \psi + \frac{\mathrm{d}}{\mathrm{d}x}\left(\alpha\frac{\mathrm{d}\psi}{\mathrm{d}x}\right) = f \text{ in } (0,s), \quad \psi(0)=0,\ \ \psi(s)=\psi_d(s)\right\}.$$

To discretize the problem, take I mesh points x_i, call $\psi_i \equiv \psi(x_i)$, $\delta x_{i+1/2} = x_{i+1} - x_i$, $\delta x_i = (x_{i+1} - x_{i-1})/2$ $x_I = s$ and consider

$$\min_{\delta\mathbf{x_i}\in\mathcal{D}\mathcal{S}_d}\left\{\sum_{i\subset[i_a,i_b]} |\psi_i-\psi_{d_i}|^2\delta x_i : \psi_i + \frac{1}{\delta x_i}\left[\alpha_{i+1/2}\frac{\psi_{i+1}-\psi_i}{\delta x_{i+1/2}} - \alpha_{i-1/2}\frac{\psi_i-\psi_{i-1}}{\delta x_{i-1/2}}\right] = f_i\right\}$$

with $\psi_0 = 0$ $\psi_I = \psi_{d_I}$, i.e., with $\Psi = \psi - \psi_d$,

$$\min_{\vec{\delta}x\in D}\{\Psi^{\mathrm{T}}B(\vec{\delta}x)\Psi : A(\vec{\alpha},\vec{\delta}x)\Psi = F\}.$$

It is seen here that the problem is akin to control in the coefficient of PDEs [8], that the discrete problem has a greater unknown space than the continuous problem because the mesh comes also as a degree of freedom.

Nevertheless the problem is differentiable and so gradient methods should work. For this we will need the derivatives of the cost function with respect to all the unknowns, $s, \alpha, \delta x_i$. This can be a momentous task and so whenever possible automatic differentiation is of great help (see Refs. [6,5], and the bibliography therein).

2. Principle of automatic differentiation

Consider the problem of finding $J'(u)$ when $j(u)$ is given by a computer program. Because the program is made of differentiable lines, J' can be computed by differentiating every line and adding them to the computer program immediately above each line. For instance

Program for J	**lines to add to the program**
$x = 2u(u+1)$	$\mathrm{d}x = 2 * u * \mathrm{d}u + 2 * \mathrm{d}u * (u+1)$
$y = x + \sin(u)$	$\mathrm{d}y = \mathrm{d}x + \cos(u) * \mathrm{d}u$
$J = x * y$	$\mathrm{d}J = \mathrm{d}x * y + x * \mathrm{d}y$

If this new program is run with u=u0, du=1, dx=0, dy=0, dJ=0, then dJ is the derivative of J with respect to u at $u0$.

2.1. Automatic AD

However, differentiating each line can be long and tedious. It can be done by the compiler by overloading the operators of arithmetic and the functions in the standard C-library. Operator overloading is available in Fortran 90 and C++. We present briefly an implementation in C++ because it seems difficult to make an efficient implementation in Fortran 90 [16]. Consider the following procedures:

- Step 1: if the program is in FORTRAN use f2C from http://www.netlib.org/f2c/
- Step 2: change float and double into `ddouble` and add `#include ddouble.h`

2.2. The library ddouble

Each variable has now two fields: its value and the value of its derivative. So we define a class of differentiable variables and stores these values in v. Every time an arithmetic operation is done, the corresponding operation on the derivatives must be done too. For instance below we give the overloading of the multiplication and of the addition:

```
class ddouble{ public:
  float v[2];
  ddouble(double a, double b = 0)
    { v[0] = a; v[1] = b;}
    // intialise la derivee a 0 sauf si b! = 0
  friend ddouble operator *
          (const ddouble& a, const ddouble& b)
      { ddouble c;
        c.v[1] = a.v[1] * b.v[0] + a.v[0] * b.v[1];
    // (fg)' = f'g + fg'
        c.v[0] = a.v [0] * b.v[0];
        return c;
      }
```

```
  friend ddouble operator +
    (const ddouble& a, const ddouble& b)
      { ddouble c;
        c.v[1] = a.v[1] + b.v[1]; // (f + g)' = f' + g'
        c.v[0] = a.v[0] + b.v[0];
        return c;
      }
// ...
} ;
```

3. Well posedness

Consider the academic problem of designing a wind tunnel with required flow properties in a region of space D (see Fig. 3). With a stream function formulation this would be

$$\min_{S\in\mathcal{S}_d}\left\{\int_D |\psi-\psi_d|^2:\ -\Delta\psi=0,\quad \psi|_S=0\ \psi|_{\partial C}=\psi_d\right\}. \tag{1}$$

Before attempting any numerical simulation we may study the existence of solutions. These are by no means nonpractical questions because many of these optimal shape design problems do not have solutions. For example the optimization of a hook, clamped to the wall on the left and pulled so by weight on the right. With respect to weight under a given max constraints so that the structure does not break: optimal structures are of composite materials; there is no simply connected solution to this problem.

Although existence can be studied directly by using continuity results with respect to domain boundaries, one may also map the unknown domain from a fixed domain and consider that the unknown is now $T: C\to\Omega$,

$$\min_{T\in\mathcal{T}_d}\left\{\int_{\hat{D}} |\psi-\psi_d|^2:\ \nabla\cdot[A\nabla\psi]-0 \text{ in } C\quad \psi|_{\partial C}-\psi_d,\ A-T'^{-1^{T}}T'^{-1}\det T'\right\} \tag{2}$$

or extend the operators by zero in S and take the characteristic function for unknown:

$$\min_{\chi\in X_d}\left\{\int_D |\psi-\psi_d|^2:\ -\nabla\cdot[\chi\nabla\psi]=0,\quad \psi(1-\chi)=0\ \psi|_{\partial C}=\psi_d\right\}. \tag{3}$$

This last approach, suggested in Ref. [15] has led to *topological optimization.*

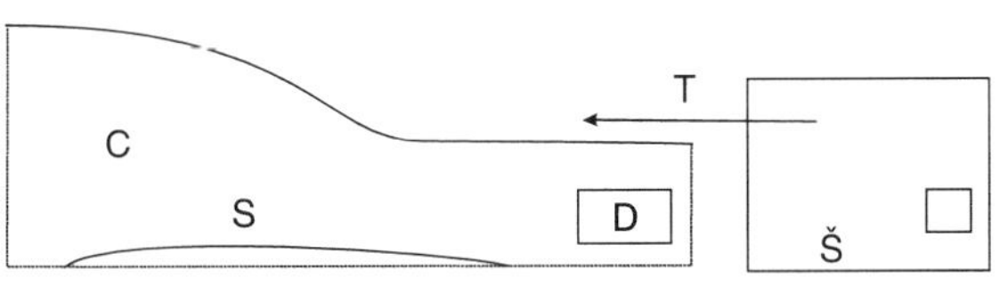

Fig. 3. Inverse design for a wind tunnel with desired properties ψ_d in D.

3.1. Results

Most results are obtained by considering minimizing sequence S^n and (in the case of our academic example) show that $\psi^n \to \psi$ for some ψ, which is the solution of the PDE.

By using directly regularity with respect to the domain, Chenais [3] (see also Ref. [10]) showed that in the class of all S uniformly Lipschitz, problem (1) has a solution.

Similarly Murat-Simon [9] working with (2) showed that in the class of $T \in W^{1,\infty}$ uniformly, the solution exists.

However, working with (3) generally leads to weaker results because if $\chi^n \to \chi$, χ may not be a characteristic function; one is lead to a *relaxed problem*.

In two-dimensional and for the Dirichlet problem there is a very elegant result due to Sverak [14]: *assume all admissible domains Ω are included in a compact set D and the number connected components of $D \setminus \Omega$ is less than a given number, then, in this class of sets, there exists a solution.*

3.2. Well-posedness by regularization

Another way to insure well posedness is to *regularize* the problem by changing the criteria and adding a “cost” to the control. For problem (1)

$$J(\Omega) = \int_D (\psi - \psi_d)^2 + \varepsilon \int_S \mathrm{d}x$$

insures existence. More generally one may consider working with

$$J(\Omega) = \int_D (\psi - \psi_d)^2 + \varepsilon \|S\|_?^2$$

but the choice of norm is a delicate one. In general for second-order problem anything related to the second derivatives would be likely to work, but it is not known if weaker norms would work too.

4. Sensitivity analysis

Even though AD can solve the problem it is wise to check differentiability analytically. This can be done by using normal variation on a reference shape (see Fig. 4)

$$\partial\Omega^\alpha = \{x + \alpha\mathbf{n} : \quad x \in \partial\Omega\}. \tag{4}$$

Following Cea [2], and Delfour-Zolezio [4] one may consider a velocity of deformation $V(x)$ and define a time dependent shape

$$\Omega(t) = \{x + V(x)t : x \in \Omega\}$$

and compute $\mathrm{d}J/\mathrm{d}t$, known as the *material derivative* of J.

Lately, for Neumann problems, the concept of topological derivative was introduced in Ref. [13]. One digs a small circular hole of center $x - 0$ in the domain and study the limit of

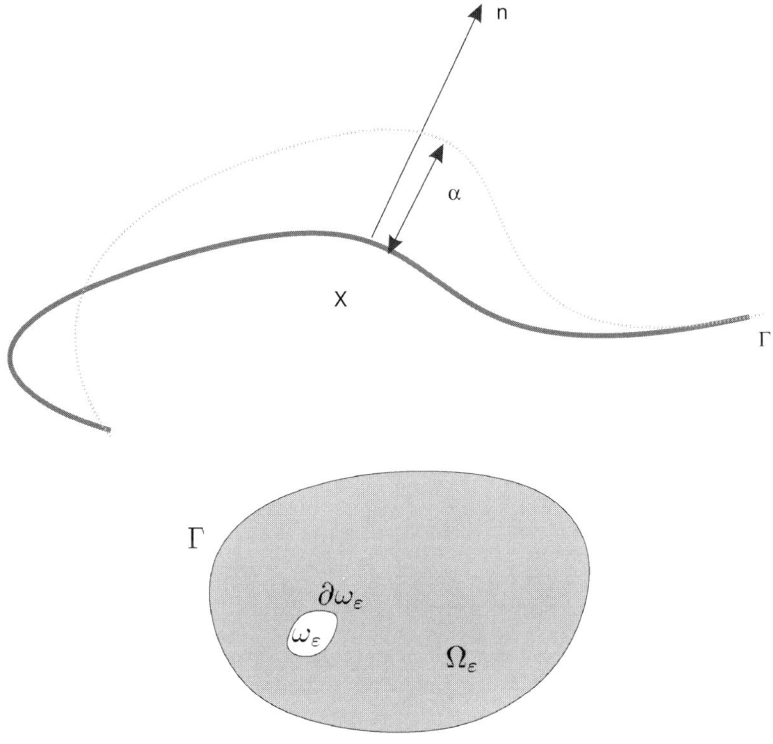

Fig. 4. Normal variations on a reference shape (top). Topological variation on the shape (bottom).

$(1/\varepsilon)(\psi^\varepsilon - \psi)$ where ψ^ε is the solution of the PDE with the hole and ψ the solution without the hole (see Fig. 4).

4.1. Sensitivity: example

For the Laplace equation with Dirichlet conditions

$$-\Delta\psi^{\varepsilon\alpha} = f \text{ in } \Omega^{\varepsilon\alpha} \quad \psi^{\varepsilon\alpha} = 0 \text{ on } \Gamma^{\varepsilon\alpha} = \{x + \varepsilon\alpha\mathbf{n} : x \in \Gamma\},$$

where $\Omega^{\varepsilon\alpha}$ is obtained by (4) the derivative with respect to α is calculated by assuming

$$\psi^{\varepsilon\alpha} = \psi + \varepsilon\psi'_\alpha + \frac{\varepsilon^2}{2}\psi''_\alpha.$$

By linearity ψ' and ψ'' satisfy the PDE with zero rhs. By Taylor expansion:

$$0 = \psi^{\varepsilon\alpha}(x + \varepsilon\alpha n) = \psi^{\varepsilon\alpha}(x) + \varepsilon\alpha\frac{\partial\psi^{\varepsilon\alpha}}{\partial n}(x) + \frac{\varepsilon^2\alpha^2}{2}\frac{\partial^2\psi}{\partial n^2}(x) + \cdots .$$

Therefore

$$-\Delta\psi'_\alpha = -\Delta\psi''_\alpha = 0 \qquad \psi'_\alpha|_\Gamma = -\alpha\frac{\partial\psi}{\partial n} \qquad \psi''_\alpha|_\Gamma = -\alpha\frac{\partial\psi'_\alpha}{\partial n} - \frac{\alpha^2}{2}\frac{\partial^2\psi}{\partial n^2}.$$

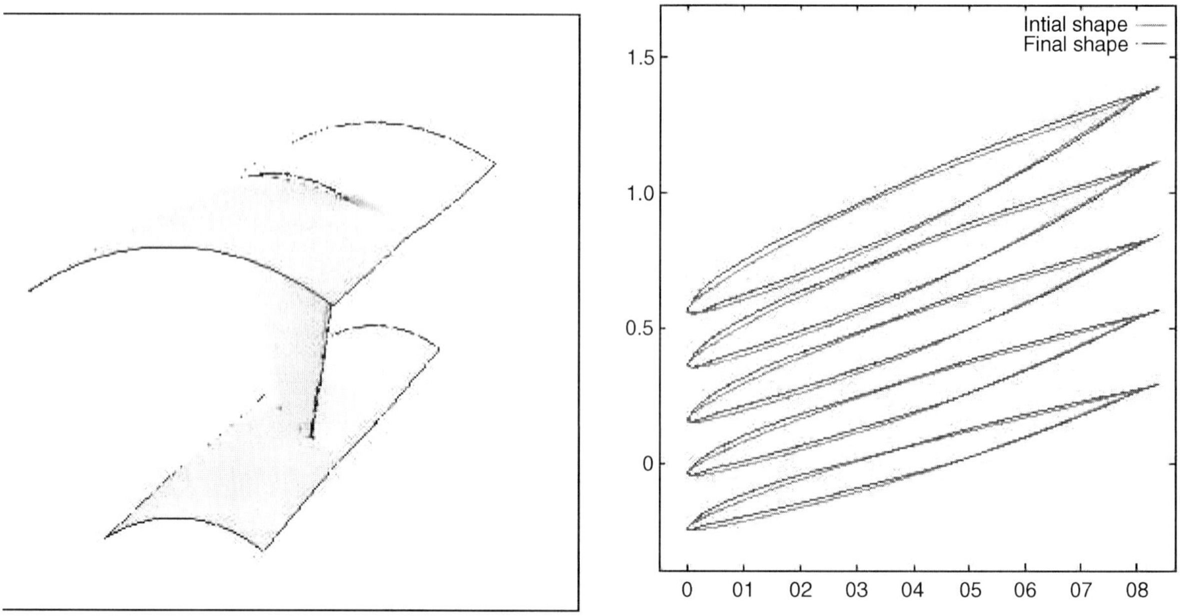

Fig. 5. Optimization of a cooling fan for a car engine. This three-dimensional optimization improved the design by 10%. The picture displays the final shape and the change at some cross section from the original hand optimized original design.

4.2. Navier–Stokes equations

4.2.1. The minimum drag problem

$$E(\Omega) \equiv \min_{\Omega \in C} \int_\Omega \frac{1}{2} \|\nabla u\|^2 \mathrm{d}x: \quad u|_{\partial\Omega} = g$$

$$u \nabla u + \nabla p - v\Delta u = 0, \quad \nabla \cdot u = 0,$$

Sensitivity analysis by local variations gives

$$\partial\Omega' = \{x + \alpha n(x) \in \partial\Omega\} \Rightarrow \delta E = \int_{\partial\Omega} \chi\alpha \, \mathrm{d}s + \mathrm{o}(|\alpha|).$$

$$\delta E = \int_\Omega \nabla u \cdot \nabla \delta u + \frac{1}{2} \int_{\partial\Omega} \alpha |\nabla u|^2 = \frac{1}{2} \int_{\partial\Omega} \alpha \partial_n u \cdot (\partial_n u + 2\underline{\partial_n w}) + \mathrm{o}(|\alpha|)$$

$$-\mathbf{u} \nabla \mathbf{w} + \mathbf{w} \nabla \mathbf{u}^{\mathrm{T}} + \nabla \mathbf{q} - \mathbf{v} \Delta \mathbf{w} = \mathbf{u} \nabla \mathbf{u}, \quad \nabla \cdot \mathbf{w} = \mathbf{0}, \quad \mathbf{w}|_{\partial\boldsymbol{\Omega}} = \mathbf{0}.$$

More can be found in Refs. [11,12]. See also Fig. 5.

4.3. Gradient methods

So one starts with a smooth shape, moves each point in its normal direction by

$$\alpha = \partial_n u \cdot (\partial_n u + 2\partial_n w).$$

But will the new shape have the same regularity? In general the answer is no, and this loss of regularity is numerically dangerous and prone to generation of oscillations. The cure is to use a *smother* which mathematically amounts to change the norm of the gradient method. For instance if the shape is moved by β solution of

$$-\frac{d^2\beta}{ds^2} = -\alpha$$

implies then

$$j(S(\beta)) - j(S) = \frac{1}{2}\int_{\partial\Omega} \beta\partial_u \cdot (\partial_n u + 2\partial_n w)$$

$$= -\frac{1}{2}\int_{\partial\Omega} \beta\alpha = -\int_{\partial\Omega} \left|\frac{d\alpha}{ds}\right|^2.$$

4.4. Discretization

Consider again the academic problem (1). It can be discretized by

$$\min_{q_h} \left\{ j_h(q_h) = \int_D |\psi - \psi_d|^2 : \int_\Omega \nabla\psi_h \nabla w_h + \frac{1}{\varepsilon}\int_\Gamma (\psi_h - \psi_d) w_h = 0 \quad \forall w_h \in V_h \right\},$$

where V_h is the finite element space of piecewise linear continuous functions.

Calculus of variation is possible but the degree of freedoms are now the node motion $q_h \in V_h$, a piecewise linear continuous function built from its values at the vertices, namely the motions of the same (see Fig. 6). Let

$$\int_\Omega \nabla p_h \nabla w_h + \frac{1}{\varepsilon}\int_\Gamma p_h w_h = 2\int_D (\psi_h - \psi_d) w_h \quad \forall w_h \in V_h.$$

Finally

$$\delta J_h = \int_\Omega \nabla\psi_h (\nabla\delta q_h + \nabla\delta q_h - \nabla\cdot\delta q_h)^{\mathrm{T}} \nabla p_h \approx \int_\Gamma \nabla\psi_h \nabla p_h \delta q_h \cdot n.$$

Recognize here the linearization of $\nabla Q_h \nabla Q_h^{\mathrm{T}} \det \nabla Q_h$ at $x + q_h(x)$, therefore

Corollary 1. *The change of cost function due to inner nodes is* $O(h^2)$

4.5. Guidelines

To our experience success in solving an OSD problem depends on the following.

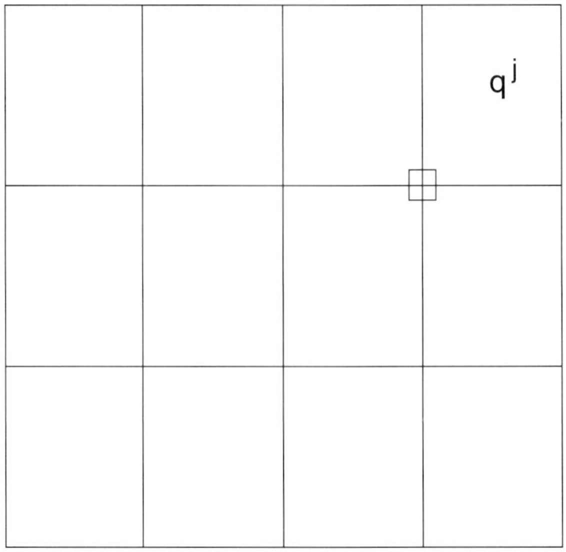

Fig. 6. After discretization, not only the boundary vertices but also the inner vertices are the degrees of freedom of the optimization problem.

- Whenever possible second-order optimization methods (BFGS for instance) should be used because the problems are stiff.
- Compute derivatives with respect to boundary nodes only and apply the theory of approximate gradients to combine mesh refinement with optimization (see Ref. [7] for example).
- Use a smoother, i.e. do not work with the $L^2(\Gamma)$ norm of the node displacements. Use it also to move the inner vertices.
- Experience also shows that

$$\delta_S j(\Omega) = \delta_S \int_S F(\psi) \cdot n \quad \approx \int_S F(\psi) \cdot \delta n + \cdots .$$

  Namely the most important variation is due to the change of the normal in the case of surface integrals.

5. Implementation issues and results

5.1. Link with CAD

In industries, shapes are stored in Computed Aided Design data bases as a set of Bezier patches or others with infinite details such as screws and bolts which are not relevant to a finite element calculation in aerodynamics for instance. Furthermore the CAD system is proprietary.

Therefore it is convenient to abstract the optimization from the CAD system and ask the engineer for any triangulation of the surface and use it as our initial design. The strategy is then what we

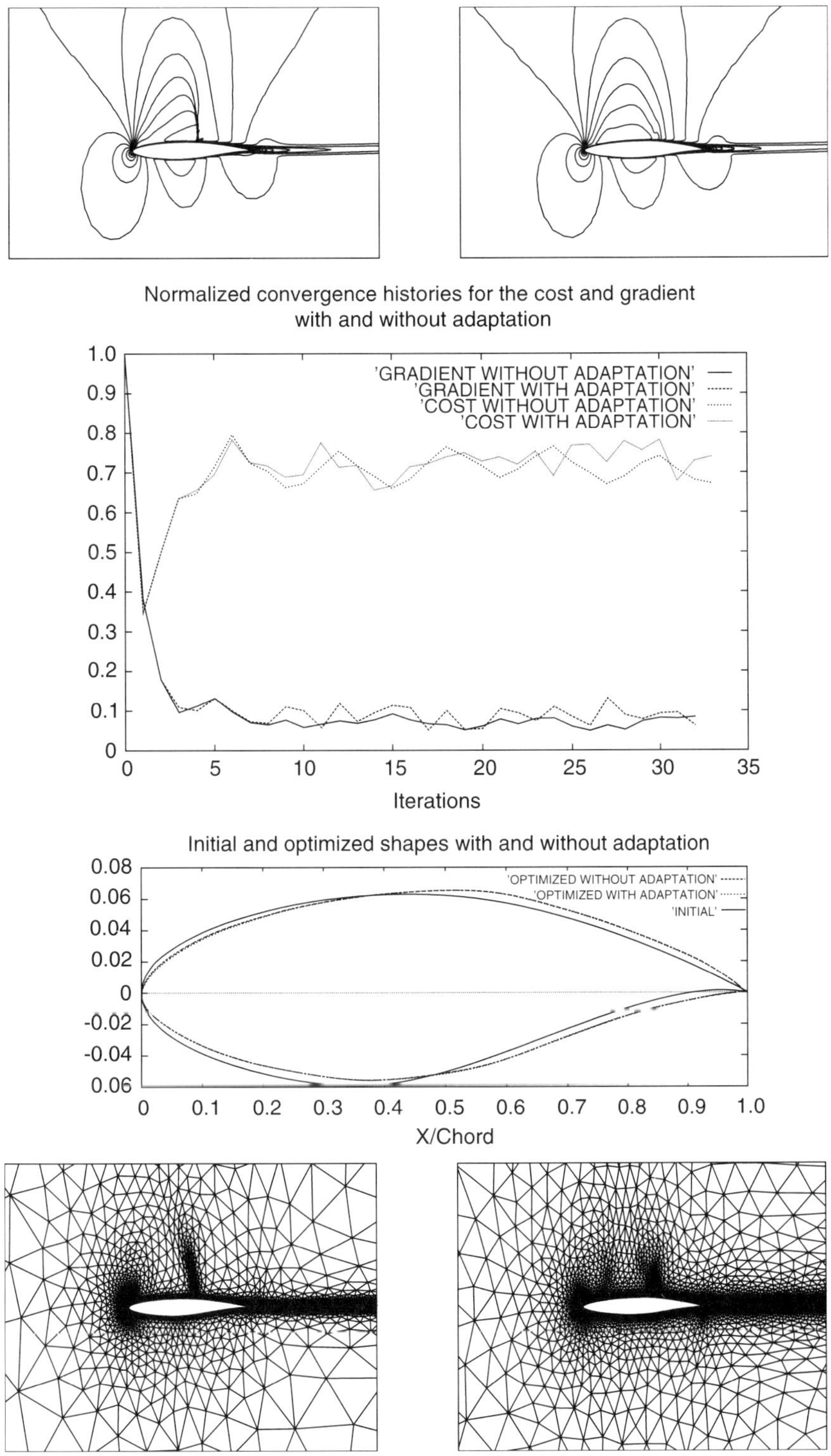

Fig. 7. Optimization of a wing profile.

call a *CAD-free optimization platform*:

(1) Generate any surface mesh from the CAD data.
(2) Apply a C^1+ edges recognition software for surface mesh refinement.
(3) Apply a three-dimensional volumic automatic mesh generator from the surface mesh.
(4) Do the optimization with mesh refinement using the same module as in step 2 but coupled with the PDE solver.
(5) Feed back the result into the CAD system.

5.2. Optimization of a wing profile

See Fig. 7.

Drag is mostly pressure drag due to the shock (pressure drag). The lift and area are imposed by a penalty method with parameters ε, β.

$$J(u, p, \theta) = F \cdot u_\infty + \frac{1}{\varepsilon}|F \times u_\infty - C_l|^2 + \frac{1}{\beta}\left(\int_S \mathrm{d}x - a\right)^2$$

with $F = \int_S (p\mathbf{n} + (\mu \nabla u + \nabla u^{\mathrm{T}}))$ and a Navier–Stokes $+ k - \varepsilon +$ wall laws flow solver.

6. Prospective

Recently the optimization of a complete aircraft (a business jet) was done with this method (even on a workstation when incomplete gradients are used) and a 10% improvement obtained after a few iterations.

However OSD is still a difficult and computer intensive task. There is a good prospect for global nondifferentiable optimization because there are often many local minima and because the flow solver is often available in binary format only (such would be the case if a commercial software was used). However, genetic algorithms are still slow and difficult to couple with gradient methods.

Incomplete gradients is also a good field of research and awaits mathematical proofs.

References

[1] A. Baron, Optimisation multi-physique, Thesis, University of Paris 6, 1998.
[2] J. Cea, Conception optimale ou identification de forme: calcul rapide de la dérivée directionelle de la fonction cout, Modélisation Math Anal, AFCET, Dunod, 1986.
[3] D. Chenais, Shape optimization in shell theory, Eng. Opt. 11 (1987) 289–303.
[4] M. Delfour, J.P. Zolezio, Surface Derivatives SIAM Series, SIAM, Philadelphia, PA, 2001.
[5] N. Di Cesare, O. Pironneau, Shock sensitivity analysis, Comput. Fluid Dyn. J. 9 (2) (2000).
[6] A. Griewank, Principles and Algorithms of Algorithmic Differentiation, SIAM, 2000.
[7] G. Lemarchand, O. Pironneau, E. Polak, Incomplete gradients, in: N. Debit, M. Garbey, J. Periauk, D. Case, Y. Kuznetsov (Eds.), Domain Decomposition Method in Science and Engineering, Proceedings of 13th Domain Decomposition Method Conference, Lyon, 2000, DDM, Bergen, 2001.
[8] J.L. Lions, Contrôle Optimal de systèmes gouvernés par des équations aux dérivées partielles, Dunod-Gauthier Villars, 1968.

[9] F. Murat, J. Simon, Etude de problèmes d'optimum design, Proceedings of the Seventh IFIP Conference, Lecture Notes in Computer Sciences, Vol. 41, 1976, pp. 54–62.
[10] P. Neittaanmäki, Computer aided optimal structural design, Surv. Math. Ind. 1 (1991) 173–215.
[11] O. Pironneau, On optimal shapes for Stokes flow, J. Fluid Mech. 70 (2) (1973) 331–340.
[12] O. Pironneau, Optimal Shape Design for Elliptic Systems, Springer, Berlin, 1984.
[13] J. Sokolowski, J.P. Zolezio, Introduction to Shape Optimization, Springer Series in Computational Mathematics, Springer, New York, 1991.
[14] A. Sverak, On existence of solution for a class of optimal shape design problems, Note CRAS, 1992.
[15] L. Tartar, in: A. Bensoussan (Ed.), Control Problems in the Coefficients of PDE, Lecture Notes in Economics and Math Systems, Springer, Berlin, 1974.
[16] J. Tiihonen, Private communication, 2001. tiihonen@mit.jyu.fi.

Further reading

B. Mohammadi, O. Pironneau, Applied Optimal Shape Design, Oxford University Press, Oxford, 2001.

ELSEVIER

Journal of Computational and Applied Mathematics 149 (2002) 207–219

JOURNAL OF COMPUTATIONAL AND APPLIED MATHEMATICS

www.elsevier.com/locate/cam

Subdivision as a fundamental building block of digital geometry processing algorithms

Peter Schröder

Department of Computer Science, California Institute of Technology, 1200 E. California Blvd., Pasadena, CA 91125, USA

Received 6 June 2001; received in revised form 19 January 2002

Abstract

Digital Geometry Processing (DGP) is a newly emerging discipline which aims to build the mathematical and algorithmic foundations for the next generation of digital multimedia. Digital *signal* processing was highly successful for earlier generations of digital multimedia such as sound, image, and video. Now that 3D scanning technologies make sampled surfaces broadly available a similar apparatus is needed to deal with this new multimedia datatype. Due to the inherent curvature of such surfaces though traditional tools are not immediately applicable. In this paper we give a brief overview of some recent developments in subdivision surfaces and how these help to build a foundation for DGP algorithms.

Keywords: Subdivision; Digital signal processing; Approximation theory; Geometric modeling; Multiscale algorithms

1. Introduction

Multimedia data types such as digital sound, images, and video are now ubiquitous in all areas of computing and daily life. This wide impact was made possible by a number of factors. A key factor in the wide use of a given data type is the ease and economy of acquiring it. Using a rough time line one can observe that this was true for sound in the 1970s, images in the 1980s, and finally video in the 1990s, roughly following the development of computing hardware with its ever increasing cpu and memory resources (Fig. 1). Another key factor in the wide use of a given data type is the existence of efficient algorithms for creation, storage, transmission, editing and other manipulations of the data. The mathematical foundation for these algorithms has for a very long time rested on sampling and associated Fourier techniques. Even more recent developments, such as the use of wavelets for image and video compression still rest upon the foundation laid by

E-mail address: ps@cs.caltech.edu (P. Schröder).

PII: S0377-0427(02)00530-7

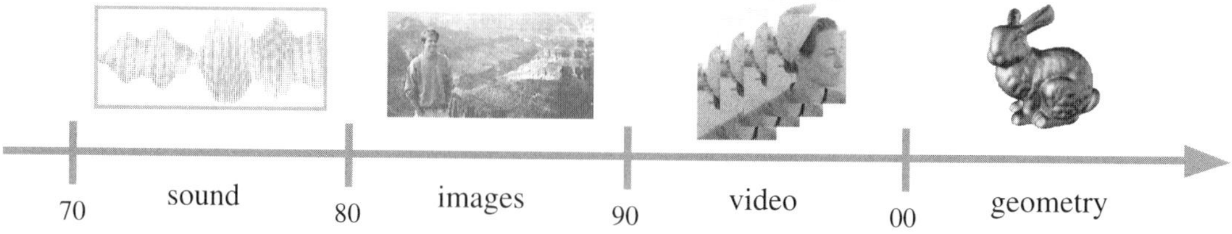

Fig. 1. The development of earlier generations of digital multimedia can be seen as an increase in the dimensionality of the data, enabled by simultaneous advances in acquisition, computation and underlying theory. Digital geometry is now entering a similar phase. © Peter Schröder, 2001.

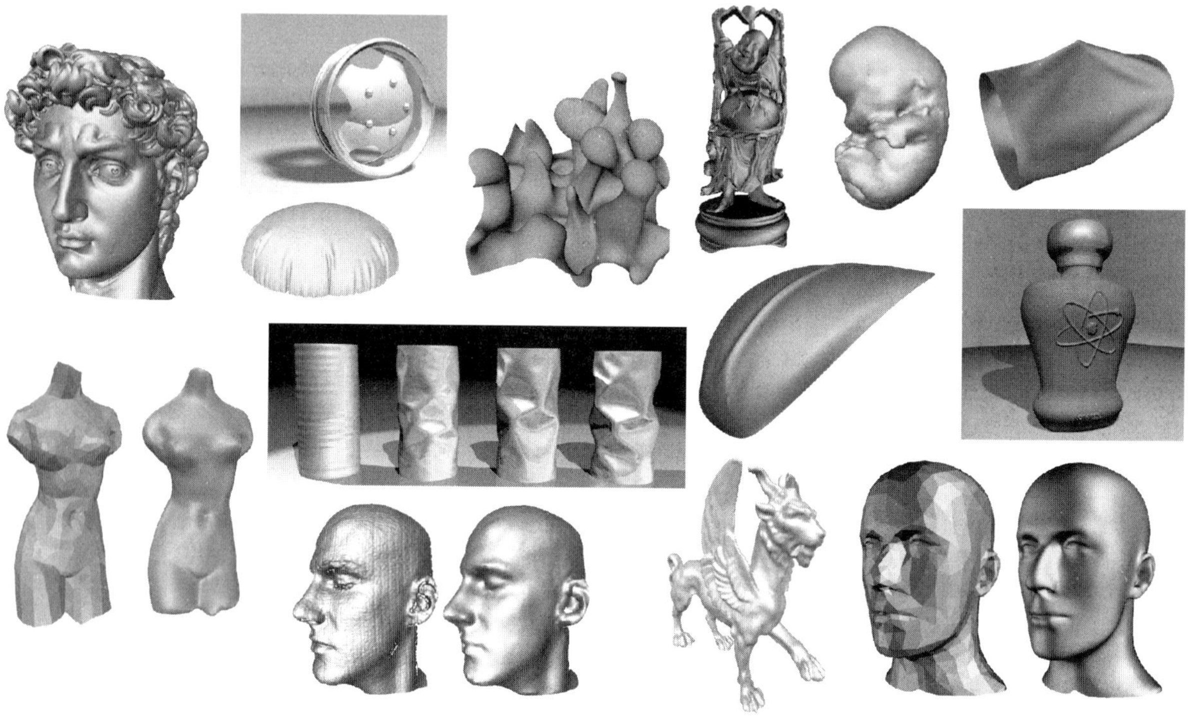

Fig. 2. Examples of digital geometry from many different application areas. © Peter Schröder, 2001.

Fourier analysis. As such, these methods now codified as "digital signal processing" (DSP) have been extraordinarily successful impacting areas ranging from cheap consumer devices such as cell phones and MP3 players to high end scientific computing applications solving some of today's most demanding PDEs, for example.

We are now witnessing the arrival of a new multimedia data type: *digital geometry*. As with the earlier waves of multimedia this development parallels increasing cpu and memory resources on modern PCs as well as the availability of increasingly cheap and reliable acquisition devices. The latter includes laser range scanners, 3D photography systems based on stereo matching, contact digitizers, as well as volumetric imaging techniques such as industrial and medical MRI and CT scanners (Fig. 2). These systems range from low-cost consumer level devices all the way to extremely high-resolution military and scientific systems. Once again, we need corresponding developments in algorithms and computational representations to help us realize the potential this data presents. Unfortunately, earlier DSP techniques cannot be immediately brought to bear on this new data type. Instead, we need to develop a new toolbox of mathematics, computational representations and algorithms for *digital geometry processing* (DGP) [37].

2. The need for new tools

Earlier generations of digital multimedia can all be treated as functions of a Euclidian domain. In the case of sound we have a univariate function of time, for images a bivariate function of a section of the image plane, and video may be thought of as a trivariate function of the image plane and time. This is not true anymore for surfaces: They carry intrinsic curvature and their essential 2-manifold (with boundary) structure cannot be ignored. For example, in the Euclidian setting it is possible to use regular sampling to uniquely represent bandlimited functions. There are no comparable uniform, i.e., equispaced, sampling patterns on general 2-manifolds. Since the entire mathematical machinery for digital signal processing is based on translation invariance, none of the DSP tools developed so far are immediately applicable. A possible solution to this problem is the use of local parameterizations on a given 2-manifold for which one could then use regular sampling and apply (windowed) Fourier techniques. However, since there are many possible parameterizations for a given smooth manifold it is not clear which one to use and, in general, the result of the computation will depend on the particular parameterization chosen. Instead, we seek to find sampling patterns for geometry which are as regular as possible (Fig. 3).

As an additional ingredient we are also looking for representations which are *hierarchical.* Hierarchical methods in the classical setting add the idea of scale invariance to the already present

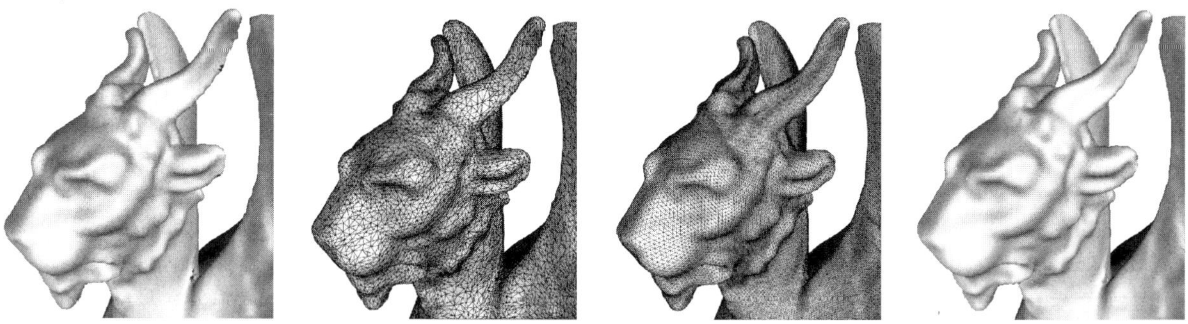

Fig. 3. A surface is typically represented by samples (points) and their connectivity (often triangles). Many sampling technologies result in irregular meshes, i.e., those in which each vertex may have any number of neighbors and triangle sizes do not vary smoothly. Much more attractive are smooth, semi-regular samplings such as the one on the right. There each point has six neighbors almost everywhere and sample spacing varies smoothly. In both cases the geometry itself is the same (leftmost and rightmost). © Peter Schröder, 2001.

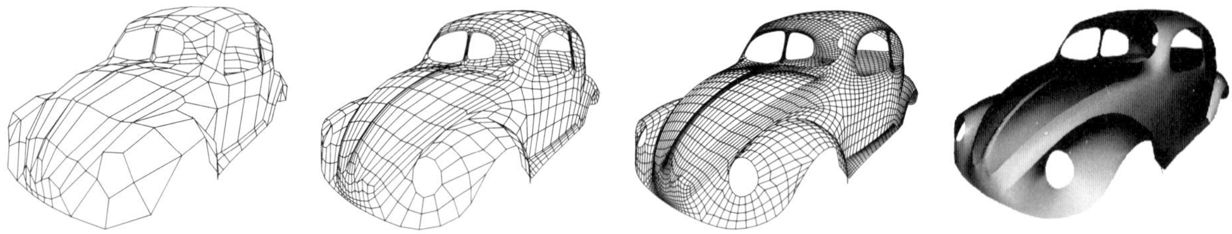

Fig. 4. Subdivision surfaces are constructed through a process of repeated refinement applied to a coarse control mesh. Topologic refinement splits each face in four while the geometric part of refinement assigns spatial positions to all new vertices as well as moves old vertices to new positions. A (piecewise) smooth surface results in the limit. While quadrilaterals are used in this case other schemes based on triangles or even hexagons are possible. © Peter Schröder, 2001.

translation invariance of the Fourier setting and lead to wavelet approaches. Multiresolution of this type, be it called wavelets, or multigrid, or filterbanks, has become a critical ingredient in building highly efficient, scalable algorithms for many processing and numerical computing tasks.

The task then is to rebuild a signal processing like toolbox of mathematical theory and numerical algorithms which replaces regular sampling with *semi-regular* sampling and enables multiresolution algorithms. The latter requires three ingredients: up/down sampling, smoothing, and detail computations. The surface representation, commonly called *Subdivision*, provides these elements and can be used to build the foundations of Digital Geometry Processing algorithms and theory (Fig. 4).

3. Subdivision surfaces

Subdivision surfaces were first introduced by Catmull and Clark [4] and Doo and Sabin [12] to address the problem of constructing free-form, smooth surfaces of arbitrary topology. Instead of "glueing" together individual tensor product spline patches, they used the fact that a spline patch can be produced in the limit of repeated uniform knot insertion ("control mesh refinement") and generalized these refinement rules from the regular setting to the arbitrary topology setting (Fig. 5).

For quadrilateral schemes vertices with valence $k = 4$ are *regular*—corresponding to the regular quadrangulation of the plane—while those with valence $k \neq 4$ are called *irregular*. In the case of triangle schemes a regular vertex has $k = 6$. The Euler–Poincaré formula for 2-manifolds (with boundary) implies that only a topologic torus, infinite plane, or infinite cylinder admit entirely regular control meshes, indicating that the problem of irregular control points is fundamental and cannot be circumvented.

Early subdivision methods were based on generalization of spline knot insertion [23] and the so-called "Oslo algorithm" [9]. Later on, subdivision was studied in its own right [5]. However, only recently have subdivision surfaces received broad attention and development in the computer graphics community (see [46] and the references therein). In particular, a broad theoretical understanding of the analytic properties of subdivision surfaces was missing for a long time. Early attempts at analysis of the convergence of surface subdivision schemes and the smoothness of the resulting

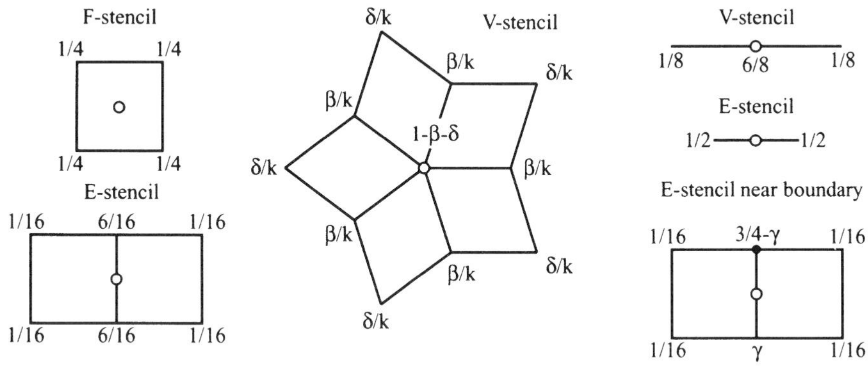

Fig. 5. Refinement rules of Catmull and Clark showing the weights ("stencils") used to compute new control points as averages of control points in the coarser mesh. These rules—face, edge, vertex—are indexed based on the topologic association of the new vertex with the coarser mesh. On the right the boundary rules including one special rule for edges adjacent to a boundary vertex. Here, k denotes the number of faces incident on a vertex, $\beta = 3/(2k)$, $\delta = 1/(4k)$, and $\gamma = \frac{3}{8} - \frac{1}{4}\cos(\pi/k)$ where the latter k is taken to be the valence of the boundary vertex. For more details see [3]. © Peter Schröder, 2001.

limit functions was based on spectral analysis around irregular vertices [2], but was incomplete. Only very recently did this situation change when an essentially complete theoretical understanding of subdivision surfaces was achieved in the works of Reif [34,36] and Zorin [50,44,43].

On the practical side, many recent algorithmic developments have moved applications of subdivision surfaces rapidly forward. Examples include, interactive multiresolution editing [49], reconstruction of sampled data with subdivision surfaces [18], direct evaluation at arbitrary parameter values [38,45], inclusion of boundary conditions and smoothness constraints [3,24], trimming [26], approximation with subdivision surfaces [25], simulation of the mechanics of surfaces [7] for engineering design [8], nonmanifold subdivision [42], and many more. In fact, techniques are mature enough for deployment in movie production [10] and many geometric modeling packages (e.g., Mirai, Maya, 3DMax, etc.).

A prototypical example of a DGP application enabled by subdivision is the compression and progressive transmission of geometry. Fig. 6 illustrates this idea. A surface is encoded into a bitstream which allows partial reconstructions of the original geometry as soon as bits arrive at the receiver. Initially, the geometry is "blurry" but improves rapidly. Such constructions are well known from the image setting where wavelets together with zero-tree coders are used for progressive encodings and transmission. Using subdivision basis functions as scaling functions and building associated wavelets [28] one can build compression algorithms for surfaces [19,6], which are very similar in construction to the image case.

4. The flavors of subdivision

As mentioned above, subdivision schemes were originally derived from B-spline knot refinement rules [9] to address the challenge of building smooth surfaces of arbitrary topology [4,12].

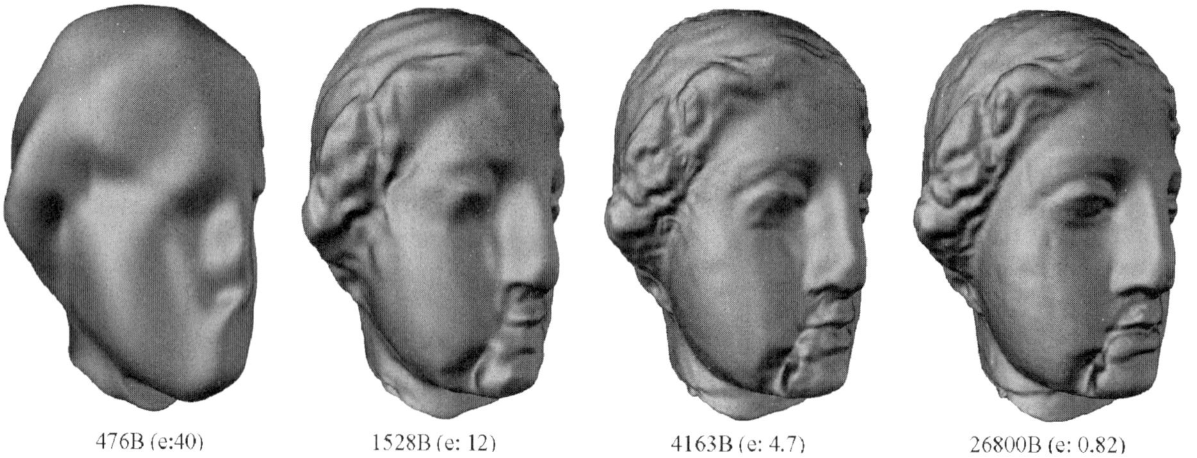

Fig. 6. Partial bit-stream reconstructions from a progressive encoding of the Venus head model. File sizes are given in bytes and relative L^2 reconstruction error in multiples of 10^{-4}. The rightmost reconstruction is indistinguishable from the original. © Peter Schröder, 2001.

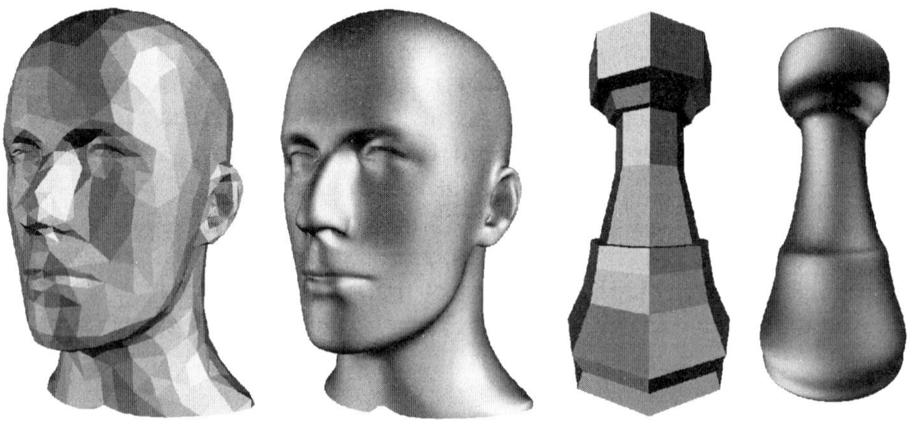

Fig. 7. Examples of subdivision based on triangles and quadrilaterals respectively. On the left, application of Loop's scheme and on the right, application of the scheme of Catmull and Clark. © Peter Schröder, 2001.

Beginning with an arbitrary connectivity control mesh, which forms a topologic 2-manifold possibly with boundary, the surface is constructed through a limiting process of repeated refinement. Fig. 7 shows examples for both quadrilateral and triangle based schemes, giving the control mesh and the resulting limit surface.

Subdivision consists of two components, a *topologic split* rule describing how the connectivity of the control mesh is refined, and a *geometric rule* which determines the new control point positions from the old positions. For the latter, most constructions use smoothing filters of (small) finite support and local definition with constant coefficients depending only on the valence of vertices or faces in the support of the filter. These schemes can be grouped according to a number of basic

Table 1
Classification of major subdivision schemes

Scheme	Approx.	Interpol.	Quad	Triangle	Primal	Dual
Midedge [17,32]	*		*			*
Doo–Sabin [12]	*		*			*
Catmull–Clark [4]	*		*		*	
Loop [27]	*			*	*	
Butterfly [14,48]		*		*	*	
Kobbelt quad.[20]		*	*		*	
Kobbelt $\sqrt{3}$ [21]	*			*	*	
Oswald $\sqrt{3}$ [30]	*			*		*

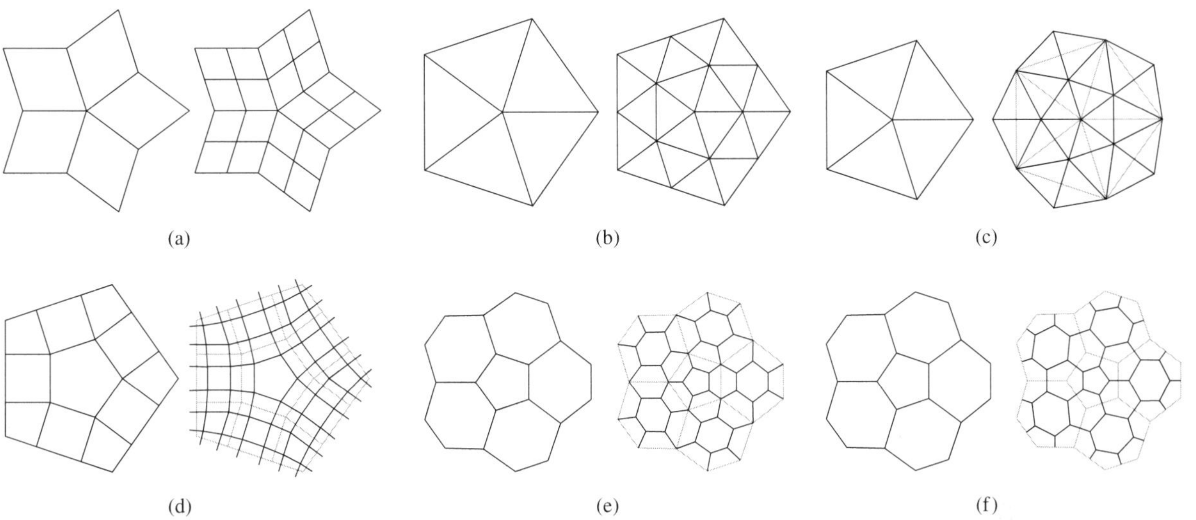

Fig. 8. Topologic styles of subdivision. For each pair the coarser level is shown on the left with the finer level on the right. (a) Primal quadrilateral quadrisecion. (b) Primal triangle quadrisection. (c) Primal $\sqrt{3}$ trisection subdivision. (d) Dual quadrilateral quadrisection. (e) Dual triangle quadrisection. (f) Dual $\sqrt{3}$ subdivision. © Peter Schröder, 2001.

criteria (Table 1) indicating whether they are

- approximating or interpolating with respect to the original control mesh point positions;
- based on quadrilateral, triangle, or hexagon faces as basic primitive;
- of primal or dual type depending on the split rule (faces or vertices, respectively).

Among approximating subdivision schemes, those based on quadrilaterals assume a special role since there exist both primal and dual schemes based on quadrilaterals [12,4,17,32,40,41] (in the case of triangles, dual schemes are based on hexagons). Typically, the topologic step of a primal scheme is described as a *face split* while dual schemes employ *vertex splits*. Primal schemes which quadrisect are shown in Fig. 8(a) for quadrilaterals and Fig. 8(b) for triangles. The topologic split

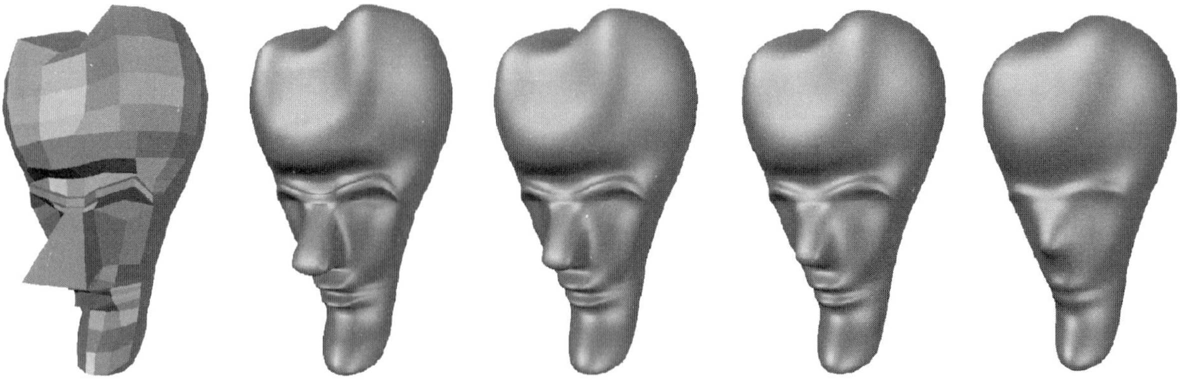

Fig. 9. A more complex example showing the initial control mesh on the left followed by the results of applying subdivision of degree two (Doo–Sabin), three (Catmull–Clark), four, and nine. © Peter Schröder, 2001.

step for dual schemes which quadrisect, is illustrated in Fig. 8(d) for quadrilaterals and Fig. 8(e) for dual triangles, i.e., hexagons. The Doo–Sabin scheme [12], for example, is a dual quadrilateral scheme, while half box splines [33]—being dual triangle schemes based on vertex quadrisection—lead to hexagonal tilings.

Given these basic schemes one may ask for interrelationships between them both in terms of geometric and topologic rules. In the case of univariate splines, for example, the Lane–Riesenfeld algorithm [23] establishes an interlaced ladder of increasing order B-splines through repeated averaging. Here a single geometric rule, which averages back and forth between primal and dual meshes after upsampling, generates B-splines of arbitrary order. For quadrilateral schemes this relationship continues to hold in the arbitrary topology setting [47,39] (see Fig. 9).

Even more interesting are schemes in which the topologic subdivision rule calls for *joining* primal and dual meshes in some fashion. An example are the primal $\sqrt{3}$-schemes [15,21,22] in which the refined mesh contains all vertices of the primal and dual mesh (Fig. 8(c)). Dual schemes of this type are possible as well (Fig. 8(f)). Just as in the case of quadrilateral subdivision, primal and dual $\sqrt{3}$-schemes can also be built based on repeated averaging [30]. Examples of primal $\sqrt{3}$-schemes are shown in Fig. 10. Dual schemes, i.e., those based on hexagons and trisection are illustrated in Fig. 11.

Many more constructions are possible and the area is still quite active. However, in practice, Catmull–Clark and Loop subdivision are the most frequently deployed methods.

5. Open research questions

The last few years have seen rapid development of fairly comprehensive theory and algorithms for basic surface subdivision, reaching a certain level of maturity. However, the area remains active and fertile for ongoing research. In the following paragraphs we lay out a few of the interesting open questions in the area of subdivision and more generally in DGP.

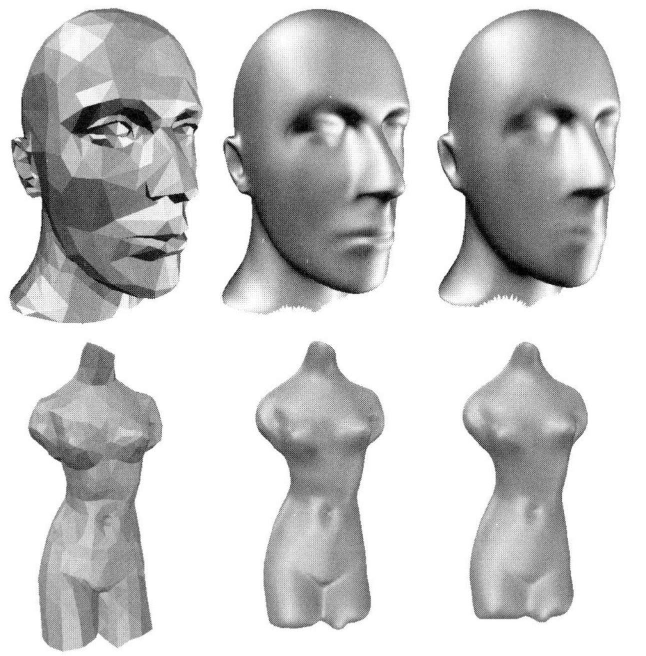

Fig. 10. Examples of repeated averaging based $\sqrt{3}$ subdivision with two respectively four averaging cycles for the mannequin and venus models with the respective control meshes on the left.

Higher order smoothness at irregular points: While C^1 smoothness at irregular points is fairly easy to achieve, no *practical* subdivision schemes which achieve (flexible) C^2 at irregular points are currently known. Lower bound estimates [35] suggest that such schemes with small stencils are unlikely. However, it may be possible to achieve higher smoothness through nonstationary schemes, i.e., schemes which change their stencil (number of coefficients, or coefficient values with fixed stencil) at each level of subdivision.

Unified treatment of subdivision schemes: Repeated averaging which alternates between primal and dual meshes has recently emerged as a general primitive for the construction of broad classes of subdivision. This is true for quadrilateral schemes [39,47] as well as certain triangle based schemes [39,30], although the pattern is not as elegant in the latter case. An open question is whether in the case of repeated averaging there exist simple sufficient conditions which ensure the smoothness of the resulting method at irregular points.

Subdivision in higher dimensions: So far subdivision has been mostly employed for curves and surfaces. Many application domains, especially in physical modeling, require treatment of higher dimensional domains, in particular 3D. Some generalizations of subdivision to 3D [29,1,13] and higher dimensions [31] have been pursued, but little is as yet known about them. It appears that the number of cases to distinguish when dealing with the smoothness analysis of such schemes is much larger, considerably complicating the picture in this setting.

Shape insensitive and irregular subdivision: An annoying problem in practical applications of subdivision schemes is the fact that all established schemes are specific to a particular element type.

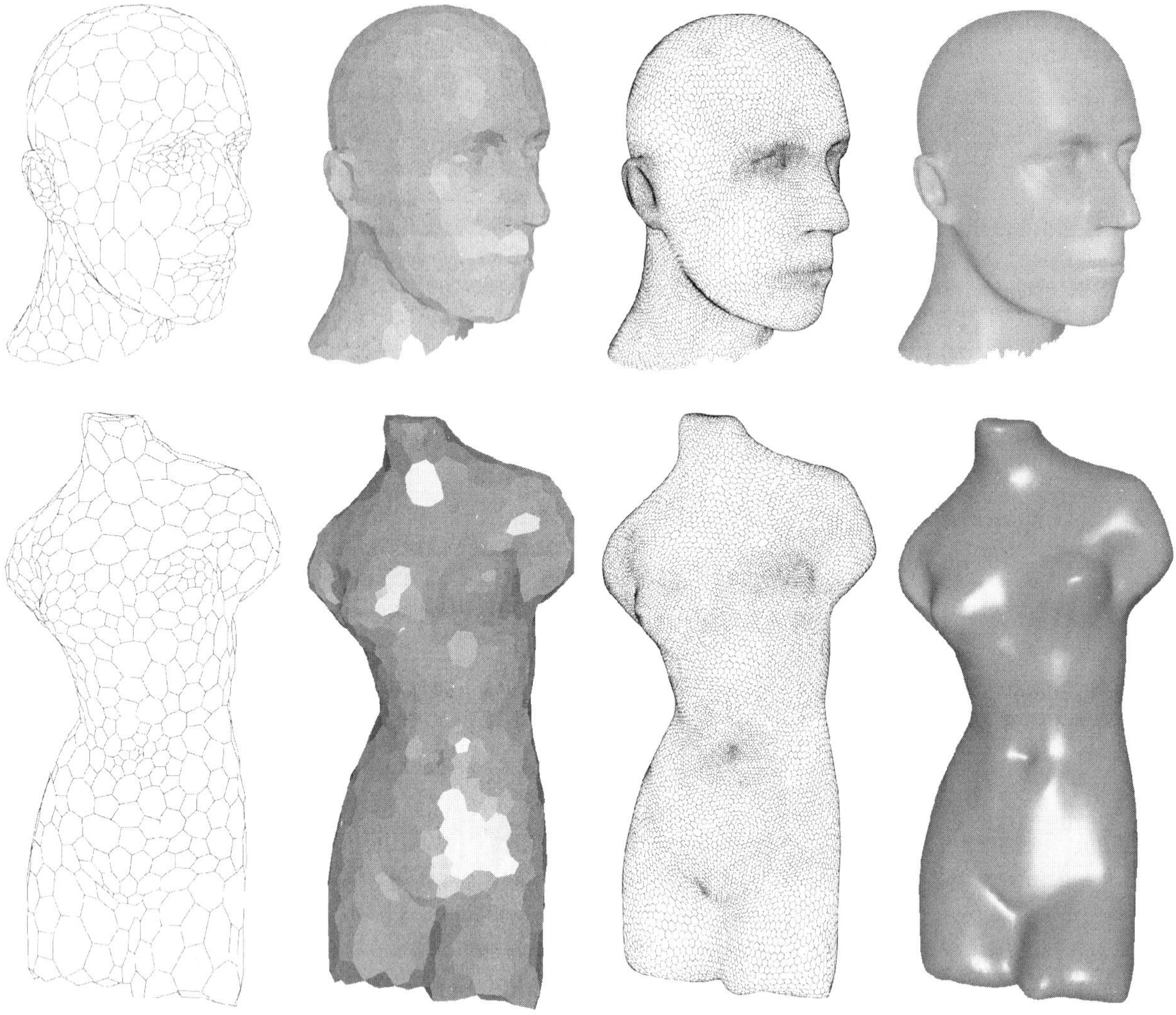

Fig. 11. Dual control mesh and surface after four levels of dual $\sqrt{3}$ subdivision. © Peter Schröder, 2001.

For example, the Loop scheme for triangle meshes and the Catmull–Clark scheme for quadrilaterals. While the latter can be applied to triangle meshes, the resulting surfaces exhibit subtle artifacts such as undulations. In practice, it is desirable to work with surfaces which locally have two preferred directions, favoring schemes such as Catmull–Clark. However, when building control meshes for such surfaces it is generally difficult to ensure that they contain quadrilaterals only. Finding a subdivision scheme which performs well, i.e., produces "fair" surfaces independent of the control mesh face types would be highly desirable. A much more general instance of this problem is the design of fully irregular subdivision schemes [16,11], i.e., those which admit arbitrary control point insertion locations in meshes with irregular connectivity everywhere. Very little is known about such schemes.

6. Summary

Digital geometry processing is a newly emerging research area which aims to make manipulation of geometry as simple and efficient as methods currently in place for image processing, for example. Subdivision surfaces and their underlying machinery and algorithms are a key component to making such algorithms a reality.

Here, we aimed to give a short overview of the variety of methods available and provide pointers to the literature for readers interested in more details.

Acknowledgements

This work was supported in part by NSF (DMS-9874082, ACI-9721349, DMS-9872890, ACI-9982273), Toyota, Intel, Alias|Wavefront, Pixar, Microsoft, and the Packard Foundation.

References

[1] C. Bajaj, J. Warren, G. Xu, A smooth subdivision scheme for hexahedral meshes, The Visual Computer (2002), to appear.

[2] A.A. Ball, D.J.T. Storry, Conditions for tangent plane continuity over recursively generated B-spline surfaces, ACM Trans. Graphics 7 (2) (1988) 83–102.

[3] H. Biermann, A. Levin, D. Zorin, Piecewise smooth subdivision surfaces with normal control, Proceedings of SIGGRAPH 2000, 2000, pp. 113–120.

[4] E. Catmull, J. Clark, Recursively generated B-spline surfaces on arbitrary topological meshes, Comput. Aided Des. 10 (6) (1978) 350–355.

[5] A.S. Cavaretta, W. Dahmen, C.A. Micchelli, Stationary subdivision, Mem. Amer. Math. Soc. 93 (1991) 453.

[6] A. Certain, J. Popovic, T. DeRose, T. Duchamp, D. Salesin, W. Stuetzle, Interactive multiresolution surface viewing, Proceedings of SIGGRAPH 96, 1996, pp. 91–98.

[7] F. Cirak, M. Ortiz, P. Schröder, Subdivision surfaces: a new paradigm for thin-shell finite-element analysis, Internat. J. Numer. Methods Eng. 47 (2000) 2039–2072.

[8] F. Cirak, M. Scott, E. Antonsson, M. Ortiz, P. Schröder, Integrated modeling, simulation, and design for thin-shell structures using subdivision, Comput. Aided Des. 34 (2002) 137–148.

[9] E. Cohen, T. Lyche, R. Riesenfeld, Discrete B-splines and subdivision techniques in computer aided geometric design and computer graphics, Comput. Graphics Image Process. 14 (2) (1980) 87–111.

[10] T. DeRose, M. Kass, T. Truong, Subdivision surfaces in character animation, Proceedings of SIGGRAPH 98, 1998, pp. 85–94.

[11] M. Desbrun, M. Meyer, P. Schröder, A.H. Barr, Implicit fairing of irregular meshes using diffusion and curvature flow, Proceedings of SIGGRAPH 99, 1999, pp. 317–324.

[12] D. Doo, M. Sabin, Analysis of the behaviour of recursive division surfaces near extraordinary points, Comput. Aided Des. 10 (6) (1978) 356–360.

[13] T. Duchamp, Construction of 3D subdivision schemes, personal communication, June 2001.

[14] N. Dyn, D. Levin, J.A. Gregory, A butterfly subdivision scheme for surface interpolation with tension control, ACM Trans. Graphics 9 (2) (1990) 160–169.

[15] I. Guskov, Irregular subdivision and its applications, Ph.D. Thesis, Princeton University, 1999.

[16] I. Guskov, W. Sweldens, P. Schröder, Multiresolution signal processing for meshes, Proceedings of SIGGRAPH 99, 1999, pp. 325–334.

[17] A. Habib, J. Warren, Edge and vertex insertion for a class of C^1 subdivision surfaces, Comput. Aided Geom. Des. 16 (4) (1999) 223–247 (previously available as a TR, Rice University, August 1997).
[18] H. Hoppe, T. DeRose, T. Duchamp, M. Halstead, H. Jin, J. McDonald, J. Schweitzer, W. Stuetzle, Piecewise smooth surface reconstruction, Computer Graphics (SIGGRAPH '94 Proceedings), 1994, pp. 295–302.
[19] A. Khodakovsky, P. Schröder, W. Sweldens, Progressive geometry compression, Proceedings of SIGGRAPH 00, 2000, pp. 271–278.
[20] L. Kobbelt, Interpolatory subdivision on open quadrilateral nets with arbitrary topology, in: Proceedings of Eurographics 96, Computer Graphic Forum, 1996, pp. 409–420.
[21] L. Kobbelt, $\sqrt{3}$ subdivision, Proceedings of SIGGRAPH 2000, 2000, pp. 103–112.
[22] U. Labsik, G. Greiner, Interpolatory $\sqrt{3}$-subdivision, Comput. Graphics Forum 19 (3) (2000) 131–138.
[23] J. Lane, R. Riesenfeld, A theoretical development for the computer generation and display of piecewise polynomial surfaces, IEEE Trans. Pattern Anal. Mach. Intell. 2 (1) (1980) 35–46.
[24] A. Levin, Interpolating nets of curves by smooth subdivision surfaces, Proceedings of SIGGRAPH 1999, 1999, pp. 57–64.
[25] N. Litke, A. Levin, P. Schröder, Fitting subdivision surfaces, in: Visualization '01, 2001, pp. 319–324.
[26] N. Litke, A. Levin, P. Schröder, Trimming for subdivision surfaces, Comput. Aided Geom. Des. 18 (5) (2001) 463–481.
[27] C. Loop, Smooth subdivision surfaces based on triangles, Master's Thesis, Department of Mathematics, University of Utah, 1987.
[28] M. Lounsbery, T.D. DeRose, J. Warren, Multiresolution analysis for surfaces of arbitrary topological type, ACM Trans. Graphics 16 (1) (1997) 34–73 (originally available as TR-93-10-05, October, 1993, Department of Computer Science and Engineering, University of Washington).
[29] R. MacCracken, K.I. Joy, Free-form deformations with lattices of arbitrary topology, Proceedings of SIGGRAPH 96, 1996, pp. 181–188.
[30] P. Oswald, P. Schröder, Composite primal/dual $\sqrt{3}$-subdivision schemes, October 2001, submitted for publication.
[31] V. Pascucci, Slow growing subdivision in any dimension: towards removing the curse of dimensionality, Technical Report UCRL-JC-140582, Lawrence Livermore National Lab, 2001.
[32] J. Peters, U. Reif, The simplest subdivision scheme for smoothing polyhedra, ACM Trans. Graphics 16 (4) (1997) 420–431.
[33] H. Prautzsch, W. Boehm, The Handbook of Computer Aided Geometric Design, Elsevier, Amsterdam, 2000 (Chapter Box Splines).
[34] U. Reif, A unified approach to subdivision algorithms near extraordinary points, Comput. Aided Geom. Des. 12 (1995) 153–174.
[35] U. Reif, A degree estimate for polynomial subdivision surfaces of higher regularity, Proc. Amer. Math. Soc. 124 (1996) 2167–2174.
[36] U. Reif, Analyse und Konstruktion von Subdivisionsalgorithmen für Freiformflächen beliebiger Topologie, Shaker Verlag, 1999, Habilitationsschrift.
[37] P. Schröder, W. Sweldens, Digital geometry processing, Proceedings of the Sixth Annual Symposium on Frontiers of Engineering, 2001, pp. 41–44.
[38] J. Stam, Exact evaluation of Catmull–Clark subdivision surfaces at arbitrary parameter values, Proceedings of SIGGRAPH 98, 1998, pp. 395–404.
[39] J. Stam, On subdivision schemes generalizing uniform B-spline surfaces of arbitrary degree, Comput. Aided Geom. Des. 18 (5) (2001) 383–396.
[40] L. Velho, Quasi 4-8 subdivision, Comput. Aided Geom. Des. 18 (4) (2001) 345–357.
[41] L. Velho, D. Zorin, 4-8 subdivision, Comput. Aided Geom. Des. 18 (5) (2001) 397–427.
[42] L. Ying, D. Zorin, Nonmanifold subdivision, Proceedings of Visualization 2001, 2001, pp. 325–332.
[43] D. Zorin, A method for analysis of C^1-continuity of subdivision surfaces, SIAM J. Numer. Anal. 37 (5) (2000) 1677–1708.
[44] D. Zorin, Smoothness of subdivision on irregular meshes, Const. Approx. 16 (3) (2000) 359–397.
[45] D. Zorin, D. Kristjansson, Evaluation of piecewise smooth subdivision surfaces, Visual Computer (2002).
[46] D. Zorin, P. Schröder (Eds.), Subdivision for Modeling and Animation, Course Notes, ACM Siggraph, 2000.

[47] D. Zorin, P. Schröder, A unified framework for primal/dual quadrilateral subdivision schemes, Comput. Aided Geom. Des. 18 (5) (2001) 429–454.

[48] D. Zorin, P. Schröder, W. Sweldens, Interpolating subdivision for meshes with arbitrary topology, Proceedings of SIGGRAPH 96, 1996, pp. 189–192.

[49] D. Zorin, P. Schröder, W. Sweldens, Interactive multiresolution mesh editing, Proceedings of SIGGRAPH 97, 1997, pp. 259–268.

[50] D.N. Zorin, Subdivision and multiresolution surface representations, Ph.D. Thesis, Caltech, Pasadena, CA, 1997.

ELSEVIER Journal of Computational and Applied Mathematics 149 (2002) 221–225

JOURNAL OF COMPUTATIONAL AND APPLIED MATHEMATICS

www.elsevier.com/locate/cam

Nonconforming finite element methods

Zhong-Ci Shi

Institute of Computational Mathematics, Chinese Academy of Sciences, P.O. Box 2719, 100080 Beijing, China

Received 12 December 2001; received in revised form 23 January 2002

Abstract

Some interesting and important nonconforming finite elements for the second- and fourth-order elliptic problems are briefly described and analyzed.

The finite element method has achieved great success in many fields of science and technology since it was first suggested in elasticity in the fifth decade of the 20th century. Today it has become a powerful tool for solving partial differential equations [7,42].

The key issue of the finite element method is using a discrete solution on the finite element space, usually consisting of piecewise polynomials, to approximate the exact solution on the given space according to a certain kind of variational principle.

When the finite element space is a subspace of the solution space, the method is called CONFORMING. It is known that in this case, the finite element solution converges to the true solution provided the finite element space approximates the given space in some sense [7].

In general, for a $2m$ order elliptic boundary value problem, the conforming finite element space is a C^{m-1} subspace. It means that the shape function in this conforming finite element space is continuous together with its $m-1$ order derivatives. That is, for a second-order problem, the shape function is continuous and for a fourth-order problem, the shape function and its first derivatives are continuous. It is a rather strong restriction put on the shape functions in the latter case.

It is proved that to build up a conforming finite element space with C^1 continuity for a two-dimensional fourth-order problem, like the plate bending problem in elasticity, at least a quintic polynomial with 18 parameters is required for a triangular element [7], and a bi-cubic polynomial with 16 parameters for a rectangular element. It causes some computational difficulties because the dimension of the related finite element space is fairly large and its structure is rather complicated.

E-mail address: shi@sec.cc.ac.cn (Z.-C. Shi).

0377-0427/02/$ - see front matter © 2002 Published by Elsevier Science B.V.
PII: S0377-0427(02)00531-9

Therefore, it is desirable to use low-order polynomials with few parameters while keeping the required continuity property. A genius approach is to subdivide a given element into several small subelements and then use low-order polynomials on each subelement so that the C^1 continuity on the entire element is achieved. It is called the macro-element method [7]. There appeared a great deal of literature in this direction. However, due to the complexity of formulation of relevant finite element spaces, this method seems not so popular in finite element method calculations.

Another approach is to relax directly the C^{m-1} continuity of the finite element space. It comes to the so-called NONCONFORMING finite element method which had and still has a great impact on the development of finite element methods [7]. However, it was found shortly that some nonconforming elements converge and some do not. The convergence behavior sometimes depends on the mesh configuration.

Therefore, it is important to have some criterion to verify which nonconforming element is convergent and which is not. Irons proposed the Patch Test [9] based on some mechanical consideration and computational experiences. The idea of the patch test is that each element should solve the problem accurately for any constant strain field. The test is very simple and easy to implement in engineering applications [40].

Unfortunately, it was found in mathematics that the patch test is neither necessary nor sufficient [37,19–21,27]. A precisely convergence condition, namely the generalized patch test, was suggested by Stummel [36] from a rigorous viewpoint of mathematics. Many nonconforming finite elements in applications can be checked with this test [22]. A simplified version of the generalized patch test was proposed by the author [25]. However, either the patch test or the generalized patch test is only an analysis tool for an assessment of the convergence of nonconforming elements. How to build up a good element is another or even more important issue for solving a real-life problem.

Up to now there has been proposed a vast number of engineer devices based on different mechanical interpretations, like unconventional elements [1], energy-orthogonal elements with free formulation [8,5,33], quasi-conforming elements [4,38,26,28,31], generalized conforming elements [14,28,32] and many others [35]. The approximate spaces related to all these elements mentioned above are not included in the given solution space. Hence they are simply called NONSTANDARD finite elements [35]. A unified mathematical treatment for analysis of these nonstandard finite elements is proposed by the author and his co-worker as the double set parameter method [6].

In this comment, we will briefly describe and analyze some interesting and important nonconforming finite elements for the second- and fourth-order elliptic problems. Some of these elements are old and well known, some are quite new, but all of them are useful in applications.

For the second-order problem, we will discuss the following three low-order nonconforming elements:

1. *P_1 nonconforming* [7]. This is a triangular element, not C^0. The shape function is a linear polynomial with three nodal parameters at mid-points of three edges of the triangle. In contrast, the simplest and oldest conforming C^0 element, the so-called Courant element, uses the same linear polynomial but with three nodal parameters at the vertices of the triangle.
2. *Wilson element* [39]. This is an old nonconforming rectangular element. The shape function is a quadratic polynomial with six parameters, four at vertices of the rectangle and two internal degrees of freedom, like the second-order derivatives. This element converges for rectangular meshes [11], but does not converge for arbitrary quadrilaterals [12]. Some mesh conditions have

to be added [20]. It is interesting to mention that the behavior of the Wilson element is better than the corresponding bilinear Q_1 conforming element as many engineering examples have indicated [42]. But this fact has not yet been proved mathematically [30,34].

3. *The rotated Q_1 element* [18]. This is a newly established nonconforming rectangular element. The shape function consists of four terms as $[1, x, y, x^2 - y^2]$. There are two versions of choosing nodal parameters. The first one uses four function values at the mid-point of each edge of the rectangle. The second version uses four mean values of the shape function along edges. Both versions are convergent for rectangular meshes. However, the first version is not convergent for arbitrary quadrilaterals unless certain mesh conditions are satisfied like the nonconforming Wilson element. Meanwhile, the second version is convergent for quadrilaterals without any restriction on meshes [16]. This kind of new elements has recently found many important applications [18,13,15].

For the fourth-order problem, we will show three interesting nonconforming elements as follows:

1. *Morley element* [17]. This is an old and simplest triangular plate element. The shape function is a quadratic polynomial with six nodal parameters. They are three function values at vertices and three normal derivatives at mid-points of three edges. This element does not even belong to C^0 class, nevertheless it is convergent for the fourth-order problem [10,29]. Surprisingly, it is recently proved [41] that the Morley element is divergent for the second-order elliptic problem. In contrast, it is well known that there exists for a long time the P_2 conforming element for the second-order problem. Its shape function is again a quadratic polynomial with six parameters as three function values at vertices and three function values at mid-points of edges. This quadratic C^0 element is convergent for the second-order problem, but divergent for the fourth-order problem.
2. *Zienkiewicz incomplete cubic triangular element* [2]. The shape function consists of incomplete cubic polynomials with specially chosen nine terms. The nine nodal parameters are three function values together with six first partial derivatives with respect to x and y at three vertices. This is a C^0 element, but not C^1. From the viewpoint of continuity of the shape function, the Zienkiewicz element is better than the Morley element. However, it is proved that this element is convergent only for very special meshes, namely, all edges of triangles are parallel to the three given directions [10,27] while the Morley element converges for arbitrary triangulations.
   As for other meshes, like the cross diagonal mesh, Zienkiewicz et al. have noticed in their early numerical experiments [2] that the finite element solution always has some discrepancy within 10% from the true solution no matter how fine the mesh is. Later, it was proved mathematically [3] that the Zienkiewicz element using the cross-diagonal mesh actually tends to a limit, but it is not the true solution of the given problem, rather of another problem. It is a very interesting phenomenon in the nonconforming finite element analysis. This strange wrong limit behavior was also observed in other nonconforming elements for the second-order problem [23]. So care must be taken in using nonconforming finite elements. We cannot rely solely on the computer output which, in general, cannot make differences between the true solution and a wrong limit.
3. *Argyris' unconventional triangular element* (*TRUNC*) [1]. Since the Zienkiewicz element is not always convergent as mentioned above, it is desirable to have a cubic C^0 element with nine parameters converging for arbitrary meshes. There are many different devices proposed by

engineers. Among them the Argyris approach, the so-called unconventional method, deserves special attention due to its creativity both in mechanics and mathematics. Argyris actually does use the same shape function space as the Zienkiewicz element. Then he describes his TRUNC element purely on a mechanical interpretation, which turns out that in the formulation of the stiffness matrix the coupling terms between the low-order model consisting of second-order polynomials and the high-order model of cubic polynomials disappear. It seems very strange at first glance, but Argyris showed by numerous examples that the TRUNC element gives convergence results for arbitrary meshes and even better numerical accuracy than the Zienkiewicz element in convergence cases.

What is the mathematics behind this genius device? In fact, it was found by the author [24] that if instead of the exact integration in stiffness matrix calculation, we use a kind of numerical integration, then the approximate coupling terms will turn to zero. It means that a proper numerical integration gives some small perturbation to the coupling terms of the Zienkiewicz stiffness matrix that dramatically changes its convergence property from divergence to convergence. As usual, the numerical integration introduces some truncation error which has happened to cancel the inherent error of the Zienkiewicz element due to its nonconformity. Therefore, we can conclude that two NEGATIVEs get a POSITIVE.

References

[1] J.H. Argyris, M. Hasse, H.P. Mlejnek, On an unconventional but natural formation of a stiffness matrix, Comput. Methods Appl. Mech. Eng. 22 (1980) 1–22.

[2] G.P. Bazeley, Y.K. Cheung, B.M. Irons, O.C. Zienkiewicz, Triangular elements in bending–conforming and nonconforming solutions, in: Proceedings of the Conference on Matrix Methods in Structural Mechanics, Air Force Ins. Tech., Wright-Patterson A.F. Base, Ohio, 1965.

[3] W. Cai, The limitation problem for Zienkiewicz triangle elements, Math. Numer. Sinica 8 (1986) 345–353.

[4] W.J. Chen, Y.X. Liu, L.M. Tang, The formulation of quasi-conforming elements, J. Dalian Inst. Technol. 19 (1980) 37–49.

[5] S.C. Chen, Z.-C. Shi, On the free formulation scheme for construction of stiffness matrices, Math. Numer. Sinica 13 (1991) 417–424.

[6] S.C. Chen, Z.-C. Shi, Double set parameter method of constructing stiffness matrices, Math. Numer. Sinica 13 (1991) 286–296.

[7] P.G. Ciarlet, The Finite Element Method for Elliptic Problems, North-Holland, Amsterdam, NY, 1978.

[8] C.A. Felippa, P.G. Bergan, A triangular bending element based on energy-orthogonal free formulation, Comput. Methods Appl. Mech. Eng. 61 (1987) 129–160.

[9] B.M. Irons, A. Razzaque, Experience with the patch test, in: A.R. Aziz (Ed.), Proceedings of the Symposium on Mathematical Foundations of the Finite Element Method, Academic Press, New York, 1972, pp. 557–587.

[10] P. Lascaux, P. Lesaint, Some nonconforming finite elements for the plate bending problem, RAIRO Anal. Numer. R-1 (1985) 9–53.

[11] P. Lesaint, On the convergence of Wilson's nonconforming element for solving the elastic problem, Comput. Methods Appl. Eng. 7 (1976) 1–6.

[12] P. Lesaint, M. Zlamal, Convergence of the nonconforming Wilson element for arbitrary quadrilateral meshes, Numer. Math. 36 (1980) 33–52.

[13] B. Li, M. Luskin, Nonconforming finite element approximation of crystalline microstructures, Math. Comput. 65 (1996) 1111–1135.

[14] Y.Q. Long, K.Q. Xin, Generalized conforming elements, J. Civil Eng. 1 (1987) 1–14.

[15] P.B. Ming, Z.-C. Shi, Nonconforming rotated Q_1 element for Reissner–Mindlin plate, MMMAS 11 (2001) 1311–1342.

[16] P.B. Ming, Z.-C. Shi, Mathematical analysis for quadrilateral rotated Q_1 elements: (I) Convergence study, Sci. China, 2001.
[17] L.S.D. Morley, The triangular equilibrium element in the solution of plate bending problems, Aero. Quart. 19 (1968) 149–169.
[18] R. Rannacher, S. Turek, A simple nonconforming quadrilateral Stokes element, Numer. Methods PDEs 8 (1992) 97–111.
[19] Z.-C. Shi, An explicit analysis of Stummel's patch test examples, Internat. J. Numer. Methods Eng. 20 (1984) 1233–1246.
[20] Z.-C. Shi, A convergence condition for the quadrilateral Wilson element, Numer. Math. 44 (1984) 349–361.
[21] Z.-C. Shi, On the convergence properties of quadrilateral elements of Sander and Beckers, Math. Comput. 42 (1984) 493–504.
[22] Z.-C. Shi, The generalized patch test for Zienkiewicz triangles, J. Comput. Math. 2 (1984) 279–286.
[23] Z.-C. Shi, Convergence properties of two nonconforming finite elements, Comput. Methods Appl. Eng. 48 (1985) 123–137.
[24] Z.-C. Shi, A remark on the optimal order of convergence of Wilson's nonconforming element, Math. Numer. Sinica 8 (1986) 159–163.
[25] Z.-C. Shi, Convergence of the TRUNC plate element, Comput. Methods Appl. Eng. 62 (1987) 71–88.
[26] Z.-C. Shi, The F-E-M-Test for nonconforming finite elements, Math. Comput. 49 (1987) 391–405.
[27] Z.-C. Shi, On the nine degree quasi-conforming plate element, Math. Numer. Sinica 10 (1988) 100–106.
[28] Z.-C. Shi, On Stummel's examples to the patch test, Comput. Mech. 5 (1989) 81–87.
[29] Z.-C. Shi, On the accuracy of the quasi-conforming and generalized conforming finite elements, Chin. Ann. Math. 11 B (1990) 148–156.
[30] Z.-C. Shi, On the error estimates of Morley element, Math. Numer. Sinica 12 (1990) 113–118.
[31] Z.-C. Shi, S.C. Chen, Direct analysis of a nine parameter quasi-conforming plate element, Math. Numer. Sinica 12 (1990) 76–84.
[32] Z.-C. Shi, S.C. Chen, Convergence of a nine degree generalized conforming element, Math. Numer. Sinica 13 (1991) 193–203.
[33] Z.-C. Shi, S.C. Chen, F. Zhang, Convergence analysis of Bergan's energy-orthogonal plate element, MMMAS 4 (1994) 489–507.
[34] Z.-C. Shi, B. Jiang, W.M. Xue, A new superconvergence property of Wilson nonconforming finite element, Numer. Math. 78 (1997) 259–268.
[35] Z.-C. Shi, M. Wang, Mathematical theory of some nonstandard finite element methods, Contemp. Math. AMS 163 (1994) 111–125.
[36] F. Stummel, The generalized patch test, SIAM J. Numer. Anal. 16 (1979) 449–471.
[37] F. Stummel, The limitation of the patch test, Internat. J. Numer. Methods Eng. 15 (1980) 177–188.
[38] L.M. Tang, W.J. Chen, Y.X. Liu, Quasi-conforming elements in finite element analysis, J. Dalian Inst. Technol. 19 (1980) 19–35.
[39] R.L. Taylor, P.J. Beresford, E.L. Wilson, A nonconforming element for stress analysis, Internat. J. Numer. Methods Eng. 10 (1976) 1211–1219.
[40] R.L. Taylor, T.C. Simo, O.C. Zienkiewicz, A.H.C. Chan, The patch test—a condition for assessing FEM convergence, Internat. J. Numer. Methods Eng. 22 (1986) 39–62.
[41] K. Trygve, X.C. Tai, R. Winther, A robust nonconforming H^2-element, Math. Comput. 2000.
[42] O.C. Zienkiewicz, The Finite Element Method, 3rd Edition, Mcgraw-Hill, New York, 1977.

ELSEVIER

Journal of Computational and Applied Mathematics 149 (2002) 227–237

JOURNAL OF COMPUTATIONAL AND APPLIED MATHEMATICS

www.elsevier.com/locate/cam

Good approximation on the sphere, with application to geodesy and the scattering of sound

Ian H. Sloan *, Robert S. Womersley

School of Mathematics, University of New South Wales, 2052 Sydney, NSW, Australia

Received 17 December 2001; received in revised form 2 March 2002

Abstract

Advances in approximation theory are often driven by applications. This paper explores two recent developments in polynomial approximation on the sphere, and relates them to applications. The first, driven by applications in geodesy, concerns cubature rules on the sphere that are exact for polynomials of high degree. The second, driven by a problem of the scattering of acoustic waves by smooth objects, concerns the so-called 'hyperinterpolation' approach to function approximation by spherical polynomials.

Keywords: Approximation; Polynomial; Sphere; Cubature

1. Introduction

The sphere is a natural setting for many mathematical models. It is therefore not surprising that approximation problems on spheres or sphere-like surfaces are often encountered. Yet because of the more difficult geometry and topology of the sphere, there are many more gaps in our understanding than there are for Euclidean domains.

In this paper we look at two case studies, in each of which the mathematical modelling of significant applications came up against shortcomings in the known results for approximation theory on the sphere.

The first application concerns the study of the earth's gravitational potential. In this case the need was for cubature rules on the sphere that are exact for all spherical polynomials up to some high degree. The second application concerns the scattering of acoustic waves from smooth three-dimensional objects, and the related 'inverse' problems of determining the shape of the scattering object from

* Corresponding author.

E-mail address: sloan@maths.unsw.edu.au (I.H. Sloan).

0377-0427/02/$ - see front matter © 2002 Published by Elsevier Science B.V.
PII: S0377-0427(02)00532-0

measurements of the 'far-field pattern' of the scattered waves. In this case the call was for improved results for the approximation of a smooth function on the sphere by a spherical polynomial.

The two applications, and some developments in approximation theory which they provoked, are described in the next two sections.

2. Geodesy and cubature

The problem of describing the earth's gravitational field is a classical problem of formidable difficulty, because it combines both large-scale (global) features and small-scale details. On the one hand, there exist well-developed spherical harmonic expansions of the earth's gravitational potential at (say) the surface of the earth, which in effect approximate the gravitational potential as a spherical polynomial of degree as high as 360. On the other hand, near mountain ranges and mineral bodies there are sharp variations over short distances that are virtually impossible to capture by global polynomials.

Modern descriptions of the earth's gravitational potential combine both polynomials and wavelets, so as to be able to capture both global and local features; for example, see [5]. Difficult issues arise when dealing with short-range features of the description, if at the same time the global (or spherical polynomial) aspects are to be respected.

Recently Freeden, of the Geomathematics Group at the University of Kaiserslautern, presented to us a problem driven by such a need; faced with the problem of designing future satellite measurements of gravitational data, he asked: how can we construct effective numerical integration rules that integrate exactly all polynomials of high degree? The requirement of integrating exactly all polynomials up to some (high) degree is driven by the need to respect the global features of the model. For various reasons, it is desirable that the points of the cubature rule be well distributed, rather than being concentrated near the poles as happens with typical tensor product rules.

Suppose f is a continuous function on the earth's surface, which is represented in appropriate units as the unit sphere S^2 embedded in $\mathbb{R}^3$. The obvious way to construct such a cubature rule is to approximate f by an interpolating polynomial $\Lambda_n f$ of degree n that agrees with f at appropriately chosen points, and then to integrate $\Lambda_n f$ exactly.

To be explicit, let $\mathbb{P}_n$ denote the space of spherical polynomials of degree n or less, that is,

$$\mathbb{P}_n = \text{span}\{Y_{\ell k} : \ell = 0, 1, \ldots, n, \ |k| \leqslant \ell\},$$

where $\{Y_{\ell k}\}$ is an orthonormal set of spherical harmonics (see for example [10]). The dimension of $\mathbb{P}_n$ is

$$\dim \mathbb{P}_n = \sum_{\ell=0}^{n} (2n + 1) = (n + 1)^2.$$

As a prelude to defining an interpolating polynomial, we need to make a 'good' choice of $d_n{=}(n{+}1)^2$ points $x_1, \ldots, x_{d_n}$ on S^2, and then to define

$$\Lambda_n f \in \mathbb{P}_n, \quad \Lambda_n f(x_j) = f(x_j), \ j = 1, \ldots, d_n.$$

The interpolating polynomial $\Lambda_n f$ is well defined if the set $\{x_1,\ldots,x_{d_n}\}$ is a 'fundamental system', that is, if the only polynomial in $\mathbb{P}_n$ that vanishes at all the points $x_1,\ldots,x_{d_n}$ is the zero polynomial. We shall from now onwards consider only fundamental systems.

Once the fundamental system $\{x_1,\ldots,x_{d_n}\}\subseteq S^2$ has been selected, the interpolating cubature rule corresponding to these points is defined by

$$Q_n f = \int_{S^2} (\Lambda_n f)(x)\,\mathrm{d}s(x), \tag{1}$$

where $\mathrm{d}s(x)$ denotes surface measure on S^2. To make the right-hand side look like a cubature rule, we introduce the fundamental Lagrange polynomials corresponding to the given fundamental system, defined (just as in the one-dimensional case) by

$$\ell_j \in \mathbb{P}_n, \quad \ell_j(x_k) = \delta_{jk}, \ j,k = 1,\ldots,d_n. \tag{2}$$

In terms of the fundamental Lagrange polynomials we may write the Lagrange formula for $\Lambda_n f$,

$$\Lambda_n f = \sum_{j=1}^{d_n} f(x_j)\ell_j,$$

which substituted into (1) gives

$$Q_n f = \sum_{j=1}^{d_n} w_j f(x_j),$$

with

$$w_j = \int_{S^2} \ell_j(x)\,\mathrm{d}s(x), \quad j = 1,\ldots,d_n.$$

The rule Q_n has the property that it integrates exactly all spherical polynomials of degree $\leqslant n$, that is

$$Q_n p = \int_{S^2} p(x)\,\mathrm{d}s(x) \quad \text{if } p \in \mathbb{P}_n.$$

This gives a convenient way of computing the weights w_j : let $\{\phi_1,\ldots,\phi_{d_n}\}$ denote any basis of $\mathbb{P}_n$ (for example, a spherical harmonic basis); then the weights $w_1,\ldots,w_{d_n}$ may be obtained by solving the square linear system

$$\sum_{j=1}^{d_n} w_j \phi_k(x_j) = \int_{S^2} \phi_k(x)\,\mathrm{d}s(x), \quad k = 1,\ldots,d_n. \tag{3}$$

What can go wrong? The general experience is that solving (3) is easy for small values of n, but that for n greater than say 30, the matrix $\{\phi_k(x_j)\}$ is so badly conditioned as to make a meaningful computation of the weights impossible.

A second difficulty is that for most choices of the points $x_1,\ldots,x_{d_n}$ some of the weights $w_1,\ldots,w_{d_n}$ will turn out to be negative, especially when n is large. Negative weights (unless they are relatively very small in magnitude) are never welcome, and can lead to a serious loss of stability if

$$\sum_{j=1}^{d_n} |w_j| \gg \sum_{j=1}^{d_n} w_j = 4\pi.$$

Many ways of choosing point systems on the sphere have been proposed in the literature, all having features that are attractive from one point of view or another. Point sets with good distribution properties and simple formulas for their generator have been given in [4,9] ('generalized spiral points').

A variety of other point systems have been considered as candidates for polynomial interpolation in [18], with computed results that turn out to vary enormously between one choice of points and another.

Intuitively appealing are the 'minimum energy' point systems which minimize

$$\psi(x_1,\ldots,x_m) = \sum_{i=1}^{m} \sum_{j=i+1}^{m} \frac{1}{|x_i - x_j|}, \tag{4}$$

over all choices of the points $x_j \in S^2$, $j = 1,\ldots,m$, where $|x_i - x_j|$ denotes the Euclidean distance in $\mathbb{R}^3$ between the points x_i and x_j. This and related potentials for selecting the points have been considered by many people. Detailed calculations of minimum energy systems have been presented in [3] for systems of $d_n = (n+1)^2$ points and values of n up to 29.

The minimum energy points of [3] were found in paper [18] to lead to remarkably poorly conditioned interpolation matrices: condition numbers as large as 10^8 were reported even for relatively small values of n. The quality of the interpolating polynomials $\Lambda_n f$ obtained with these points was also found to be remarkably poor. One way to measure the quality of the interpolation operator Λ_n is to compute its uniform norm (or 'Lebesgue constant')

$$\begin{aligned} \|\Lambda_n\|_\infty &= \sup_{f \in C(S^2)} \frac{\|\Lambda_n f\|_\infty}{\|f\|_\infty} \\ &= \max_{x \in S^2} \sum_{j=1}^{d_n} |\ell_j(x)|. \end{aligned} \tag{5}$$

The values of $\|\Lambda_n\|_\infty$ reported in [18] were typically several orders of magnitude larger than those for better choices of the points.

While it is true that the point systems in [3] were intended for interpolatory cubature rather than interpolation, in fact they are not entirely successful for this role either, for the cubature weights reported in [3] generally take both positive and negative values, except for the smallest values of n. In any event, the poor conditioning of the matrices makes this an impossible approach for large values of n.

In the next section we describe very recent results for cubature formulas based on so-called 'extremal' systems, which are found to have excellent properties (including positive weights) even for very large values of n.

3. Cubature with extremal systems

We begin with the observation that if $\phi_1, \ldots, \phi_{d_n}$ is any basis for $\mathbb{P}_n$, then the fundamental Lagrange polynomial ℓ_j may be written as the ratio of two determinants,

$$\ell_j(x) = \frac{\det \begin{bmatrix} \phi_1(x_1) \cdots & \phi_1(x) \cdots & \phi_1(x_{d_n}) \\ \phi_2(x_1) \cdots & \phi_2(x) \cdots & \phi_2(x_{d_n}) \\ \vdots & \vdots & \vdots \\ \phi_{d_n}(x_1) \cdots & \phi_{d_n}(x) \cdots & \phi_{d_n}(x_{d_n}) \end{bmatrix}}{\det \begin{bmatrix} \phi_1(x_1) \cdots & \phi_1(x_j) \cdots & \phi_1(x_{d_n}) \\ \phi_2(x_1) \cdots & \phi_2(x_j) \cdots & \phi_2(x_{d_n}) \\ \vdots & \vdots & \vdots \\ \phi_{d_n}(x_1) \cdots & \phi_{d_n}(x_j) \cdots & \phi_{d_n}(x_{d_n}) \end{bmatrix}},$$

since the right-hand side clearly satisfies (2).

An extremal system is one that maximises the absolute value of the denominator in the above expression,

$$\det \begin{bmatrix} \phi_1(x_1) \cdots & \phi_1(x_{d_n}) \\ \vdots & \vdots \\ \phi_{d_n}(x_1) \cdots & \phi_{d_n}(x_{d_n}) \end{bmatrix}. \tag{6}$$

The motivation is that if the system of points maximises the magnitude of (6) then immediately we have

$$|\ell_j(x)| \leqslant 1, \quad j = 1, \ldots, d_n, \ x \in S^2,$$

and through (5) this property then gives an upper bound on the Lebesgue constant,

$$\|\Lambda_n\|_\infty \leqslant d_n = (n+1)^2. \tag{7}$$

This construction appears to traces back to Fekete and Auerbach, and for the particular case of the sphere to be due to Reimer [10].

It is known that pairs of points of an extremal system can never be very close together, in that from a result of [10] the minimum angle subtended at the centre of the sphere by the arc between two distinct points of an extremal system is $\pi/2n$. On the other hand from [17] the 'mesh norm'

$$h(x_1, \ldots, x_{d_n}) = \max_{x \in S^2} \min_{j=1,\ldots,d_n} \cos^{-1}(xx_j)$$

(which may be thought of as the largest angle between a point on the sphere and the nearest mesh point) cannot be too large, being bounded asymptotically by $2j_0/n$, where $j_0 \approx 2{\cdot}4048$ is the smallest positive zero of the Bessel function J_0.

A recent study [16] of extremal systems on the sphere has shown that the above estimates are not sharp. The minimum angle between pairs of points asymptotically appears to approach π/n rather than $\pi/2n$. The computed mesh norm is found to be roughly half of the theoretical bound above. And from the numerical computations in [18] it is known that the Lebesgue constant is more like $n+1$ then $(n+1)^2$.

In the paper [16] extremal systems are computed for values of n as large as 128. The first surprise is that this is possible at all, given the comments about ill-conditioning made in the preceding section. The next is that *all of the cubature weights turned out to be positive*, even for the largest values of n. Indeed, even more is true: all weights turn out to lie between 1/2 and 3/2 of the average weight $4\pi/d_n$. Although positivity of the weights has so far been proved only for $n=2$ (see [11]), the computed weights in [16] lead to the conjecture that the cubature weights of the interpolatory cubature rule associated with an extremal system has positive weights for *all n*.

In any event, it seems clear that computed cubature rules based on extremal systems can have a role in practical computations on the sphere. Full listings of the computed points and weights are given at the web site http://www.maths.unsw.edu.au/~rsw/Sphere for all values of n up to 50, then for $n=64$, 96 and 128. The paper [16] gives details of the calculations, and additional theoretical results. Note that for the latter case the number of cubature points and weights is $129^2 = 16\,641$. Since their computation involves optimising a dense matrix of size $129^2 \times 129^2$ as a function of over 33 000 variables, high demands are placed on computing resources, especially memory. At the moment memory limitations make it unfeasible for us to go to values of n much beyond $n = 128$.

4. Scattering of sound by a smooth object

The second application that challenges the present state of knowledge of approximation on the sphere concerns the scattering of sound by a smooth 'ball-like' object. By this we mean that the surface $\partial\Omega$ of the object can be mapped smoothly and invertibly to the unit sphere S^2.

In this presentation we want to emphasise only the central ideas, and especially the link to approximation theory. For all technical details we refer freely to the literature.

Given a smooth ball-like object with surface $\partial\Omega$, our problem is to solve the wave equation for the air pressure u in the region exterior to $\partial\Omega$, subject to the boundary condition that $u=0$ on $\partial\Omega$ (the 'sound soft' condition), and that at infinity the solution takes the form of an incident plane wave with vector $\mathbf{k}$ plus an outgoing scattered wave ('radiation condition').

The 'direct' problem is: given $\partial\Omega$, find the far-field pattern, which is the angular distribution at infinity of the scattered-wave part of the solution. Even more difficult, and the ultimate goal of most studies, is the 'inverse' problem, in which the goal is to infer the scattering surface $\partial\Omega$ from observations of the far-field pattern.

By looking for a solution that is harmonic in time, the problem reduces to the Helmholtz equation in the exterior region, with appropriate boundary conditions.

It is by now very well known that one can replace the problem of solving the boundary value problem over the exterior region by solving instead a boundary integral equation on $\partial\Omega$ for $v=\partial u/\partial n$, the outward normal derivative of u on $\partial\Omega$ (see [2] for full details). The boundary integral equation takes the form

$$v + \mathscr{M}v = f,$$

where $\mathscr{M}$ is a linear combination of the single-layer potential for the Helmholtz equation, and the normal derivative of the single-layer potential. For the present discussion the details do not matter: it suffices to know that $\mathscr{M}$ is an integral operator on $\partial\Omega$ with a weakly singular kernel (i.e. the singularity in the kernel is absolutely integrable).

We are assuming that the surface $\partial\Omega$ can be mapped smoothly to a unit sphere. More precisely, we suppose that there exists a C^∞ mapping q: $\partial\Omega \to S^2$ and a C^∞ inverse q^{-1}: $S^2 \to \partial\Omega$. In the forward problem we assume that the mapping q is known. With V denoting the function v transferred to S^2, i.e. $V(\hat{x}) = v(q^{-1}(\hat{x}))$ for $\hat{x} \in S^2$, the boundary integral equation can now be expressed as a boundary integral equation on S^2, say

$$V + MV = F, \tag{8}$$

where the Jacobian of the surface mapping is incorporated into the new boundary integral operator M.

Now that the problem has been converted to one lying on the sphere, we are ready to think about approximation. There are many approaches to the approximate solution of boundary integral equations, including many that rely on dividing the surface into small patches, and then approximating the solution on a patch mapped to some reference region by a low-degree polynomial. In the present discussion, however, the focus is on spectral methods: in other words, the exact solution V: $S^2 \to \mathbb{C}$ is approximated by a spherical polynomial $V_n \in \mathbb{P}_n$.

There are different ways of selecting the particular polynomial approximation V_n. To motivate the discussion, it is useful to consider first the Galerkin approximation,

$$V_n^{\mathrm{G}} \in \mathbb{P}_n, \quad (V_n^{\mathrm{G}}, \chi) + (MV_n^{\mathrm{G}}, \chi) = (F, \chi) \quad \forall \chi \in \mathbb{P}_n, \tag{9}$$

where

$$(f, g) = \int_{S^2} f(x)\overline{g(x)}\,\mathrm{d}s(x). \tag{10}$$

Atkinson [1] has given a comprehensive stability and error analysis for the Galerkin approximation in the space $C(S^2)$.

The trouble with the Galerkin method is that it is virtually impossible to implement exactly: for if we introduce a basis for $\mathbb{P}_n$ into (9), then computing the matrix elements in the linear system is seen to involve two levels of integration—one for the operation of M, the other for the inner product integral. Moreover, each of these is a surface integral. In practice some form of numerical integration is needed, but in that case the analysis of the Galerkin method does not strictly apply.

That brings us to a fully discrete spectral method for (8), which is the real object of our discussion. This method is essentially similar to one studied in [17], but Wienert was not able to prove stability of the method or to obtain a complete error analysis. The missing ingredient, as we shall see, was a knowledge of the properties of a certain polynomial approximation on the sphere. (Wienert's actual method is different from that indicated below, but an analysis of the present method is easily mapped to an analysis of Wienert's method. For the precise relationship between the methods and the respective analyses, see [8].)

The method we actually discuss here has the further simplified form

$$V_n^s \in \mathbb{P}_n, \quad (V_n^s, \chi)_n + (MV_n^s, \chi)_n = (F, \chi)_n \quad \forall \chi \in \mathbb{P}_n, \tag{11}$$

where $(f,g)_n$ is a cubature approximation to the integral in (10),

$$(f,g)_n = \sum_{k=1}^{m} w_k f(x_k)\overline{g(x_k)},$$

where the rule

$$Q_n h := \sum_{k=1}^{m} w_k h(x_k)$$

is such that the weights w_k are all positive, the points x_k lie on S^2, and the rule is exact for all $p \in \mathbb{P}_{2n}$, i.e. for all polynomials of degree at most $2n$. In particular, this holds if the cubature rule is the $2(n+1)^2$ point product Gauss rule; see [8] for details.

Method (11) is semidiscrete, in that the integral involved in the operator M is still supposed to be done exactly. The fully discrete variant that truly corresponds to Wienert's method is obtained by replacing the integral in MV_n^s in (11) by a cubature sum which takes account of the special nature of the weak singularity in the surface integral in MV_n^s. For a detailed discussion see [8]. For the present it suffices to consider (11), which is already difficult enough.

How could the stability and convergence properties of the method in (11) be established? An obvious idea was to try to mimic the successful proof in [1] for the Galerkin approximation, but there is a problem with this, which we now try to draw out. The Galerkin analysis used the function space setting $C(S^2)$ for the analysis. In that space the operator M is a compact operator with some smoothing effect, in that it maps continuous functions on S^2 into Lipschitz continuous functions on S^2. The analysis of [1] proceeded by rewriting (9) in the equivalent form

$$V_n^{\mathrm{G}} + P_n M V_n^{\mathrm{G}} = P_n F, \tag{12}$$

where $P_n: C(S^2) \to \mathbb{P}_n$ is the orthogonal projection operator defined by

$$P_n f \in \mathbb{P}_n, \quad (P_n f, \chi) = (f, \chi) \quad \forall \chi \in \mathbb{P}_n. \tag{13}$$

We do not repeat the analysis here, but note only the crucial point in the analysis, which is that

$$\|P_n\|_\infty := \sup_{\substack{f \in C(S^2) \\ \|f\|_\infty \leqslant 1}} \|P_n f\|_\infty$$

grows only slowly with n: in fact it is known that

$$\|P_n\|_\infty \leqslant C n^{1/2}, \tag{14}$$

where C is a known constant. This rate of growth is in fact slow enough to allow the existence and the uniform boundedness of $(I + P_n M)^{-1}$ in the space $C(S^2)$ to be proved, the central point being that the growth in $\|P_n\|_\infty$ is slow enough to be compensated by the smoothing property of M.

By analogy with (12), the semidiscrete equation (11) can be written as

$$V_n^s + L_n M V_n^s = L_n F,$$

where $L_n : C(S^2) \to \mathbb{P}_n$ is the projection operator defined by the discrete version of (13),

$$L_n f \in \mathbb{P}_n, \quad (L_n f, \chi)_n = (f, \chi)_n \quad \forall \chi \in \mathbb{P}_n. \tag{15}$$

The problem was that until recently a result for L_n analogous to (14) was not known. The weaker result

$$\|L_n\|_\infty \leqslant C_1 n^2$$

was obtained as a special case of a more general result in [6], but this was not good enough to make the stability proof go through: in fact, for Atkinson's proof a bound of the form $Cn^{1-\varepsilon}$ for some $\varepsilon > 0$ is needed. Since such a result was not available, the convergence problem was solved in [7] by a different but ingenious approach, in which stability and convergence is achieved for an augmented finite dimensional space, obtained by adding to the space $\mathbb{P}_n$ a roughly equal number of nonpolynomial basis functions.

The operator L_n defined by (15) is sometimes called the 'hyperinterpolation' operator, see [13]; in that paper good properties of L_n as a map from $C(S^2)$ to $L^2(S^2)$ were proved. More recently our colleagues Graham and Ganesh encouraged us to study L_n as a map from $C(S^2)$ to $C(S^2)$, in the knowledge that a result analogous to (14) would enable the stability and error analysis of the semidiscrete equation (11) to be completed. Eventually we were able to show, in [15], that the analogue of (14) does indeed hold. Our proof placed some restrictions on the cubature rule, but these restrictions were later removed [12].

Theorem 4.1 (Reimer [12], Sloan and Womersley [15]). *Let Q_n be a positive weight cubature rule on S^2 which is exact for all spherical polynomials of degree $\leqslant 2n$. Then the corresponding hyperinterpolation operator L_n defined by* (15) *satisfies*

$$\|L_n\|_\infty \leqslant C' n^{1/2},$$

where C' is independent of n.

The proof, in essence, follows the same line as that in [1] for the classical proof for $\|P_n\|_\infty$, but with the integrals replaced by sums. The key step is to show that the sums are in turn bounded by integrals.

The result above allowed Graham and Sloan in [8] to complete an analysis of both (11) and its fully discrete analogue, and thence to obtain a complete stability and error analysis for Wienert's method. We omit the details.

We should not finish by leaving the impression that the theoretical results are complete. One obvious lack is that as yet we do not have any convergence results for derivatives of $L_n f$, or convergence results in Sobolev spaces. The present result allows only the simplest (second-kind) operator equations to be analysed. Fortunately, that is all that is needed for the application to the scattering of sound.

5. Open questions

These two applications and the numerical investigations in [18,16] give rise to a number of interesting open questions.

For extremal systems of degree n on S^2,

- are the cubature weights w_j always positive for $n \geqslant 3$?
- is the minimum distance between points asymptotically π/n?
- is the Lebesgue constant $O(n)$?

More generally

- do there exist systems of interpolation points on the sphere S^2 for which the Lebesgue constant is $O(n)$ or $o(n)$?

Also, given that it is known [14] that the norm of the interpolation operator Λ_n as a map from $C(S^2)$ to $L_2(S^2)$ is strictly greater than $\sqrt{4\pi}$,

- does $\|\Lambda_n\|_{C\to L_2} \to \infty$ as $n \to \infty$?

Other more general questions concern the characterisation of good systems of points:

- what geometrical or analytical properties characterise extremal systems? (That is, is there some way of characterising global extrema, as distinct from local extrema, of the determinant?)

Finally for the scattering of sound, can we still use an analysis similar to that in [8] if instead of hyperinterpolation we use interpolation with respect to a suitable point system? (The present analysis would require the uniform norm to be $O(n^{1-\varepsilon})$ for some $\varepsilon > 0$.)

Acknowledgements

The support of the Australian Research Council is gratefully acknowledged.

References

[1] K.E. Atkinson, Numerical integration on the sphere, J. Austral. Math. Soc. Ser. B 23 (1982) 332–347.
[2] D. Colton, R. Kress, Inverse Acoustic and Electromagnetic Scattering Theory, Springer, Berlin, 1998.
[3] J. Fliege, U. Maier, The distribution of points on the sphere and corresponding cubature formulae, IMA J. Numer. Anal. 19 (1999) 317–334 `http://www.mathematik.uni-dortmund.de/lsx/fliege/nodes.html`.
[4] W. Freeden, T. Gervens, M. Schreiner, Constructive Approximation on the Sphere, Clarendon Press, Oxford, 1998.
[5] W. Freeden, U. Windheuser, Combined spherical harmonics and wavelet expansion—a future concept in Earth's gravitational determination, Appl. Comput. Harmon. Anal. 4 (1997) 1–37.
[6] M. Ganesh, I. Graham, J. Sivaloganathan, A pseudospectral three-dimensional boundary integral method applied to a nonlinear model problem from finite elasticity, SIAM J. Numer. Anal. 31 (1994) 1378–1414.
[7] M. Ganesh, I. Graham, J. Sivaloganathan, A new spectral boundary integral collocation method for three-dimensional potential problems, SIAM J. Numer. Anal. 35 (1998) 778–805.

[8] I. Graham, I.H. Sloan, Fully discrete spectral boundary integral methods on smooth closed surfaces in $\mathbb{R}^3$, Numer. Math., published on-line December 18, 2001.

[9] E.A. Rahkmanov, E.B. Saff, Y.M. Zhou, Minimal discrete energy on the sphere, Math. Res. Lett. 1 (1994) 647–662.

[10] M. Reimer, Constructive Theory of Multivariate Functions, BI Wissenschaftsverlag, Mannheim, Wien, Zürich, 1990.

[11] M. Reimer, Quadrature rules for the surface integral of the unit sphere based on extremal fundamental systems, Math. Nachr. 169 (1994) 235–241.

[12] M. Reimer, Hyperinterpolation on the sphere at the minimal projection order, J. Approx. Theory 104 (2000) 272–286.

[13] I.H. Sloan, Polynomial interpolation and hyperinterpolation over general regions, J. Approx. Theory 83 (1995) 238–254.

[14] I.H. Sloan, Interpolation and hyperinterpolation on the sphere, in: W. Haußmann, K. Jetter, M. Reimer, (Eds.), Multivariate Approximation: Recent Trends and Results, Akademie Verlag GmbH, Berlin (Wiley-VCH), 1997, pp. 255–268.

[15] I.H. Sloan, R.S. Womersley, Constructive polynomial approximation on the sphere, J. Approx. Theory 103 (2000) 91–118.

[16] I.H. Sloan, R.S. Womersley, Extremal systems of points and numerical integration on the sphere, Adv. Computational Math., to appear.

[17] L. Wienert, Die numerische Approximation von Randintegraloperatoren für die Helmholtzgleichung in $\mathbb{R}^3$, Ph.D. Thesis, University of Göttingen, 1990.

[18] R.S. Womersley, I.H. Sloan, How good can polynomial interpolation on the sphere be? Adv. Comput. Math. 14 (2001) 195–226.

JOURNAL OF COMPUTATIONAL AND APPLIED MATHEMATICS

ELSEVIER Journal of Computational and Applied Mathematics 149 (2002) 239–250

www.elsevier.com/locate/cam

Double exponential transformation in the Sinc-collocation method for two-point boundary value problems ☆

Masaaki Sugihara

Department of Computational Science and Engineering, Graduate School of Engineering, Nagoya University, Furo-cho Chikusa, Nagoya 464-8603, Japan

Received 17 December 2001; received in revised form 2 March 2002

Abstract

In 1974, Takahasi and Mori proposed the double exponential transformation for efficient evaluation of integrals of an analytic function with end point singularity. Recently, it has turned out that the double exponential transformation is also useful for various kinds of Sinc methods. In the present paper, we consider the Sinc-collocation method for two-point boundary value problems to show that the method can be considerably enhanced when incorporated with the double exponential transformation technique.

MSC: 65L10; 65L60

Keywords: Double exponential transformation; Sinc methods; Sinc-collocation method; Two-point boundary value problem

1. Introduction

The double exponential formula (DE formula), which is a quadrature formula based on the double exponential transformation (DE transformation), was first proposed by Takahasi and Mori [13] in 1974 in order to compute integrals with end point singularity such as

$$I=\int_{-1}^{1}\frac{\mathrm{d}x}{(2-x)(1-x)^{1/4}(1+x)^{3/4}}. \tag{1.1}$$

☆ This work is partially supported by a Grant-in-Aid for Scientific Research of the Ministry of Education, Science, Sports and Culture of Japan, and by "Research for the Future Program" of Japan Society for the Promotion of Science.

E-mail address: sugihara@na.cse.nagoya-u.ac.jp (M. Sugihara).

PII: S0377-0427(02)00533-2

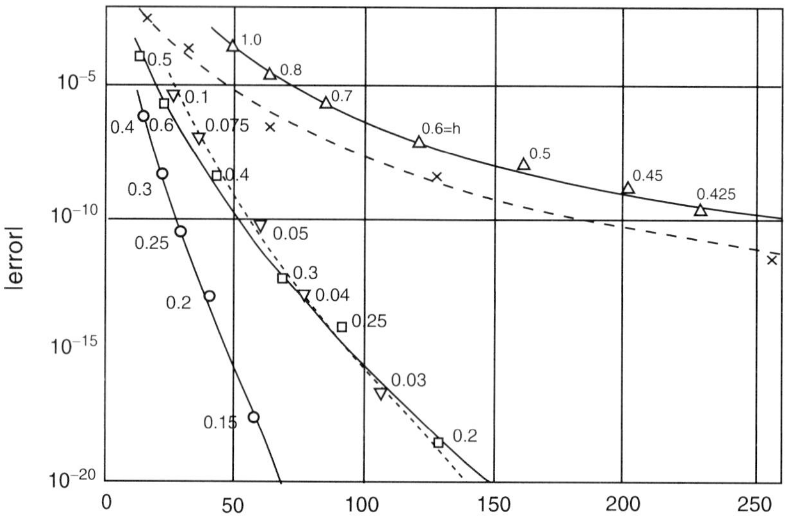

Fig. 1. Error of various formulas when applied to $\int_{-1}^{1} \frac{dx}{(2-x)(1-x)^{1/4}(1+x)^{3/4}}$.

For an integral over $(-1, 1)$,

$$\int_{-1}^{1} f(x)\,dx,$$

the DE transformation and the DE formula are given by

$$x = \psi(t) = \tanh\left(\frac{\pi}{2}\sinh t\right),$$

and

$$h \sum_{k=-N_-}^{N_+} f(\psi(kh))\psi'(kh)$$

$$= h \sum_{k=-N_-}^{N_+} f\left(\tanh\left(\frac{\pi}{2}\sinh kh\right)\right) \frac{(\pi/2)\cosh kh}{\cosh^2((\pi/2)\sinh kh)},$$

respectively. The DE formula has been widely used in the last three decades and is now recognized to be one of the most efficient quadrature formulas [6,7]. To illustrate how efficient the DE formula is, we cite the numerical result from [13] (see Fig. 1), where a variety of quadrature formulas obtained by variable transformation including the DE formula are applied to problem (1.1).

The effectiveness of the DE transformation technique in numerical integration naturally suggests that the DE transformation technique could be useful in other numerical methods. Indeed, it has been demonstrated in [7,12] that the use of the DE transformation technique in the *Sinc methods* developed by Stenger and his school [3,5,9,10] yields highly efficient numerical methods for interpolation, quadrature, approximation of transforms, and solution of integral, differential, and partial differential equations.

In the present paper, we consider the Sinc-collocation method for two-point boundary value problems, and show that the Sinc-collocation method gets considerably enhanced when incorporated with the DE transformation technique. To be specific, we consider the Sinc-collocation method for the following simple two-point boundary problem with the zero Dirichlet boundary condition:

$$y''(x) + \mu(x)y'(x) + \nu(x)y(x) = \sigma(x), \quad x \in (a,b),$$
$$y(a) = y(b) = 0. \tag{1.2}$$

It is known that, for this problem, the standard Sinc-collocation method [1] with n collocation points converges at the rate of $\exp(-\kappa\sqrt{n})$ with some $\kappa > 0$ under certain mild conditions. We here show that the Sinc-collocation method incorporated with the DE transformation converges at the rate of $\exp(-\kappa' n/\log n)$ with some $\kappa' > 0$, though under rather stringent conditions.

The paper is organized as follows. In Section 2, we describe the Sinc-collocation method for the two-point boundary problem on the entire interval $(-\infty,\infty)$ together with known error analysis. In Section 3, we deal with the two-point boundary problem on the general interval $[a,b]$, for which the Sinc-collocation method is incorporated with variable transformation techniques. In Section 4, we demonstrate via numerical examples that the Sinc-collocation method is remarkably enhanced when combined with the DE transformation technique. Final comments are made in Section 5.

2. Sinc-collocation method on the entire interval $(-\infty, \infty)$

In this section, we treat the Sinc-collocation method for the approximate solution of the following two-point boundary value problem on the interval $(-\infty,\infty)$,

$$Ly(x) \equiv y''(x) + \mu(x)y'(x) + \nu(x)y(x) = \sigma(x), \quad x \in (-\infty,\infty),$$
$$\lim_{x\to\pm\infty} y(x) = 0. \tag{2.1}$$

2.1. Description of the Sinc-collocation method

Following [5,9], we describe the Sinc-collocation method for the two-point boundary value problem (2.1).

We assume an approximate solution $y_n(x)$ to (2.1) of the form

$$y_n(x) \equiv \sum_{j=-N}^{N} w_j S(j,h)(x), \quad n = 2N+1, \tag{2.2}$$

where $S(j,h)(x)$ designates the Sinc function:

$$S(j,h)(x) \equiv \frac{\sin[(\pi/h)(x-jh)]}{(\pi/h)(x-jh)} \tag{2.3}$$

[1] By the "standard Sinc-collocation method" we mean the Sinc-collocation method developed by Stenger and his school.

Note that $y_n(x)$ satisfies the boundary condition in (2.1) because

$$\lim_{x\to\pm\infty} S(j,h)(x) = 0.$$

The coefficients $\{w_j\}$ are determined from the requirement:

$$Ly_n(kh) = \sigma(kh), \quad k = -N, -N+1, \ldots, N-1, N, \tag{2.4}$$

which yields a system of linear equations in $\{w_j\}$. The explicit form of the system of equations is obtained from the substitution of (2.1) and (2.2) into (2.4) as

$$\sum_{j=-N}^{N} \{\delta_{jk}^{(2)}/h^2 + \mu(kh)\delta_{jk}^{(1)}/h + v(kh)\ \delta_{jk}^{(0)}\} w_j = \sigma(kh),$$
$$k = -N, -N+1, \ldots, N-1, N, \tag{2.5}$$

where

$$\delta_{jk}^{(0)} \equiv S(j,h)(kh) = \begin{cases} 1 & \text{if } j = k, \\ 0 & \text{if } j \neq k, \end{cases}$$

$$\delta_{jk}^{(1)} \equiv hS'(j,h)(kh) = \begin{cases} 0 & \text{if } j = k, \\ \dfrac{(-1)^{k-j}}{(k-j)} & \text{if } j \neq k, \end{cases}$$

$$\delta_{jk}^{(2)} \equiv h^2 S''(j,h)(kh) = \begin{cases} -\dfrac{\pi^2}{3} & \text{if } j = k, \\ \dfrac{-2(-1)^{k-j}}{(k-j)^2} & \text{if } j \neq k. \end{cases}$$

We can solve (2.5) by the Gaussian elimination.

2.2. Convergence analysis

Before stating theorems, let us introduce notation and definitions.

Definition 2.1. Let $\mathscr{D}_d$ denote the infinite strip region of width $2d$ ($d > 0$) in the complex plane:

$$\mathscr{D}_d \equiv \{z \in \mathbb{C} \mid |\mathrm{Im}\, z| < d\}.$$

For $0 < \varepsilon < 1$, let $\mathscr{D}_d(\varepsilon)$ be defined by

$$\mathscr{D}_d(\varepsilon) = \{z \in \mathbb{C} \mid |\mathrm{Re}\, z| < 1/\varepsilon, |\mathrm{Im}\, z| < d(1-\varepsilon)\}.$$

Let $H^1(\mathscr{D}_d)$ be the set of functions analytic in $\mathscr{D}_d$ such that

$$\lim_{\varepsilon\to 0} \int_{\partial \mathscr{D}_d(\varepsilon)} |f(z)|\,|\mathrm{d}z| < \infty.$$

The following theorem shows that the Sinc-collocation method converges at the rate of $\exp(-\kappa\sqrt{n})$, if the solution of problem (2.1) decays single exponentially.

Theorem 2.2 (Bialecki [1], Stenger [9]). *Assume that problem* (2.1) *has a unique solution* $y(x)$, *and that the solution* $y(x)$ *is analytic on the real line. Furthermore assume with positive constants* α, β *and* d *that*

(1) μ *and* ν *are analytic and bounded in the strip region* $\mathscr{D}_d$;
(2) μ *takes real values on the real line*;
(3) $\mathrm{Re}\{2\nu(x) - \mu'(x)\} \leqslant 0$ *for all* $x \in \mathbb{R}$;
(4) σ *belongs to* $H^1\ (\mathscr{D}_d)$ *and decays single exponentially on the real line, that is*,

$$|\sigma(x)| \leqslant \alpha \exp(-\beta|x|) \quad \textit{for all } x \in \mathbb{R};$$

(5) y *belongs to* $H^1\ (\mathscr{D}_d)$ *and decays single exponentially on the real line, that is*,

$$|y(x)| \leqslant \alpha \exp(-\beta|x|) \quad \textit{for all } x \in \mathbb{R}.$$

Then we have

$$\sup_{-\infty<x<\infty} |y(x) - y_n(x)| \leqslant cN^{5/2} \exp[-(\pi d\beta N)^{1/2}],$$

where the mesh size h *in the* Sinc-*collocation method is taken as*

$$h = \left(\frac{\pi d}{\beta N}\right)^{1/2}.$$

In the case where the solution decays double exponentially, we can prove the following theorem, which shows that the Sinc-collocation method converges at the rate of $\exp(-\kappa' n/\log n)$.

Theorem 2.3. *Assume that problem* (2.1) *has a unique solution* $y(x)$, *and that the solution* $y(x)$ *is analytic on the real line. Furthermore assume with positive constants* $A, B, \alpha, \beta, \gamma$ *and* d *that*

(1) μ, μ' *and* ν *are analytic in the strip region* $\mathscr{D}_d$, *and their absolute values on the real line are bounded from above as follows*:

$$|\mu(x)|,\ |\mu'(x)|,\ |\nu(x)| \leqslant A \exp(B|x|) \quad \textit{for all } x \in \mathbb{R};$$

(2) $\mu y, \mu' y$ *and* νy *belong to* $H^1(\mathscr{D}_d)$;
(3) μ *takes real values on the real line*;
(4) $\mathrm{Re}\{2\nu(x) - \mu'(x)\} \leqslant 0$ *for all* $x \in \mathbb{R}$;
(5) σ *belongs to* $H^1\ (\mathscr{D}_d)$ *and decays double exponentially on the real line, that is*,

$$|\sigma(x)| \leqslant \alpha \exp(-\beta \exp(\gamma|x|)) \quad \textit{for all } x \in \mathbb{R};$$

(6) y *belongs to* $H^1\ (\mathscr{D}_d)$ *and decays double exponentially on the real line, that is*,

$$|y(x)| \leqslant \alpha \exp(-\beta \exp(\gamma|x|)) \quad \textit{for all } x \in \mathbb{R}.$$

Then we have

$$\sup_{-\infty<x<\infty} |y(x) - y_n(x)| \leqslant c(\log N)N^{B/\gamma+3/2} \exp\left[\frac{-\pi d\gamma N}{\log(\pi d\gamma N/\beta)}\right],$$

where the mesh size h *in the* Sinc-*collocation method is taken as*

$$h = \frac{\log(\pi d\gamma N/\beta)}{\gamma N}.$$

Remark 2.1. The double exponential decay treated in Theorem 2.3 is extremal in that a function that decays more rapidly must vanish. To be more precise, the following theorem is known, which has recently been obtained in establishing the optimality of the double exponential quadrature formulas [11].

Theorem 2.4. *If a function* $f(z)$ *satisfies the following two conditions*:

(1) $f(z)$ *belongs to* H^1 $(\mathscr{D}_d)$;
(2) *the decay rate on the real line satisfies*

$$f(x) = O(\exp(-\beta \exp(\gamma|x|))) \quad \text{as } |x| \to \infty,$$

where $\beta > 0$ *and* $\gamma > \pi/(2d)$,

then $f \equiv 0$.

3. Sinc-collocation method on a general interval [*a*, *b*]

The two-point boundary value problem (1.2) on a general interval $[a,b]$ is now treated.

3.1. *Description of the Sinc-collocation method*

The basic idea of the Sinc-collocation method on the interval $[a,b]$ is to transform the problem, with a properly selected variable transformation, to the interval $(-\infty,\infty)$ and then to solve the transformed problem on $[a,b]$ by the Sinc-collocation method on the interval $(-\infty,\infty)$ described in Section 2.2. A more formal description of the Sinc-collocation method on the interval $[a,b]$ follows.

Step 1: With an adroitly selected variable transformation $x = \psi(\xi)$ such that

$$\psi : (-\infty,\infty) \to (a,b),$$

transform problem (1.2) to that on the interval $(-\infty,\infty)$; the resulting problem is given by

$$\tilde{y}''(\xi) + \tilde{\mu}(\xi)\tilde{y}'(\xi) + \tilde{v}(\xi)\tilde{y}(\xi) = \tilde{\sigma}(\xi), \quad \xi \in (-\infty,\infty),$$

$$\lim_{\xi\to\pm\infty} \tilde{y}(\xi) = 0, \tag{3.1}$$

where $\tilde{y}(\xi) = y(\psi(\xi))$, and

$$\tilde{\mu}(\xi) \equiv \psi'(\xi)\mu(\psi(\xi)) - \psi''(\xi)/\psi'(\xi),$$

$$\tilde{v}(\xi) \equiv (\psi'(\xi))^2 v(\psi(\xi)),$$

$$\tilde{\sigma}(\xi) \equiv (\psi'(\xi))^2 \sigma(\psi(\xi)).$$

Step 2: Apply the Sinc-collocation method on the interval $(-\infty, \infty)$ to the transformed problem (3.1).

Remark 3.1. In the self-adjoint case of (1.2):

$$y''(x) + v(x)y(x) = \sigma(x), \quad x \in (a,b),$$
$$y(a) = y(b) = 0, \tag{3.2}$$

the transformed problem (3.1) is not self-adjoint, which gives rise to a non-symmetric system of linear equations. To circumvent this unwelcome situation, a symmetrization technique is developed [2]. By considering $Y(\xi) \equiv (\psi'(\xi))^{-1/2} y(\psi(\xi))$ instead of $y(\psi(\xi))$ in step 1, we can get a self-adjoint boundary problem:

$$Y''(\xi) + N(\xi)Y(\xi) = S(\xi), \quad \xi \in (-\infty, \infty),$$
$$\lim_{\xi \to \pm\infty} Y(\xi) = 0, \tag{3.3}$$

where

$$N(\xi) \equiv (\psi'(\xi))^{1/2} \frac{\mathrm{d}}{\mathrm{d}\xi} \left\{ \frac{1}{\psi'(\xi)} \frac{\mathrm{d}}{\mathrm{d}\xi} ((\psi'(\xi))^{1/2}) \right\} + (\psi'(\xi))^2 v(\psi(\xi)),$$

$$S(\xi) \equiv (\psi'(\xi))^{3/2} \sigma(\psi(\xi)).$$

This yields a symmetric system of linear equations.

Remark 3.2. Closely related to the above remark, it should be mentioned that Ng has recently developed fast iterative solvers of the symmetric system of linear equations obtained by the above symmetrization of the Sinc-collocation method [8]. He employs the preconditioned conjugate gradient method with banded matrices as preconditioners.

3.2. Convergence analysis

For the convergence of the Sinc-collocation method on the interval $[a, b]$, combined with a variable transformation, the following two theorems immediately follow from Theorems 2.2 and 2.3.

Theorem 3.1 (Bialecki [1], Stenger [9]). *Assume that problem* (1.2) *has a unique solution* $y(x)$, *and that* $y(x)$ *is analytic on the interval* (a,b). *Furthermore, assume that there exists a* *variable transformation* $z = \psi(\zeta)$ *such that the transformed problem satisfies assumptions* (1)–(5) *in Theorem* 2.2 *with some* α, β *and* d. *Then we have*

$$\sup_{a<x<b} |y(x) - \tilde{y}_n(\psi^{-1}(x))| \leqslant cN^{5/2} \exp[-(\pi\, \mathrm{d}\beta N)^{1/2}].$$

where $\tilde{y}_n(\xi)$ *is the approximate solution to the transformed problem obtained by the Sinc-collocation method on the interval* $(-\infty, \infty)$ *with mesh size*

$$h = \left(\frac{\pi d}{\beta N}\right)^{1/2}.$$

Theorem 3.2. *Assume that problem* (1.2) *has a unique solution* $y(x)$, *and that* $y(x)$ *is analytic on the interval* (a,b). *Furthermore, assume that there exists a variable transformation* $z = \psi(\zeta)$ *such that the transformed problem satisfies assumptions* (1)–(6) *in Theorem* 2.3 *with some* $A, B, \alpha, \beta, \gamma$ *and* d. *Then we have*

$$\sup_{a<x<b} |y(x) - \tilde{y}_n(\psi^{-1}(x))| \leqslant c(\log N) N^{B/\gamma + 3/2} \exp\left[\frac{-\pi\, \mathrm{d}\gamma N}{\log(\pi \mathrm{d}\gamma N/\beta)}\right],$$

where $\tilde{y}_n(\xi)$ *is the approximate solution to the transformed problem obtained by the Sinc-collocation method on the interval* $(-\infty, \infty)$ *with mesh size*

$$h = \frac{\log(\pi\, \mathrm{d}\gamma N/\beta)}{\gamma N}.$$

The Sinc-collocation method based on Theorem 3.1 is what we call the standard Sinc-collocation method, which is dealt with in textbooks on Sinc methods [5,9]. On the other hand, the variable transformation assumed in Theorem 3.2 is what we call a DE transformation, and Theorem 3.2 affords a foundation for the Sinc-collocation method incorporated with the DE transformation.

4. Numerical examples

We show two examples on the interval [0, 1] to illustrate how the Sinc-collocation method is enhanced when incorporated with the DE transformation technique. The formulas required for the assembly of the transformed differential equation (3.1) are summarized in Table 1.

Example 4.1. We first consider the problem

$$y''(x) - \frac{3}{4x^2} y(x) = -3\sqrt{x}, \quad x \in (0,1),$$
$$y(0) = y(1) = 0, \tag{4.1}$$

which appears in [4, Example 3.1]. This problem has a regular singular point at $x = 0$, and has the exact solution $y(x) = x^{(3/2)}(1-x)$. We apply both the standard Sinc-collocation method and the Sinc-collocation method incorporated with DE transformation.

Table 1
Components for the transformed differential equation (3.1):

$$\tilde{y}''(\xi) + \left\{ \psi'(\xi)\mu(\psi(\xi)) - \frac{\psi''(\xi)}{\psi'(\xi)} \right\} \tilde{y}'(\xi) + (\psi'(\xi))^2 v(\psi(\xi))\tilde{y}(\xi) = (\psi'(\xi))^2 \sigma(\psi(\xi))$$

i	$\psi_i\,(\xi)$	$\psi_i'\,(\xi)$	$-\dfrac{\psi_i''(\xi)}{\psi_i'(\xi)}$
1	$\frac{1}{2}\tanh\frac{\xi}{2} + \frac{1}{2}$	$\psi_1(\xi)(1-\psi_1(\xi))$	$2\psi_1(\xi) - 1$
2	$\frac{1}{2}\tanh\left(\frac{\pi}{2}\sinh\xi\right) + \frac{1}{2}$	$(\pi\cosh\xi)\psi_2(\xi)(1-\psi_2(\xi))$	$(\pi\cosh\xi)(2\psi_2(\xi)-1) - \tanh\xi$

The standard Sinc-collocation method: With some calculation we can verify that this problem meets all the assumptions in Theorem 3.1. In fact, with the variable transformation:

$$x = \psi_1(\xi) \equiv \frac{1}{2}\tanh\frac{\xi}{2} + \frac{1}{2}, \tag{4.2}$$

the transformed problem, which is given by

$$\begin{aligned} &\tilde{y}''(\xi) + (2\psi_1(\xi) - 1)\tilde{y}'(\xi) - \tfrac{3}{4}(1-\psi_1(\xi))^2\tilde{y}(\xi) \\ &\quad = -3(\psi_1(\xi))^{5/2}(1-\psi_1(\xi))^2, \quad \xi \in (-\infty, \infty), \\ &\lim_{\xi\to\pm\infty} \tilde{y}(\xi) = 0, \end{aligned}$$

satisfies all the assumptions in Theorem 2.2 with $\beta = 1$ and $d < \pi$. (We here omit the value of α, for it is irrelevant to both the order of magnitude of the error estimate and the selection of the mesh size h. For the same reason we will do so in the following.) Using $\beta = 1$, $d = \pi/2$, we apply the Sinc-collocation method on the interval $(-\infty, \infty)$ to the transformed problem. Fig. 2 shows the error in the approximate solution, which is denoted by Standard-Sinc. (We estimate the error $\sup_{0<x<1} |y(x) - \tilde{y}_n(\psi^{-1}(x))|$ by computing the absolute error between $y(x)$ and $\tilde{y}_n(\psi^{-1}(x))$ at 1000 equally spaced points in $(0,1)$.) In Fig. 2 we observe that the error behaves like $\exp(-\kappa\sqrt{n})$, as expected from Theorem 3.1.

The Sinc-collocation method with the DE transformation: With elaborate calculation we can also show that problem (4.1) meets all the assumptions in Theorem 3.2. In fact, with the variable transformation:

$$x = \psi_2\,(\xi) \equiv \frac{1}{2}\tanh\left(\frac{\pi}{2}\sinh\xi\right) + \frac{1}{2} \tag{4.3}$$

the transformed problem, which is given by

$$\begin{aligned} &\tilde{y}''(\xi) + \{(\pi\cosh\xi)(2\psi_2(\xi) - 1) - \tanh\xi\}\tilde{y}'(\xi) - \tfrac{3}{4}(\pi\cosh\xi)^2(1-\psi_2(\xi))^2\tilde{y}(\xi) \\ &\quad = -3(\pi\cosh\xi)^2(\psi_2(\xi))^{5/2}(1-\psi_2(\xi))^2, \qquad \xi \in (-\infty, \infty), \\ &\lim_{\xi\to\pm\infty} \tilde{y}(\xi) = 0, \end{aligned}$$

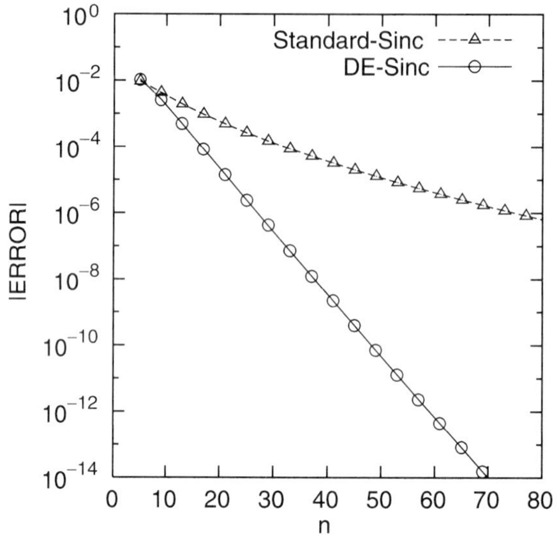

Fig. 2. Errors in the Sinc-collocation solutions of the problem (4.1).

satisfies all the assumptions in Theorem 2.3 with $B=2, \beta=\pi/2, \gamma=1$ and $d < \pi/2$. Using $\beta=\pi/2, \gamma=1, d=\pi/4$, we apply the Sinc-collocation method on the interval $(-\infty,\infty)$ to the transformed problem. The error in the approximate solution is denoted by DE-Sinc in Fig. 2. We there observe that the error converges to zero like $\exp(-\kappa' n/\log n)$, as expected from Theorem 3.2.

Example 4.2. Consider the problem

$$\begin{aligned} &y''(x) + \frac{1}{6x} y'(x) - \frac{1}{x^2} y(x) = -\frac{19}{6}\sqrt{x}, \quad x \in (0,1), \\ &y(0) = y(1) = 0, \end{aligned} \tag{4.4}$$

which appears in [5, Chapter 4]. This problem has a regular singular point at $x=0$, and has the exact solution $y(x)=x^{3/2}(1-x)$, which is the same as in Example 4.1. The exact solution is approximated by both the standard Sinc-collocation method and the Sinc-collocation method incorporated with DE transformation.

The standard Sinc-*collocation method*: It is proved that the problem transformed by the variable transformation (4.2), which is given by

$$\begin{aligned} &\tilde{y}''(\xi) + \frac{11\psi_1(\xi) - 5}{6} \tilde{y}'(\xi) - (1-\psi_1(\xi))^2 \tilde{y}(\xi) \\ &\quad = -\frac{19}{6} (\psi_1(\xi))^{5/2} (1-\psi_1(\xi))^2, \quad \xi \in (-\infty,\infty), \\ &\lim_{\xi \to \pm\infty} \tilde{y}(\xi) = 0, \end{aligned}$$

satisfies all the assumptions except (3) in Theorem 2.2 with $\beta = 1$ and $d < \pi$. The Sinc-collocation method on the interval $(-\infty,\infty)$ is applied to the transformed problem, where the parameters β

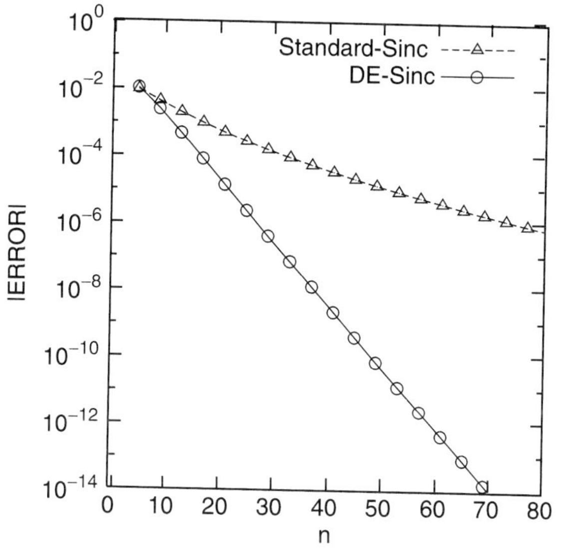

Fig. 3. Errors in the Sinc-collocation solutions of the problem (4.4).

and d are chosen as $\beta = 1$, $d = \pi/2$. The error in the approximate solution is shown in Fig. 3 by Standard-Sinc. It is observed that the error converges to zero at the rate of $\exp(-\kappa\sqrt{n})$, though the transformed problem does not meet all the assumptions in Theorem 2.2.

The Sinc-collocation method with the DE transformation: It is shown that the problem transformed by the variable transformation (4.3), which is given by

$$\tilde{y}''(\xi) + \left\{(\pi\cosh\xi)\left(\frac{11\psi_2(\xi) - 5}{6}\right) - \tanh\xi\right\}\tilde{y}'(\xi) - (\pi\cosh\xi)^2(1-\psi_2(\xi))^2\tilde{y}(\xi)$$

$$= -\frac{19}{6}(\pi\cosh\xi)^2(\psi_2(\xi))^{5/2}(1-\psi_2(\xi))^2, \quad \xi \in (-\infty, \infty),$$

$$\lim_{\xi\to\pm\infty} \tilde{y}(\xi) = 0,$$

satisfies all the assumptions except (4) in Theorem 2.3 with $B = 2$, $\beta = \pi/2$, $\gamma = 1$ and $d < \pi/2$. With $\beta = \pi/2$, $\gamma = 1$, $d = \pi/4$, the Sinc-collocation method on the interval $(-\infty, \infty)$ is applied to the transformed problem. The error in the approximate solution is shown in Fig. 3 by DE-Sinc. It is observed that the convergence rate of $\exp(-\kappa' n/\log n)$ is achieved, though the transformed problem does not meet all the assumptions in Theorem 2.3.

5. Final comments

A variety of numerical experiments, not included in the present paper, suggest that assumption (3) in Theorem 2.2 and assumption (4) in Theorem 2.3 are not mandatory. To establish the convergence theorems without these assumptions is an important future work.

The Sinc-collocation method incorporated with the DE transformation proposed in this paper can be extended not only to boundary-value problems for higher-order ordinary differential equations, but also to boundary-value problems for partial differential equations. This is left for future investigations.

Acknowledgements

The author is grateful to the referees for their valuable comments.

References

[1] B. Bialecki, Sinc-collocation methods for two-point boundary value problems, IMA J. Numer. Anal. 11 (1991) 357–375.
[2] N. Eggert, M. Jarrat, J. Lund, Sinc function computation of the eigenvalues of Sturm–Liouville problems, J. Comput. Phys. 69 (1987) 209–229.
[3] M.A. Kowalski, K.A. Sikorski, F. Stenger, Selected Topics in Approximation and Computation, Oxford University Press, Oxford, 1995.
[4] J. Lund, Symmetrization of the Sinc–Galerkin method for boundary value problems, Math. Comput. 47 (1986) 571–588.
[5] J. Lund, K.L. Bowers, Sinc Methods for Quadrature and Differential Equations, SIAM, Philadelphia, PA, 1992.
[6] M. Mori, Quadrature formulas obtained by variable transformation and DE rule, J. Comput. Appl. Math. 12& 13 (1985) 119–130.
[7] M. Mori, M. Sugihara, The double exponential transformation in numerical analysis, in: Numerical Analysis in the 20th Century Vol. V, W. Gautschi, F. Marcellán, L. Reichel (Eds.), Quadrature and Orthogonal Polynomials, J. Comput. Appl. Math. 127 (2001) 287–296.
[8] M.K. Ng, Fast iterative methods for symmetric Sinc–Galerkin systems, IMA J. Numer. Anal. 19 (1999) 357–373.
[9] F. Stenger, Numerical Methods Based on Sinc and Analytic Functions, Springer, New York, 1993.
[10] F. Stenger, Summary of Sinc numerical methods, in: Numerical Analysis in the 20th Century, Vol. I, L. Wuytack, J. Wimp (Eds.), Approximation Theory, J. Comput. Appl. Math. 121 (2000) 379–420.
[11] M. Sugihara, Optimality of the double exponential formula—functional analysis approach, Numer. Math. 75 (1997) 379–395.
[12] M. Sugihara, Near optimality of the Sinc approximation, Math. Comput., to appear.
[13] H. Takahasi, M. Mori, Double exponential formulas for numerical integration, Publ. Res. Inst. Math. Sci. 9 (1974) 721–741.

Journal of Computational and Applied Mathematics 149 (2002) 251–265

JOURNAL OF COMPUTATIONAL AND APPLIED MATHEMATICS

www.elsevier.com/locate/cam

Efficient and reliable iterative methods for linear systems

Henk A. van der Vorst

Department of Mathematics, Utrecht University, P.O. Box 80010, NL-3508 TA, Utrecht, The Netherlands

Received 27 October 2001; received in revised form 4 January 2002

Abstract

The approximate solutions in standard iteration methods for linear systems $Ax = b$, with A an n by n nonsingular matrix, form a subspace. In this subspace, one may try to construct better approximations for the solution x. This is the idea behind Krylov subspace methods. It has led to very powerful and efficient methods such as conjugate gradients, GMRES, and Bi-CGSTAB. We will give an overview of these methods and we will discuss some relevant properties from the user's perspective view.

The convergence of Krylov subspace methods depends strongly on the eigenvalue distribution of A, and on the angles between eigenvectors of A. Preconditioning is a popular technique to obtain a better behaved linear system. We will briefly discuss some modern developments in preconditioning, in particular parallel preconditioners will be highlighted: reordering techniques for incomplete decompositions, domain decomposition approaches, and sparsified Schur complements.

Keywords: Iterative methods; Krylov methods; Conjugate gradients; GMRES; Bi-CGSTAB; Preconditioning; Domain decomposition; Incomplete Cholesky

1. Krylov subspace methods

Krylov methods started in the early 1950s with the introduction of the conjugate gradients methods. These methods are designed to construct approximate solutions in the so-called Krylov subspace. Given a linear system $Ax = b$, with a large, usually sparse, nonsingular n by n matrix, then the standard Richardson iteration

$$x_k = (I - A)x_{k-1} + b$$

generates approximate solutions in shifted Krylov subspaces

$$x_0 + K^k(A; r_0) = x_0 + \{r_0, Ar_0, \ldots, A^{k-1}r_0\}$$

with $r_0 = b - Ax_0$, for some given initial vector x_0.

E-mail address: vorst@math.uu.nl (H.A. van der Vorst).

0377-0427/02/$ - see front matter
PII: S0377-0427(02)00534-4

With relatively little additional work, we can often construct much better approximate solutions from these Krylov subspaces, which leads to Krylov subspace projection methods. These methods fall in three different classes:

1. The *Ritz–Galerkin approach*: Construct the x_k for which the residual is orthogonal to the current subspace:

   $b - Ax_k \perp K^k(A; r_0)$.

   This approach leads to such popular and well-known methods as conjugate gradients and FOM.
2. The *minimum residual approach*: Identify the x_k for which the Euclidean norm $\|b - Ax_k\|_2$ is minimal over $K^k(A; r_0)$.
   The minimum residual approach leads to methods such as GMRES and MINRES.
3. The *Petrov–Galerkin approach*: Find an x_k so that the residual $b - Ax_k$ is orthogonal to some other suitable k-dimensional subspace.
   If we select the k-dimensional subspace in the third approach as $K^k(A^{\mathrm{T}}; s_0)$ for some s_0, then we obtain the Bi-CG, and QMR methods. With $AK^k(A; r_0)$ for the subspace, we obtain SYMMLQ and GMERR.

More recently, hybrids of the three approaches have been proposed, like CGS, TFQMR, FGMRES, GMRESR, Bi-CGSTAB, and BiCGSTAB(ℓ). The first of the Bi-CG hybrids was CGS, published in 1989 by Sonneveld [36], and followed by Bi-CGSTAB, by van der Vorst in 1992 [41], and others. The hybrid variants of GMRES are flexible GMRES and GMRESR. In this approach GMRES is combined with some other iteration scheme, for instance GMRES itself or Bi-CGSTAB.

2. The Krylov subspace solvers

In this section we will review three of the most popular iterative methods, namely conjugate gradients (CG) [19], GMRES [34], and Bi-CGSTAB [41].

2.1. The conjugate gradient method

The conjugate gradients method was initially seen as a direct method, because of its property to return the solution within n steps (in exact arithmetic). It took a few years to realize that it was more fruitful to consider the conjugate gradients method as a truly iterative method for large classes of linear systems and Reid, in 1972 [32], was one of the first to point in this direction. Meanwhile, analysis had shown already that the convergence of this method was dictated by a factor involving the ratio of the largest and smallest eigenvalue of A, independent of the dimension of the matrix. About the same time, it was recognized that good approximations K for A could be constructed with the property that the eigenvalues of $K^{-1}A$ were clustered around 1, which implied that the ratio of these eigenvalues was moderate and hence led to fast convergence of conjugate gradients, when applied to $K^{-1}Ax = K^{-1}b$ (under the condition that K is also symmetric positive definite). This process is referred to as *preconditioned conjugate gradients*. The algorithm can be described by the scheme in Fig. 1.

```
Compute r_0 = b − Ax_0 for some initial guess x_0
for i = 1, 2, ....
    Solve z_{i−1} from K z_{i−1} = r_{i−1}
    ρ_{i−1} = r*_{i−1} z_{i−1}
    if i = 1
        p_1 = z_0
    else
        β_{i−1} = ρ_{i−1} / ρ_{i−2};
        p_i = z_{i−1} + β_{i−1} p_{i−1}
    endif
    q_i = A p_i
    α_i = ρ_{i−1} / (p*_i q_i)
    x_i = x_{i−1} + α_i p_i
    r_i = r_{i−1} − α_i q_i
    check convergence; continue if necessary
end;
```

Fig. 1. The CG algorithm.

Table 1

Method	Number of iterations
Gauss–Seidel	208 000
Block SOR	765
ICCG	25

In our schemes, x^*y denotes the innerproduct of two vectors x and y (complex conjugate if the system is complex).

A physicist from Livermore, Kershaw, was one of the first to experiment with the conjugate gradient method, with incomplete Cholesky factorization of A as a preconditioner, for tough problems related to fusion problems [20]. We quote some iteration numbers for the basic Gauss–Seidel iteration [43], the accelerated version SOR (in fact a slightly faster variant: Block SOR [43]), and conjugated gradients preconditioned with incomplete Cholesky (also known as ICCG [25]). The iteration numbers were necessary to reduce the norm of the initial residual by a factor 10^{-6} (Table 1):

This gives an impression of the sometimes gigantic improvements that could be obtained by the (preconditioned) conjugate gradients and results like these greatly contributed to the current popularity of the method. It also motivated the search for other powerful Krylov subspace methods for more general systems of equations.

2.2. GMRES

The currently de facto standard for unsymmetric systems is the GMRES method, proposed in 1986 by Saad and Schultz [34]. In this method the x_i, in the Krylov subspace of dimension i, is constructed for which the norm $\|b - Ax_i\|_2$ is minimal. This builds on an algorithm proposed by Arnoldi [1], for the construction of an orthonormal basis for the Krylov subspace for unsymmetric A. The price to pay for unsymmetry is that one has to store a full orthogonal basis for the Krylov

```
r = b − Ax^(0), for a given initial guess x = x^(0)
for j = 1, 2, ....
    β = ||r||_2, v^1 = r/β, b̂ = βe^1
    for i = 1, 2, ..., m
        w = Av^i
        for k = 1, ..., i
            h_{k,i} = (v^k)^* w; w = w − h_{k,i} v^k
        h_{i+1,i} = ||w||_2; v^{i+1} = w/h_{i+1,i}
        old Givens rot's on new column:
        for k = 2, .., i
            γ = h_{k−1,i}
            h_{k−1,i} = c_{k−1} γ + s_{k−1} h_{k,i}
            h_{k,i} = −s_{k−1} γ + c_{k−1} h_{k,i}
        new Givens rot for elim. of h_{i+1,i}
        δ = sqrt(h_{i,i}^2 + h_{i+1,i}^2), c_i = h_{i,i}/δ, s_i = h_{i+1,i}/δ
        h_{i,i} = c_i h_{i,i} + s_i h_{i+1,i}
        b̂_{i+1} = −s_i b̂_i, b̂_i = c_i * b̂_i
        ρ = |b̂_{i+1}| (= ||b − Ax^((j−1)m+i)||_2))
        if ρ is small enough then
            (n_r = i, goto SOL)
    end for i
    Solution of H_{n_r,n_r} y = b̂
    n_r = m, y_{n_r} = b̂_{n_r}/h_{n_r,n_r}
SOL: for k = n_r − 1, ..., 1
        y_k = (b̂_k − Σ_{i=k+1}^{n_r} h_{k,i} y_i)/h_{k,k}
    x = x + Σ_{i=1}^{n_r} y_i v^i, if ρ small enough quit
    r = b − Ax
```

Fig. 2. GMRES(m).

subspace, which means the more iterations are done the more basis vectors have to be stored. Also the work per iteration increases linearly. This makes the method only attractive if it converges really fast. For many practical problems GMRES takes indeed a few tens of iterations, for many other problems it may go in the hundreds which makes full GMRES unfeasible. We present in Fig. 2 a version of GMRES in which a restart takes place after every m iterations, in order to keep the memory requirements and the work per iteration limited. The application for a preconditioned system $K^{-1}Ax = K^{-1}b$ is straightforward.

If we do not apply the Givens rotations to the matrix H, that is, if we leave the matrix as it is after the first k-loop, then the leading k by k part of the small matrix H may help to obtain useful information on the big matrix A. Typically, if the matrix A is not too nonnormal, the eigenvalues of $H_{k,k}$ (the Ritz values of A with respect to the current Krylov subspace) approximate eigenvalues of A. In general, the eigenvalues close to the convex hull of the spectrum of A are first well approximated. By inspecting the approximated eigenvalues, one may get an idea of the spectrum of A. In particular, when the algorithm is applied with the preconditioned matrix $K^{-1}A$, one may see the effect of preconditioning on A. One would like to see some clustering of the eigenvalues around the value 1. If the approximated eigenvalues indicate that the hull of the spectrum of the

Compute $r^{(0)} = b - Ax(0)$ for some initial guess $x^{(0)}$
Choose $\widehat{r}^{(0)}$, for example $\widehat{r}^{(0)} = r^{(0)}$
for $i = 1, 2, ...$
  $\rho_{i-1} = (\widehat{r}^{(0)})^* r^{(i-1)}$
  if $\rho_{i-1} = 0$ **method fails**
  if $i = 1$
    $p^{(i)} = r^{(i-1)}$
  else
    $\beta_{i-1} = (\rho_{i-1}/\rho_{i-2})(\alpha_{i-1}/\omega_{i-1})$
    $p^{(i)} = r^{(i-1)} + \beta_{i-1}(p^{(i-1)} - \omega_{i-1}v^{(i-1)})$
  endif
  Solve $\widehat{p}$ from $K\widehat{p} = p^{(i)}$
  $v^{(i)} = A\widehat{p}$
  $\alpha_i = \rho_{i-1}/(\widehat{r}^{(0)})^* v^{(i)}$
  $s = r^{(i-1)} - \alpha_i v^{(i)}$
  if $\|s\|$ small enough then
    $x^{(i)} = x^{(i-1)} + \alpha_i \widehat{p}$ and stop
  Solve z from $Kz = s$
  $t = Az$
  $\omega_i = s^* t / t^* t$
  $x^{(i)} = x^{(i-1)} + \alpha_i \widehat{p} + \omega_i z$
  if $x^{(i)}$ is accurate enough **then quit**
  $r^{(i)} = s - \omega_i t$
  for continuation it is necessary that $\omega_i \neq 0$
end

Fig. 3. The Bi-CGSTAB algorithm.

(preconditioned) matrix A encloses the origin, then one should have very moderate expectations on the speed of convergence of any Krylov solver. Krylov solvers can be very slow for indefinite problems.

2.3. Bi-CGSTAB

The costs per iteration of GMRES has also led to a search for cheaper near-optimal methods. A famous result by Faber and Manteuffel [13] showed that it is in general not possible to construct optimal solutions in the Krylov subspace for unsymmetric A by short recurrences, as in the conjugate gradients method. The generalization of conjugate gradients for unsymmetric systems, Bi-CG, displays often a quite irregular convergence behavior. This method has the other disadvantage that per iteration step an operation with A^T is required and this additional operation does not lead to a further reduction of the residual. Sonneveld, in the mid-1980s, recognized that the A^T operation could be used for a further reduction of the residual, by a minor modification to the Bi-CG scheme, almost without additional computational costs: the CGS method. The CGS method was often faster, but often also significantly more irregular which led to a loss in precision. In 1992, van der Vorst [41] showed that Bi-CG could also be combined, at almost no additional cost, with minimal residual steps (comparable to GMRES(1) steps). This resulted in the popular Bi-CGSTAB algorithm, schematically represented in Fig. 3 (for the solution of $Ax = b$ with preconditioner K).

3. Reliable updating

For general systems, some of the popular methods often show an irregular type of convergence behavior, which deteriorates the accuracy of the method. By very irregular convergence we refer to the situation where successive residual vectors in the iterative process differ by orders of magnitude in norm, and some of these residuals may even be much larger in norm than the starting residual.

We say that an algorithm is *accurate* for a certain problem if the *updated residual* r_j and the *true residual* $b - Ax_j$ are of comparable size for the j's of interest. In most iteration schemes based on short term recurrences, such as Bi-CG, CGS, and Bi-CGSTAB, the approximation for the solution and the corresponding residual are updated independently, for instance:

$$x_{j+1} = x_j + w_{j+1},$$

$$r_{j+1} = r_j - Aw_{j+1}.$$

In exact arithmetic we have that $r_{j+1} = b - Ax_{j+1}$, but in finite precision arithmetic, there will at least be a discrepancy between the two updated quantities due to the multiplication by A.

It is shown in [35] how far the updated residual r_j may drift away from the more informative true residual $b - Ax_j$:

$$|\|r_j\| - \|b - Ax_j\|| \leqslant 2jn_A\bar{\xi}\|A\|\|A^{-1}\| \max_k \|r_k\|,$$

where n_A denotes the average number of non-zero elements per row of A, and $\bar{\xi}$ is the relative machine precision. Except for the factor j, the upper bound appears to be rather sharp. We see that an approximation with a large approximation error (and hence a large residual) may lead to inaccurate results in the remaining iteration process.

The vector $b - Ax_j$ is often not computed at each iteration since that would be too expensive. It turns out that the iterative schemes can be adapted so that they deliver, at relatively minor additional costs, updated residuals that are sufficiently close to the true residuals. This is done by replacing the updated vector r_j by $b - Ax_j$ at selected places in the iteration process, so that the recurrence relations are not too much perturbed. The resulting *reliable updating* process is robust, efficient, and easy to implement. For more details we refer to [42].

4. Preconditioning

It is difficult to make a general statement on how fast these Krylov methods converge. While they converge certainly much faster than the classical iteration schemes, and convergence takes place for a much wider class of matrices, there are still many practical systems that cannot be solved in a satisfactory way. Much depends on whether one is able to define a nearby matrix K that can be used as a preconditioner. Recent research is more oriented in that direction than in trying to further accelerate the Krylov subspace methods (although some improvements may be expected for these methods as well). The construction of effective and efficient preconditioners is largely problem dependent. A preconditioner is considered as effective if the number of iterations steps of the preconditioned Krylov subspace method is in the order of 100 or less.

```
ILU for an n by n matrix A (cf [2]):
for k = 1, 2, ..., n − 1
    d = 1/a_{k,k}
    for i = k + 1, k + 2, ..., n
      if (i, k) ∈ S
        e = d a_{i,k}; a_{i,k} = e
        for j = k + 1, ..., n
          if (i, j) ∈ S and (k, j) ∈ S
            a_{i,j} = a_{i,j} − e a_{k,j}
          end if
        end j
      end if
    end i
end k
```

Fig. 4. ILU for a general matrix A.

Many different preconditioners have been suggested over the years, each of these preconditioners is more or less successful for restricted classes of problems. Among all these preconditioners the incomplete LU factorizations [25,6] are the most popular ones.

Standard Gaussian elimination is equivalent to factoring the matrix A as $A = LU$, where L is lower triangular and U is upper triangular. In actual computations these factors are explicitly constructed. The main problem in sparse matrix computations is that the factors of A are often a good deal less sparse than A, which makes solution expensive. The basic idea in the point ILU preconditioner is to modify Gaussian elimination to allow fill-ins at only a restricted set of positions in the LU factors. Let the allowable fill-in positions be given by the index set S, i.e.,

$$\begin{aligned} l_{i,j} &= 0 \quad \text{if} \quad j > i \quad \text{or} \quad (i,j) \notin S, \\ u_{i,j} &= 0 \quad \text{if} \quad i > j \quad \text{or} \quad (i,j) \notin S. \end{aligned} \tag{1}$$

A commonly used strategy is to define S by

$$S = \{(i,j) \mid a_{i,j} \neq 0\}. \tag{2}$$

That is, the only nonzeros allowed in the LU factors are those for which the corresponding entries in A are nonzero. Let the preconditioner M be defined by the product of the resulting LU factors, i.e., $M = LU$. For M to be a good preconditioner, it must be a good approximation to A in some measure. A typical strategy is to require the entries of M to match those of A on the set S:

$$m_{i,j} = a_{i,j} \quad \text{if } (i,j) \in S. \tag{3}$$

Even though conditions (1) and (3) together are sufficient (for certain classes of matrices) to determine the nonzero entries of L and U directly, it is more natural and simpler to compute these entries based on a simple modification of the Gaussian elimination algorithm; see Fig. 4. The main difference from the usual Gaussian elimination algorithm is in the inner-most j-loop where an update to $a_{i,j}$ is computed only if it is allowed by the constraint set S.

After the completion of the algorithm, the incomplete LU factors are stored in the corresponding lower and upper triangular parts of the array A. It can be shown that the computed LU factors satisfy (3). These LU factors can be used in subsequent applications of the preconditioner; there is no need to explicitly form the product $M = LU$.

Attempts have been made to improve the ILU factorizations, for instance by including more fill [26], or by modifying the diagonal of the ILU factorization in order to force rowsum constraints [16,2,30,38,12], or by changing the ordering of the matrix [40]. A collection of experiments with respect to the effects of ordering is contained in [10]. More recently, it was discovered that a multigrid-inspired ordering can be very effective for discretized diffusion–convection equations, leading in some cases to almost grid-independent speeds of convergence [37,5], see also [7]. In these publications the ordering strategy is combined with a drop-tolerance strategy for discarding small enough fill elements.

4.1. Reordering the unknowns

A standard trick for exploiting parallelism is to select all unknowns that have no direct relationship with each other and to number them first. For the 5-point finite-difference discretization over rectangular grids, this approach is known as a *red–black ordering*. For elliptic PDEs, this leads to very parallel preconditioners. The computational performance of the preconditioning step is as high as the performance of the matrix–vector product. However, changing the order of the unknowns leads in general to a different preconditioner. Duff and Meurant [10] report on experiments that show that most reordering schemes (for example, the red–black ordering) lead to a considerable increase in iteration steps (and hence in computing time) compared with the standard lexicographical ordering. For the red–black ordering associated with the discretized Poisson equation, it can be shown that the condition number of the preconditioned system is only about one-quarter that of the unpreconditioned system for ILU, MILU and SSOR [21]. Since the speed of convergence of conjugate gradients is proportional to the square root of the condition number, this would imply asymptotically a reduction in the number of iterations by a factor of 2. However, because the costs per iteration for the preconditioned process is almost doubled, we may not expect significant reductions in CPU-time as a result of this red–black reordered ILU preconditioner.

Another approach, suggested by Meurant [27], exploits the idea of the two-sided (or twisted) Gaussian elimination procedure for tridiagonal matrices. This is generalized for the incomplete factorization. Van der Vorst [39] has shown how this procedure can be done in a nested way. For 3D finite-difference problems, twisting can be used for each dimension, which gives an increase in parallelism by a factor of two per dimension. This leads, without further computational overhead, to incomplete decompositions, as well as triangular solves, that can be done in eight parallel parts (2 in each dimension). For a discussion of these techniques see [9].

Meurant [28] reports on timing results obtained on a CRAY Y-MP/832, using an incomplete repeated twisted block factorization for two-dimensional problems. For this approach for preconditioned CG, Meurant reports a speedup of nearly 6 on an 8-processor CRAY Y-MP. This speedup has been measured relative to the same repeated twisted factorization process executed on a single processor. Meurant also reports an increase in the number of iteration steps as a result of this repeated twisting. This increase implies that the effective speedup with respect to the best nonparallel code is only about 4.

A more sophisticated approach that combines ideas from twisting, domain decomposition with overlap, and reordering, was proposed in [22–24]. We will explain this idea for the special situation

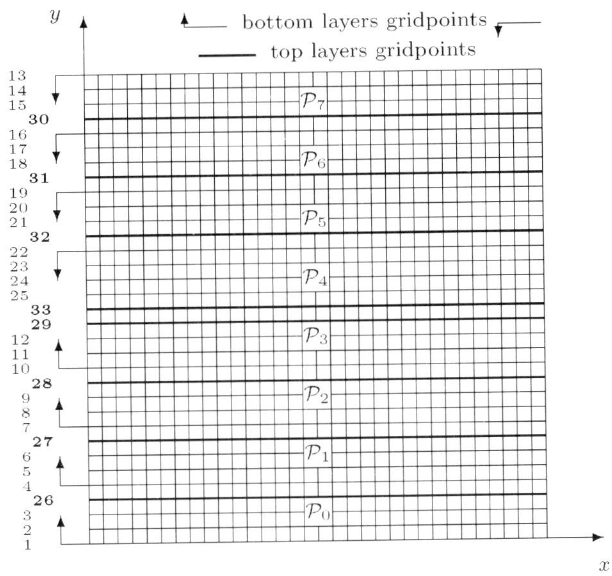

Fig. 5. Decomposition of the grid into stripes, and assignment of subdomains to processors for $p=8$. Arrows indicate the progressing direction of the line numbering per subdomain. Numbers along the y-axis give an example of global (line) ordering, which satisfy all the required conditions. Within each horizontal line, gridpoints are ordered lexicographically.

of a discretized second order elliptic PDE over a rectangular domain. The discretization has been carried out with the standard 5-point central difference stencil which, over a rectangular grid with lexicographical ordering leads to the familiar block matrix with 5 nonzero diagonals.

The first step is to split the domain in blocks, as in domain decomposition methods, and to order the unknowns lexicographically per block. This has been indicated, for the case of 8 horizontal blocks, in Fig. 5. Per block we start counting from one side ("the bottom layer"); the points on the last line ("the top layer") are ordered after all subdomains, as is indicated in Fig. 6. For instance, the lines 1, 2, 3, and 26, all belong to the block stored with processor $\mathscr{P}_0$, but in the matrix interpretation the first 3 lines are ordered first and line 26 appears in the matrix only after all other "interior" lines. This means that the matrix has the nonzero structure (we give only a relevant part of the matrix) as in Fig. 6. Note that we have already introduced another element in our ordering, namely the idea of twisting: the lines of the subdomains are ordered from bottom to top and from top to bottom in Fig. 5.

Now imagine what happens if we carry out an incomplete LU factorization with zero fill. That would create level-1 fill in the error matrix. Note that, in particular, we would introduce fill in the subblock of the matrix that connects line 26 with line 5, and note also that we would not have seen this level-1 fill if we would have selected all points lexicographically.

This means that if we want the block ordering to be at least as effective as the standard ordering, we have to remove this additional fill. This can be interpreted as permitting level-1 fill in a small overlap, and this is the reason for the name "*pseudo-overlap*" for this way of ordering. It is obvious how to generalize this idea for more arbitrary matrices: one compares the new ordering with the standard given one and one includes the possibly additional level-1 fill in the preconditioner. The idea can also be easily applied to preconditioners with a higher level fill.

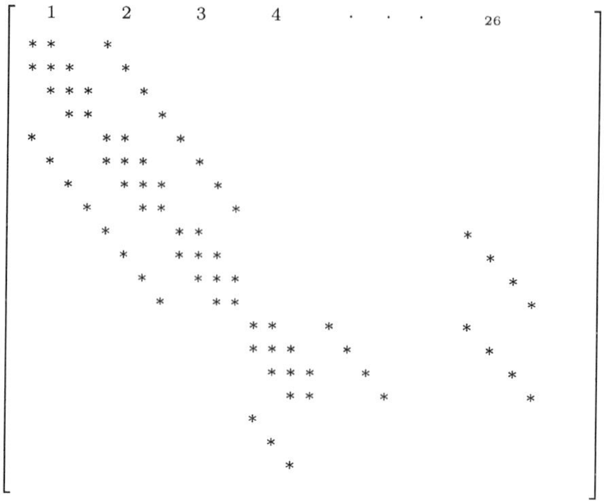

Fig. 6. The structure of the left-most part of the reordered matrix; the stars indicate the elements in the blocks corresponding to the line numbers on top.

In [23,22] it is suggested to increase the pseudo-overlap and to include also higher levels of fill that are introduced by the new block-wise ordering. For high dimensional problems and relatively low numbers of processors this leads to almost negligible overhead. It is shown by analysis in [24] and by experiments [23,22] that the block ordering with pseudo-overlap may lead to parallelizable incomplete decompositions that are almost perfectly scalable if the number of processors p is less than $\sqrt{n}$, where n denotes the order of the given linear system (the reported experiments include experiments for 16 processors, for $n \approx 260\,000$).

4.2. Hybrid iterative direct techniques

In the classical incomplete decompositions one ignores fill-in right from the start of the decomposition process. However, it might be a good idea to delay this until the matrix becomes too dense. This leads to a hybrid combination of direct and iterative techniques. One of such approaches has been described in [4]; we will describe it here in some detail.

We first permute the given matrix of the linear system $Ax=b$ to a doubly bordered block diagonal form:

$$\tilde{A} = P^{\mathrm{T}}AP = \begin{bmatrix} A_{00} & 0 & \cdots & 0 & A_{0m} \\ 0 & A_{11} & \ddots & \vdots & A_{1m} \\ \vdots & \ddots & \ddots & 0 & \vdots \\ 0 & \cdots & 0 & A_{m-1m-1} & \vdots \\ A_{m0} & A_{m1} & \cdots & \cdots & A_{mm} \end{bmatrix}. \tag{4}$$

```
Parallel_for i = 0, 1, ..., m − 1
    Decompose A_ii: A_ii = L_ii U_ii
    L_mi = A_mi U_ii^{-1}
    U_im = L_ii^{-1} A_im
    y_i = L_ii^{-1} b̃_i
    S_i = L_mi U_im
    z_i = L_mi y_i
end
S = A_mm − Σ_{i=0}^{m−1} S_i
y_m = b̃_m − Σ_{i=0}^{m−1} z_i
Solve S x_m = y_m
Parallel_for i = 0, 1, ..., m − 1
    x_i = U_ii^{-1}(y_i − U_im x_m)
end
```

Fig. 7. Parallel elimination.

Of course, the parallelism in the eventual method depends on the value of m, and some problems lend themselves more to this than others. Many circuit simulation problems can be rewritten in an effective way, as a circuit is often composed of components that are only locally coupled to others.

We permute the right-hand side b as well to $\tilde{b} = P^{\mathrm{T}}b$, which leads to the system

$$\tilde{A}\tilde{x} = \tilde{b} \tag{5}$$

with $\tilde{x} = Px$.

The parts of $\tilde{b}$ and $\tilde{x}$ that correspond to the block ordering, will be denoted by $\tilde{b}_i$ and $\tilde{x}_i$. The first step in the (parallelizable) algorithm will be to eliminate the unknown parts $\tilde{x}_0, \ldots, \tilde{x}_{m-1}$, which is done by the Algorithm in Fig. 7.

Note that S in Fig. 7 denotes the Schur complement after the elimination of the blocks $0, 1, \ldots, m-1$. In many relevant situations, direct solution of the reduced system $Sx_m = y_m$ requires the dominating part of the total computational costs, and this is where we bring in the iterative component of the algorithm.

Suppose that we solve the reduced system $Sx_m = y_m$ with an iterative method and that after termination we have the approximated solution $\hat{x}_m$, with $r_m = S\hat{x}_m - y_m$. When we take this approximated solution for the computation of the x_i in Fig. 7, then this leads to an approximated solution $\hat{x}$ for the system $Ax = b$. Then we have, in exact computation, that

$$\|A\hat{x} - b\|_2 = \|r_m\|_2. \tag{6}$$

The next step is to construct a preconditioner for the reduced system. This is based on discarding small elements in S. The elements larger than some threshold value define the preconditioner C:

$$c_{ij} = \begin{cases} s_{ij} & \text{if } |s_{ij}| > t|s_{ii}| \text{ or } |s_{ij}| > t|s_{jj}|, \\ 0 & \text{elsewhere} \end{cases} \tag{7}$$

with a parameter $0 \leqslant t < 1$. In the experiments, reported in [4] the value $t = 0.02$ turned out to be satisfactory, but this may need some experimentation for specific problems.

When we take C as the preconditioner, then we have to solve systems like $Cv = w$, and this requires decomposition of C. In order to prevent too much fill-in, it is suggested to reorder C with a minimum degree ordering. The system $Sx_m = y_m$ is then solved with, for instance, GMRES with preconditioner C. For the examples described in [4] it turns out that the convergence of GMRES was not very sensitive to the choice of t. The preconditioned iterative solution approach for the reduced system offers also opportunities for parallelism, although in [4] it is shown that even in serial mode the iterative solution (too sufficiently high precision) is often more efficient than direct solution of the reduced system.

Note that, because of (6), it is not necessary to iterate on the complete system.

In [4] heuristics are described for the decision on when the switch from direct to iterative should take place. These heuristics are based on mild assumptions on the speed of convergence of GMRES. The paper also reports on a number of experiments for linear systems, not only from circuit simulation, but also for some matrix problems taken from Matrix Market. [1] These experiments indicate that attractive savings in computational costs can be achieved, even in serial computation mode.

4.3. Some other approaches

Washio and Hayami [44] employed a domain decomposition approach for a rectangular grid in which one step of SSOR is performed for the interior part of each subdomain. In order to make this domain-decoupled SSOR more like global SSOR, the SSOR iteration matrix for each subdomain is modified. In order to further improve the parallel performance, the inverses in these expressions are approximated by low-order truncated Neumann series. A similar approach is suggested in [44] for a block modified ILU preconditioner. Experimental results have been reported for a 32-processor NEC Cenju distributed memory computer.

Radicati and Robert [31] used an algebraic version of this approach by computing ILU factors within overlapping block diagonals of a given matrix A. When applying the preconditioner to a vector v, the values on the overlapped region are taken as the average of the two values computed by the overlapping ILU factors. The approach of Radicati and Robert has been further refined by de Sturler [8], who studies the effects of overlap from the point of view of geometric domain decomposition. He introduces artificial mixed boundary conditions on the internal boundaries of the subdomains. In [8, Table 5.8], experimental results are shown for a decomposition into 20×20 slightly overlapping subdomains of a 200×400 mesh for a discretized convection–diffusion equation (5-point stencil). Using a twisted ILU preconditioning on each subdomain, it is shown that the complete linear system can be solved by GMRES on a 400-processor distributed memory Parsytec system with an efficiency of about 80% (this means that, with this domain adapted preconditioner, the process is about 320 times faster than ILU preconditioned GMRES for the unpartitioned linear system on a single processor). Since twisting leads to more parallelism, one can use bigger blocks (which usually means a better approximation). This helps to explain the good results.

Haase [17] suggests constructing an incomplete Choleski decomposition on each subdomain and modifying the decomposition using information from neighbouring subdomains. His results, for the discretized Poisson equation in 3D, show that an increase in the number of domains scarcely affects the effectiveness of the preconditioner. Experimental results for a realistic finite-element model, on

[1] Collection of testmatrices available at `ftp://ftp.cise.ufl.edu/cis/tech-reports/tr98/tr98-016.ps`

a 16-processor Parsytec Xplorer, show very good scalability of the conjugate gradient method with this preconditioner.

Heisse and Jung [18] attempt to improve the effectiveness of a domain decomposition preconditioner by using a multigrid V-cycle with only one pre- and one post-smoothing step of a parallel variant of Gauss–Seidel type to solve a coarse grid approximation to the problem. With the usual domain decomposition technique, effects of local changes in a domain that lead to global changes in the solution travel only one layer of neigbouring domains far per iteration. The coarse grid corrections are used to get this globally relevant information quicker to all domains. The combination with conjugate gradients, which is the underlying method used for the local subproblems, leads to good results on a variety of platforms, including a 64-processor.

5. Concluding remarks

In this contribution we have highlighted some of the Krylov subspace methods that have become accepted as powerful tools for the iterative solution of very large linear systems (say, of the order of millions of unknowns).

Krylov subspace methods and preconditioning have received considerable attention in literature. For a general background we refer to the textbooks published by Golub and Van Loan [14], Greenbaum [15], Saad [33], Axelsson [2], and Meurant [29]. For practical aspects, as well as descriptions of implementations we refer to the *Templates* [3], and for implementation aspects on high-performance computers see [9]. A useful discussion on parallel aspects appeared in [11].

References

[1] W.E. Arnoldi, The principle of minimized iteration in the solution of the matrix eigenproblem, Quart. Appl. Math. 9 (1951) 17–29.

[2] O. Axelsson, Iterative Solution Methods, Cambridge University Press, Cambridge, 1994.

[3] R. Barrett, M. Berry, T. Chan, J. Demmel, J. Donato, J. Dongarra, V. Eijkhout, R. Pozo, C. Romine, H. van der Vorst, Templates for the Solution of Linear Systems: Building Blocks for Iterative Methods, SIAM, Philadelphia, PA, 1994.

[4] C.W. Bomhof, H.A. van der Vorst, A parallel linear system solver for circuit-simulation problems, Numer. Linear Algebra Appl. 7 (2000) 649–665.

[5] E.F.F. Botta, A. van der Ploeg, Nested grids ILU-decomposition (NGILU), J. Comput. Appl. Math. 66 (1996) 515–526.

[6] P. Concus, G.H. Golub, D.P. O'Leary, A generalized conjugate gradient method for the numerical solution of elliptic partial differential equations, in: J.R. Bunch, D.J. Rose (Eds.), Sparse Matrix Computations, Academic Press, New York, 1976.

[7] E.F. D'Azevedo, F.A. Forsyth, W.P. Tang, Ordering methods for preconditioned conjugate gradient methods applied to unstructured grid problems, SIAM J. Matrix Anal. Appl. 13 (1992) 944–961.

[8] E. de Sturler, Iterative methods on distributed memory computers, Ph.D. Thesis, Delft University of Technology, Delft, the Netherlands, 1994.

[9] J.J. Dongarra, I.S. Duff, D.C. Sorensen, H.A. van der Vorst, Numerical Linear Algebra for High-Performance Computers, SIAM, Philadelphia, PA, 1998.

[10] I.S. Duff, G.A. Meurant, The effect of ordering on preconditioned conjugate gradient, BIT 29 (1989) 635–657.

[11] I.S. Duff, H.A. van der Vorst, Developments and trends in the parallel solution of linear systems, Parallel Comput. 25 (1999) 1931–1970.

[12] H.C. Elman, Relaxed and stabilized incomplete factorizations for non-self-adjoint linear systems, BIT 29 (1989) 890–915.

[13] V. Faber, T.A. Manteuffel, Necessary and sufficient conditions for the existence of a conjugate gradient method, SIAM J. Numer. Anal. 21 (2) (1984) 352–362.

[14] G.H. Golub, C.F. Van Loan, Matrix Computations, The Johns Hopkins University Press, Baltimore, MD, 1996.

[15] A. Greenbaum, Iterative Methods for Solving Linear Systems, SIAM, Philadelphia, PA, 1997.

[16] I. Gustafsson, A class of first order factorization methods, BIT 18 (1978) 142–156.

[17] G. Haase, Parallel incomplete Cholesky preconditioners based on the nonoverlapping data distribution, Parallel Comput. 24 (1998) 1685–1703.

[18] B. Heisse, M. Jung, Parallel solvers for nonlinear elliptic problems based on domain decomposition ideas, Parallel Comput. 22 (1997) 1527–1544.

[19] M.R. Hestenes, E. Stiefel, Methods of conjugate gradients for solving linear systems, J. Res. Natl. Bur. Standards 49 (1954) 409–436.

[20] D.S. Kershaw, The incomplete Choleski-conjugate gradient method for the iterative solution of systems of linear equations, J. Comput. Phys. 26 (1978) 43–65.

[21] J.C.C. Kuo, T.F. Chan, Two-color Fourier analysis of iterative algorithms for elliptic problems with red/black ordering, SIAM J. Sci. Statist. Comput. 11 (1990) 767–793.

[22] M. Magolu monga Made, H.A. van der Vorst, A generalized domain decomposition paradigm for parallel incomplete LU factorization preconditionings, Future Generation Comput. Systems 17 (2001) 925–932.

[23] M. Magolu monga Made, H.A. van der Vorst, Parallel incomplete factorizations with pseudo-overlapped subdomains, Parallel Comput. 27 (2001) 989–1008.

[24] M. Magolu monga Made, H.A. van der Vorst, Spectral analysis of parallel incomplete factorizations with implicit pseudo-overlap. Numer. Linear Algebra Appl. 9 (2002) 45–64.

[25] J.A. Meijerink, H.A. van der Vorst, An iterative solution method for linear systems of which the coefficient matrix is a symmetric M-matrix, Math. Comp. 31 (1977) 148–162.

[26] J.A. Meijerink, H.A. van der Vorst, Guidelines for the usage of incomplete decompositions in solving sets of linear equations as they occur in practical problems, J. Comput. Phys. 44 (1981) 134–155.

[27] G. Meurant, Numerical experiments for the preconditioned conjugate gradient method on the CRAY X-MP/2, Technical Report LBL-18023, University of California, Berkeley, CA, 1984.

[28] G. Meurant, The conjugate gradient method on vector and parallel supercomputers, Technical Report CTAC-89, University of Brisbane, July 1989.

[29] G. Meurant, Computer Solution of Large Linear Systems, North-Holland, Amsterdam, 1999.

[30] Y. Notay, DRIC: a dynamic version of the RIC method, Numer. Linear Algebra Appl. 1 (1994) 511–532.

[31] G. Radicati di Brozolo, Y. Robert, Parallel conjugate gradient-like algorithms for solving sparse non-symmetric systems on a vector multiprocessor, Parallel Comput. 11 (1989) 223–239.

[32] J.K. Reid, The use of conjugate gradients for systems of equations possessing 'Property A', SIAM J. Numer. Anal. (1972) 325–332.

[33] Y. Saad, Iterative Methods for Sparse Linear Systems, PWS Publishing Company, Boston, 1996.

[34] Y. Saad, M.H. Schultz, GMRES: a generalized minimal residual algorithm for solving nonsymmetric linear systems, SIAM J. Sci. Statist. Comput. 7 (1986) 856–869.

[35] G.L.G. Sleijpen, H.A. van der Vorst, Reliable updated residuals in hybrid Bi-CG methods, Computing 56 (1996) 141–163.

[36] P. Sonneveld, CGS: a fast Lanczos-type solver for nonsymmetric linear systems, SIAM J. Sci. Statist. Comput. 10 (1989) 36–52.

[37] A. van der Ploeg, Preconditioning for sparse matrices with applications, Ph.D. Thesis, Department of Mathematics, University of Groningen, 1995.

[38] H.A. van der Vorst, Iterative solution methods for certain sparse linear systems with a non-symmetric matrix arising from PDE-problems, J. Comput. Phys. 44 (1981) 1–19.

[39] H.A. van der Vorst, Large tridiagonal and block tridiagonal linear systems on vector and parallel computers, Parallel Comput. 5 (1987) 45–54.

[40] H.A. van der Vorst, High performance preconditioning, SIAM J. Sci. Statist. Comput. 10 (1989) 1174–1185.

[41] H.A. van der Vorst, Bi-CGSTAB: a fast and smoothly converging variant of Bi-CG for the solution of non-symmetric linear systems, SIAM J. Sci. Statist. Comput. 13 (1992) 631–644.

[42] H.A. van der Vorst, Q. Ye, Refined residual replacement techniques for subspace iterative methods for convergence of true residuals, SIAM J. Sci. Comput. 22 (2000) 835–852.

[43] R.S. Varga, Matrix Iterative Analysis, Prentice-Hall, Englewood Cliffs, NJ, 1962.

[44] T. Washio, K. Hayami, Parallel block preconditioning based on SSOR and MILU, Numer. Linear Algebra Appl. 1 (1994) 533–553.

ELSEVIER

Journal of Computational and Applied Mathematics 149 (2002) 267–277

JOURNAL OF COMPUTATIONAL AND APPLIED MATHEMATICS

www.elsevier.com/locate/cam

The next generation CIP as a conservative semi-Lagrangian solver for solid, liquid and gas

T. Yabe *, Y. Ogata, K. Takizawa, T. Kawai, A. Segawa, K. Sakurai

Department of Mechanical Sciences and Engineering, Tokyo Institute of Technology, Ohokayama, Meguro-ku, Tokyo 152-8552, Japan

Received 26 November 2001; received in revised form 13 January 2002

Abstract

We present a review of the CIP method, which is a kind of semi-Lagrangian scheme and has been extended to treat incompressible flow in the framework of compressible fluid. Since it uses primitive Euler representation, it is suitable for multi-phase analysis. The recent version of this method guarantees the exact mass conservation even in the framework of semi-Lagrangian scheme. Comprehensive review is given for the strategy of the CIP method that has a compact support and subcell resolution including front capturing algorithm with functional transformation.

Keywords: Multi-phase flow; Semi-Lagrangian; Solid; Liquid; Gas; Universal solver; CIP

1. Introduction

Recent development of simulation technology made possible the combined analysis of solid, liquid and gas. However, simultaneous treatment of materials undergoing phase state transition is still not well established. This is because the conventional combined analysis relies on the separate treatment of each phase. For example, overset grid is the easiest way to treat the structure–fluid interaction but it fails for a severely distorted structure or for melting of structure. A universal treatment of all phases by one simple algorithm is essential. In order to attack the problems mentioned above, we must first find a method to treat a sharp interface and to solve the interaction of compressible gas with incompressible liquid or solid.

Toward this goal, we take Eulerian approach based on the CIP (cubic-interpolated propagation) method (see for review, Ref. [14]) which does not need adaptive grid system and therefore removes

*Corresponding author. Tel.: +81-3-57342165; fax: +81-3-57342165.
E-mail address: yabe@mech.titech.ac.jp (T. Yabe).

0377-0427/02/$ - see front matter
PII: S0377-0427(02)00535-6

the problems of grid distortion caused by structural break up and topology change. The material surface can be captured almost by one grid throughout the computation. Furthermore, the code can treat all the phases of matter from solid state through liquid and two phase state to gas without restriction on the time step from high sound speed.

Although the CIP can give accurate results keeping conservation error in quite low level, the semi-Lagrangian form cannot guarantee the exact mass conservation by itself. Recent version of the CIP-CSL [10,13] can overcome this difficulty and provide exactly conservative semi-Lagrangian scheme. Since these schemes do not use the cubic polynomial but use different orders of polynomial, we redefine the name of these CIP families as "constrained interpolation profile" and still keep the abbreviation, CIP. This means that various constraints such as the time evolution of spatial gradient, that is used in the original CIP method, or spatially integrated conservative quantities can be used to construct the profile. In this paper, we shall give a review of the CIP method and related schemes to attack these important subjects.

2. CIP method

2.1. Advection processes

Although the nature is in a continuous world, digitization process is unavoidable in order to be implemented in numerical simulations. Primary goal of numerical algorithm will be to retrieve the lost information inside the grid cell between these digitized points. Most of numerical schemes proposed before, however, did not take care of real solution inside the grid cell and resolution has been limited to the grid size. The CIP method proposed by one of the authors tries to construct a solution inside the grid cell close enough to this real solution of the given equation with some constraints [14]. A simple advection equation

$$\frac{\partial f}{\partial t} + u\frac{\partial f}{\partial x} = 0. \tag{1}$$

is calculated by $f^{n+1} = F(x - u\Delta t), g^{n+1} = \mathrm{d}F(x - u\Delta t)/\mathrm{d}x$ in the CIP where g stands for $\partial f/\partial x$. If two values of f and g are given at two grid points, the profile between these points can be interpolated by cubic polynomial $F(x) = ax^3 + bx^2 + cx + d$. Thus, the profile at $n+1$ step can be obtained as

$$f_i^{n+1} = a_i\xi^3 + b_i\xi^2 + g_i^n\xi + f_i^n,$$

$$g_i^{n+1} = 3a_i\xi^2 + 2b_i\xi + g_i^n, \tag{2}$$

where we define $\xi = -u\Delta t$. The coefficients of polynomial a, b are given by continuity requirement [14].

It would be interesting to examine phase error of various schemes using the method proposed by Purnell [9] and Utsumi et al. [12]. Fig. 1 summarizes those results. As is well known, phase speed of conventional schemes depart from the exact one, that is shown by the solid line, around $k\Delta x = \pi/2$. Surprisingly, however, the CIP can reproduce the correct phase speed even up to $k\Delta x = \pi$. This is remarkable because $k\Delta x = \pi$ means that one wavelength is described by 2 grid size. Let us consider the case where values of the 3 points are zero. Even in this case, one wave can exist. The CIP

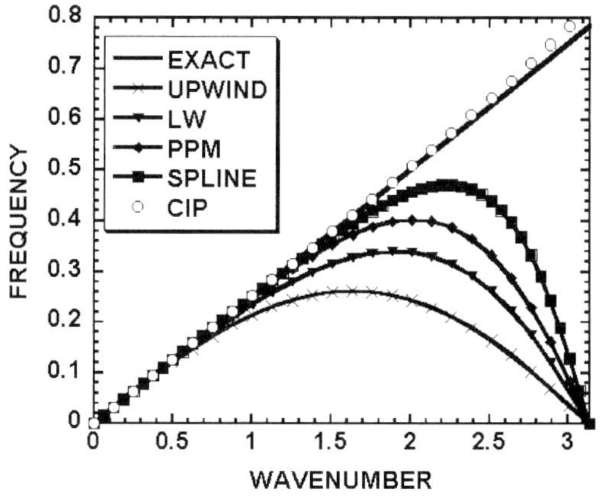

Fig. 1. Phase error of various schemes such as first order upwind, Lax–Wendroff, PPM, spline, and CIP.

gives correct spatial gradients, which are non-zero at these points, and therefore it recognize the existence of the wave inside the grid cell. Any code that uses only the information of the value, that is zero now, cannot correctly recognize the wave even if higher order polynomial is employed. The superiority of the CIP also appears in numerical damping and dissipation [12].

The importance of propagated gradients can be clearly demonstrated in comparison with the cubic spline [9]. Although the cubic spline uses the same cubic polynomial and the information of gradient as the CIP, it cannot reproduce the result of the CIP, because the gradient of the spline is determined merely from smoothness requirement. As is easily recognized, such a constraint that is independent of the original equation will not help to retrieve the profile inside the grid cell. It is important to note that the computation time does not seriously increase by the addition of spatial gradient because coefficients of cubic polynomial can be used also for gradient calculation. Actually the CPU time required for a simple advection problem in one dimension is cubic-lagrange/CIP = 1.0, spline/CIP = 1.68 (Thomas method is used to solve matrix), RCIP/CIP = 1.77, PPM/CIP = 2.31. RCIP is the rational function CIP [14].

2.2. Burgers turbulence

The CIP method is applied to Burgers equation:

$$\frac{\partial u}{\partial t} + u\frac{\partial u}{\partial x} = v\frac{\partial^2 u}{\partial x^2}. \tag{3}$$

After imposed random-noise initial perturbation, we observe the power spectrum calculated from various schemes as shown in Fig. 2 where $v = 1/250$ and velocity is $u = 20 + \delta u$ and δu is random number of ± 5 size. The CIP can reproduce the spectrum regardless of mesh size. It should be noticed that the end point of the highest k corresponds to $k\Delta x = \pi$ which corresponds to the wavelength of 2 grid size. As shown in Fig. 1, the CIP can accurately trace the phase speed of this wave and this is the reason why the spectrum does not change regardless of mesh size. Since the spectrum method does not converge in this situation, we shall compare our result with UTOPIA scheme and K–K scheme [5]. UTOPIA and K–K schemes give similar result. Although K–K scheme can produce a

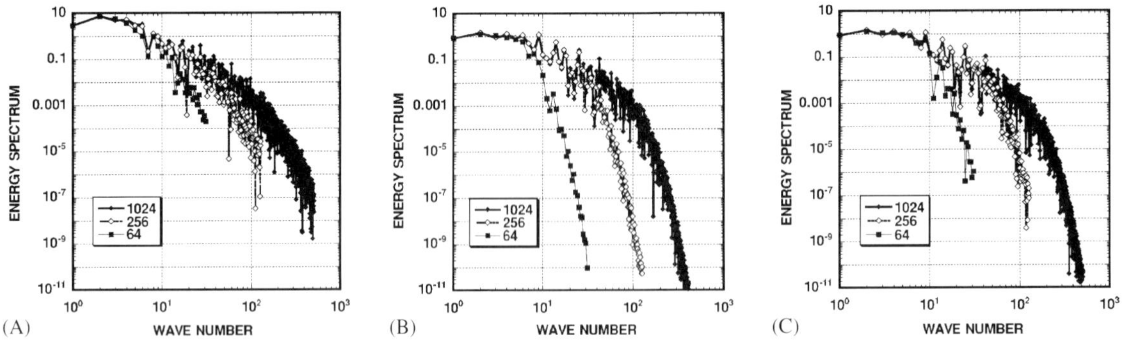

Fig. 2. Energy spectrum for Burgers turbulence (A) CIP, (B) Kawamura–Kuwahara, (C) UTOPIA for 64, 256, 1024 grids.

universal spectrum for stationary turbulence that is $u = \delta u$ without main flow [6], it suffers from significant diffusion for streaming flow because of fourth order dissipation term.

2.3. Interface tracking: a sharpness preserving method

Various kinds of methods have been developed so far to achieve a compact and correctly defined interface by introducing extra programming. Among those mostly used algorithms are the level set methods and the VOF (volume of fluid) methods [4] for front capturing, and others for front tracking [11]. Level set method that was firstly proposed by Osher and Sethian [8] gets around the computation of interfacial discontinuity by evaluating the field in higher dimensions. The interface of interest is then recovered by taking a subset of the field. Practically, the interface is defined as the zero level set of a distance function from the interface.

We devised an interface tracking technique [14] which appears efficient, geometrically faithful and diffusionless. The method is a combination of the CIP advection solver and a tangent function transformation. Let us consider the time evolution of color function which is described by Eq. (1), where the color function f represents the interior of material by 1 and the outside by 0.

Solving the above equation by finite difference schemes in an Eulerian representation will produce numerical diffusion and tend to smear the initial sharpness of the interfaces. In our method, rather than the original variable f itself, its transformation, say $F(f)$, is calculated by the CIP method. We specify $F(f)$ to be a function of f only, which means that the new function $F(f)$ is also governed by the same equation as (1). Hence, we have

$$\frac{\partial F(f)}{\partial t} + \mathbf{u} \cdot \nabla F(f) = 0, \tag{4}$$

and all the algorithms proposed for f (schemes for advection equation) can be used to $F(f)$. Hopefully, by the considerable simplicity, this kind of techniques would be very attractive for practical implementation. We here use a transformation of a tangent function for $F(f)$.

Although f experiences a rapid change from 0 to 1 at the interface, $F(f)$ shows a quite regular behavior. Because most of the values of $F(f)$ are concentrated near $f=0$ and 1, the function transformation improves locally the spatial resolution near the large gradients. Thus, the sharp discontinuity can be described quite easily. This method does not involve any interface construction procedure

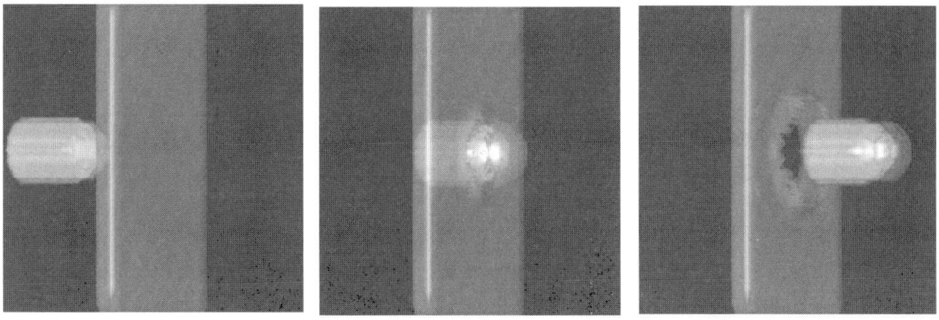

Fig. 3. A solid bullet penetrates through water curtain. The fixed uniform Cartesian grid of $80 \times 60 \times 40$ is used.

and is quite economical in computational complexity. It should be notified that the present method is more attractive in 3-D computation since the extension of the scheme to 3-D is straightforward.

Fig. 3 shows an example of this treatment. The initial sharpness of a bullet is well preserved and the discontinuities are advected with a correct speed.

2.4. Conservative semi-Lagrangian scheme

It is quite useful to look for conservative semi-Lagrangian scheme because semi-Lagrangian method can be quite effective on parallel computers, suits for multi-phase flow, and enables the advection calculation with large time step free from CFL condition. Although the semi-Lagrangian scheme has been successfully used in short-term atmospheric problems, the loss of exact conservation makes the scheme inappropriate for long-term problems and oceanic problems.

Furthermore, there exist subjects that require exact conservation of mass. One of the typical example is the black-hole formation and emission of gravity wave. In this case, small fraction of mass is converted into gravity wave and strict mass conservation is essential. Another example is plasma simulation in which Vlasov equation in six-dimensional phase space must be solved and total density of particle must be conserved, otherwise large electric field appears. The CIP method can be constructed to exactly conserve the total mass in Vlasov system of equally spaced grid in phase space [7]. However, the use of non-equally spaced grid or other general grids can save the computational cost and is worthy for investigation.

In this sense, the effort to establish exact conservation in semi-Lagrangian form would be a challenging task. Toward this goal, we proposed two schemes CIP-CSL4 [10] and CIP-CSL2 [13]. In CIP-CSL4, the fourth-order polynomial is chosen as an interpolation function and the time development of f and g is calculated simply by shifting the interpolation function $F_i(x)$ by $u\Delta t$ in the same way as Eq. (2) of the CIP method. Since the order of polynomial is increased, we have now freedom to choose another constraint to determine the polynomial. We use the constraint to the integrated value to guarantee the mass conservation. Thus this scheme is able to describe a structure of one-grid size as shown in Fig. 4 even without tangent transformation.

The CIP-CSL2 uses a set of equations:

$$\frac{\partial D}{\partial t} + u\frac{\partial D}{\partial x} = 0, \quad \frac{\partial f}{\partial t} + \frac{\partial (uf)}{\partial x} = 0. \tag{5}$$

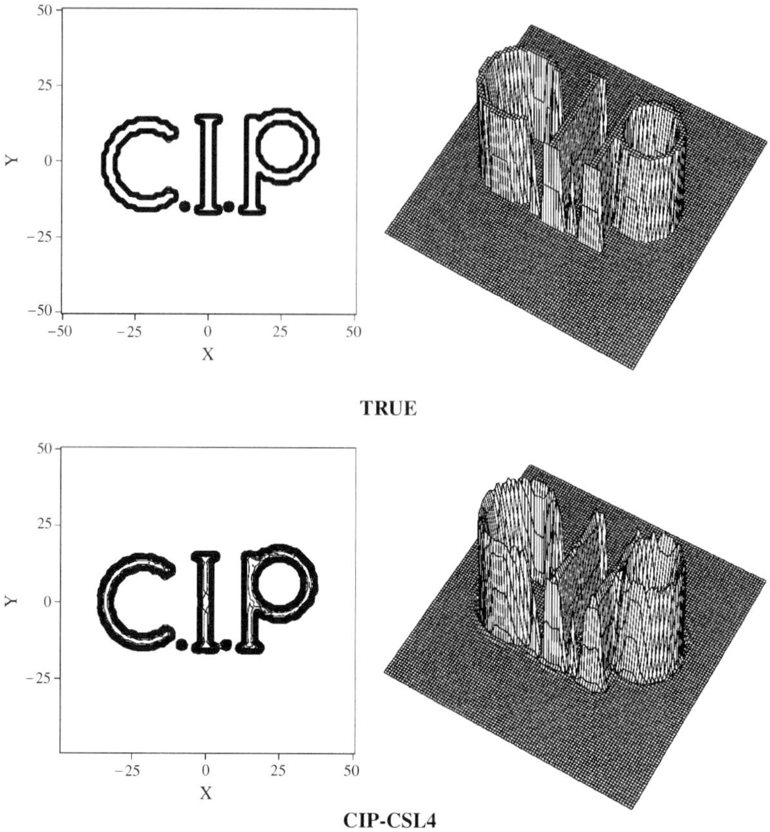

Fig. 4. Contour plots and three-dimensional views after one complete revolution of a solid-body which consists of three characters of "C.I.P" and all the lines composing the characters are thinner than 3 grid points. The dot of only one-grid size is preserved after one revolution.

These two equations are similar to the equations for f and g in Eq. (1) by replacing f by $D=\int f\,dx$ and g by $f=\partial D/\partial x$. Therefore, we can use all the procedures of the CIP method for the integral value of f. We can impose the conservation of integrated value easily by this change.

3. Semi-Lagrangian approach to hydrodynamic equations

3.1. Pressure-projection algorithm in primitive Euler scheme

Before presenting a method to solve all the phases of materials, we must at first construct a unified equation to describe all the phases. For this purpose we use the following set of hydrodynamic-type equations:

$$\frac{\partial \mathbf{f}}{\partial t} + (\mathbf{u} \cdot \nabla)\mathbf{f} = \mathbf{S}. \tag{6}$$

Here, $\mathbf{f}=(\rho,\mathbf{u},T)$, $\mathbf{S}=(-\rho\nabla\cdot\mathbf{u}+Q_{\rm m},-\nabla p/\rho+Q_u,-P_{\rm TH}\nabla\cdot\mathbf{u}/\rho C_{\rm v}+Q_E)$, and ρ is the density, $\mathbf{u}$ the velocity, p the pressure, T the temperature, $Q_{\rm m}$ represents the mass source term, Q_u represents viscosity, elastic stress tensor, surface tension, etc., and Q_E represents viscous heating, thermal conduction and heat source. Here, $C_{\rm v}$ is the specific heat for constant volume and we define $P_{\rm TH}=T(\partial p/\partial T)_\rho$ which is derived from the first principle of thermodynamics.

The underlying physics included in the above equations for continuum dynamics is complex and may include processes which possess quite different time scales of variation. It should be expedient to separate the solution procedure into several fractional steps.

The CIP method uses the primitive Euler method to solve Eq. (6) with a separate treatment of advection term like conventional semi-Lagrangian formulation, thus the formulation into a simultaneous solution of incompressible and compressible fluid is readily obtained.

Original CCUP (CIP-combined unified procedure) method was proposed only for a special equation of state, but here we rebuild it with more general EOS [14]. That is, for small change of density and temperature, the pressure change can be linearly proportional to them as

$$\Delta p=\left(\frac{\partial p}{\partial\rho}\right)_T\Delta\rho+\left(\frac{\partial p}{\partial T}\right)_\rho\Delta T, \tag{7}$$

where Δp means the pressure change $p^{n+1}-p^*$ during one time step and * is the profile after advection. This applies also to ρ, T. From this relation, once $\Delta\rho,\Delta T$ are predicted, Δp will be predicted based on Eq. (7). Needless to say, $\partial p/\partial\rho, \partial p/\partial T$ are given by EOS. Since the CIP separates the non-advection terms from the advection, we can concentrate on the non-advection terms related to sound waves which are the primary cause of the difficulty in liquid having large sound speed. Thus we obtain

$$\nabla\left(\frac{1}{\rho^*}\nabla p^{n+1}\right)=\frac{p^{n+1}-p^*}{\Delta t^2(\rho C_{\rm s}^2+(P_{\rm TH}^2/\rho C_{\rm v}T))}+\frac{\nabla\cdot\mathbf{u}^*}{\Delta t}. \tag{8}$$

This equation shows that, at the sharp discontinuity, $\mathbf{n}\cdot(\nabla p/\rho)$ is continuous. Since $\nabla p/\rho$ is the acceleration, it is essential that this term is continuous since the density changes by several orders of magnitude at the boundary between liquid and gas. In this case, the denominator of $\nabla p/\rho$ changes by several orders and pressure gradient must be calcuated accurately enough to ensure the continuous change of acceleration. The equations derived by the ICE [3] in the framework of conservative Euler seems to be quite similar to Eq. (8) but the continuity of $\nabla p/\rho$ in the formers is not guaranteed.

3.2. Application to hydro planning

Figs. 5 and 6 show two examples of fluid-structure interaction calculated by the procedure given in the previous section. A thin aluminum cylinder is thrown over water surface. As our experience tells, the cylinder slides over the surface and then jumps up. In Fig. 5, the cylinder has a round bottom shape while the bottom is flat in Fig. 6.

The present simulation code can replicate well the overall behavior of aluminum disk observed in the experiment.

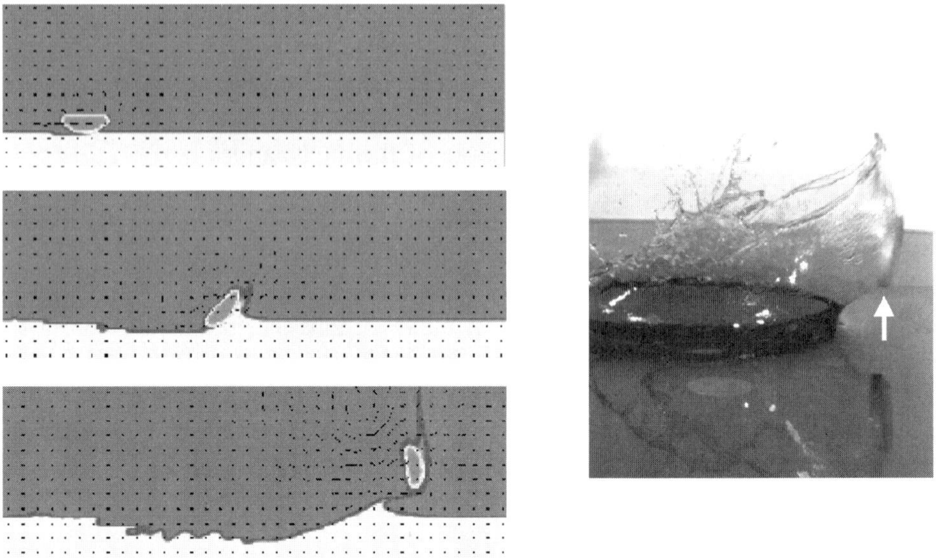

Fig. 5. Round bottom shape (Left: calculation, Right: experiment). The arrow indicates the location of aluminum cylinder.

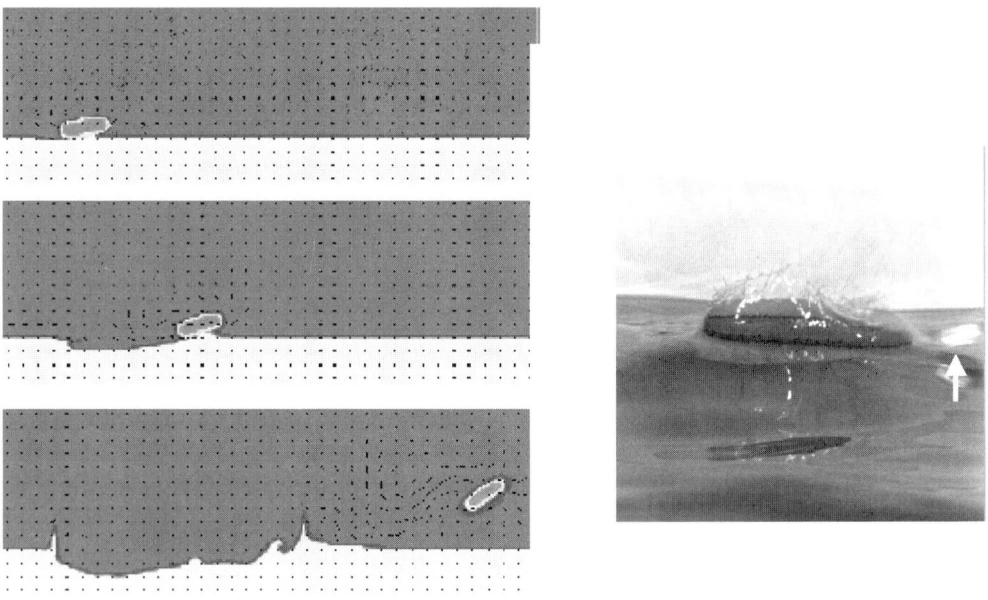

Fig. 6. Flat bottom (Left: calculation, Right: experiment). The arrow indicates the location of aluminum cylinder.

4. Matrix-free incompressible-flow calculation

Since the CIP is superior in the calculation of advection equation, we can apply the procedure to conventional characteristic equations. Then, if this equation can be solved with large CFL,

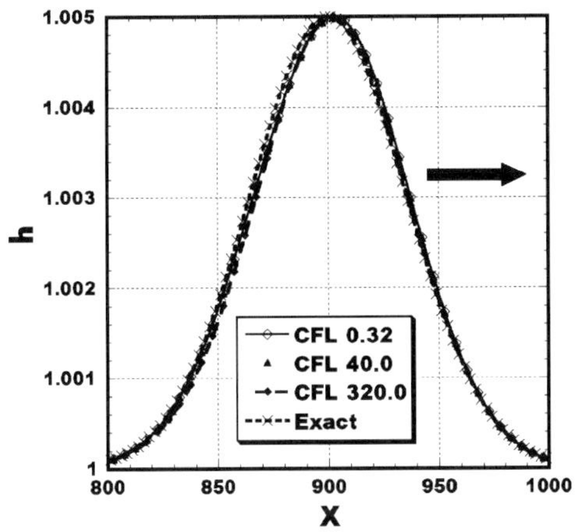

Fig. 7. The height of water wave at $t=400.0$ calculated by the CIP. The results with various CFL numbers are compared. The scheme works even for CFL = 320.

compressible and incompressible flow can be solved without matrix solution of pressure. As a simple application, we deal with shallow water equation. Let h be the height of water surface, then shallow water equations are simply written in one dimension as

$$\frac{\partial}{\partial t}\begin{pmatrix} h \\ u \end{pmatrix} + \begin{pmatrix} u & h \\ g & u \end{pmatrix}\frac{\partial}{\partial x}\begin{pmatrix} h \\ u \end{pmatrix} \equiv \frac{\partial \mathbf{W}}{\partial t} + \mathbf{A}(\mathbf{W})\frac{\partial \mathbf{W}}{\partial x} = 0. \tag{9}$$

Eq. (9) leads to two equations for the Riemann invariants,

$$\frac{D^{\pm}}{Dt}\left(\Gamma \pm \frac{u}{2}\right) = 0, \quad \frac{D^{\pm}}{Dt} \equiv \frac{\partial}{\partial t} + \lambda^{\pm}\frac{\partial}{\partial x}. \tag{10}$$

where $\Gamma \equiv \sqrt{gh}$, $\lambda^{\pm} = u \pm \Gamma$ and g represents the gravity. The form of Eq. (10) is the same as Eq. (1), so the CIP method can be applied to Eq. (10). Since the phase speed of the gravity waves Γ, which is equivalent to sound waves in hydrodynamics, is so fast, time step is restricted by the CFL ($=\lambda^{\pm}\Delta t/\Delta x$) condition of gravity wave when conventional explicit Eulerian schemes are employed. Therefore, the semi-Lagrangian approach [1,2] to the characteristic equations has been used for solving gravity wave to make time step be much longer than CFL condition. This algorithm is tested with initial perturbation of h in Gaussian form.

$$h(x,t=0) = 1.0 + 0.01\exp\left\{-\left(\frac{x - X_{\max}/2}{50.0}\right)\right\}, \tag{11}$$

The mesh size of $\Delta x = 0.25$ and the number of meshes is 4000. The numerical solution is shown in Fig. 7 and agrees well with analytical solution even for very large CFL = 320.0.

Next, we shall extend the scheme to two dimensions. We use directional splitting here and repeatedly apply the procedure of one dimension to each direction. The initial condition is set as follows.

$$h(x,y,t=0) = 1.0 + 0.001\exp\left\{-\left(\frac{x - X_{\max}/2}{50.0}\right) - \left(\frac{y - Y_{\max}/2}{50.0}\right)\right\}. \tag{12}$$

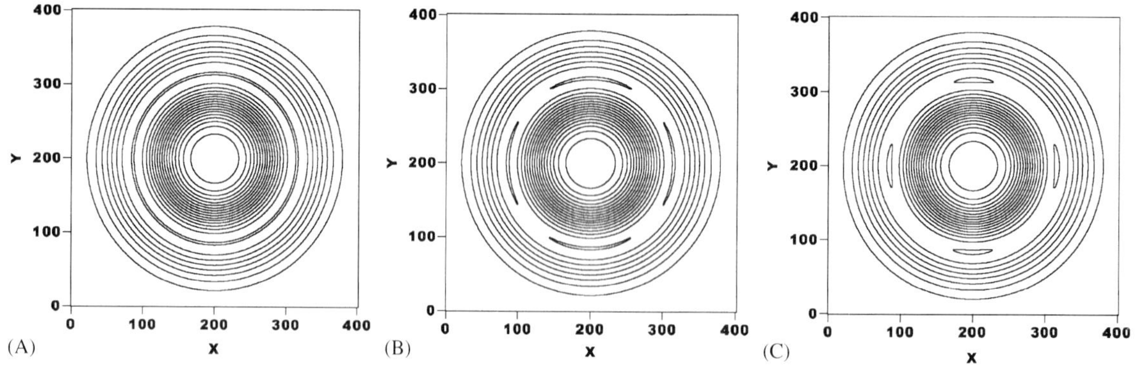

Fig. 8. The height contour of water wave at $t = 100.0$. (A) CFL $= 1.4$, (B) 5.0, (C) 14.0.

Mesh size is $\Delta x = \Delta y = 1.0$, the number of meshes is $(NX, NY) = (400, 400)$. We take CFL $= 1.4$, 5.0 and 14.0, and calculate up to $t = 100.0$. The contours for each CFL are shown in Fig. 8. Even in two dimensions, the method can give a symmetrical wave propagation under large CFL condition. Therefore we conclude that the present method is applicable to the cases in which the change of characteristic speed can be neglected for one time step, that is, perturbations are small. This does not restrict the applicability of the present method. If the speed of gravity wave is much larger than fluid velocity, variation of characteristic speed becomes smaller and large CFL can be used. The same can be said for incompressible flow because sound speed is much faster than fluid velocity.

Considering the conventional solution to incompressible flow, the use of the Poisson equation for pressure and its implicit solution do not trace the exact movement of sound wave profile, and the wave form in the solution above Lipschitz condition $(\Delta u \Delta t/\Delta x = 1)$ is smeared out by strong damping. If the phenomena in the time scale of $\Delta u \Delta t/\Delta x < 1$ is important for the system, these should be correctly treated in the time scale of $\Delta u \Delta t/\Delta x < 1$. If not, we can still use larger time step $\Delta u \Delta t/\Delta x \gg 1$ suffering the exact shape of solutions. Since the present scheme is applicable to multi-dimensional shallow water equations, it is possible to apply it to hydrodynamics directly in future.

References

[1] J.R. Bates, A. Mcdonald, Multiply-upstream, semi-Lagrangian advective schemes: analysis and application to a multi-level primitive equation model, Mon. Wea. Rev. 110 (1982) 1831.

[2] G. Erbes, A semi-Lagrangian method of characteristics for the shallow-water equations, Mon. Weather Rev. 121 (1993) 3443.

[3] F.H. Harlow, A.A. Amsden, Numerical simulation of almost incompressible flow, J. Comput. Phys. 3 (1968) 80.

[4] C.W. Hirt, B.D. Nichols, Volume of fluid (VOF) method for the dynamics of free boundaries, J. Comput. Phys. 39 (1981) 201.

[5] T. Kawamura, K. Kuwahara, Computation of high Reynolds number flow around a circular cylinder with surface roughness, AIAA Paper 84-0340, 1984.

[6] K. Kuwahara, Unsteady flow simulation and its visualization, AIAA Paper 99-3405, 1999.

[7] T. Nakamura, T. Yabe, Cubic interpolated propagation scheme for solving the hyper-dimensional Vlasov–Poisson equation in phase space, Comput. Phys. Comm. 120 (1999) 122.

[8] S. Osher, J.A. Sethian, Front propagating with curvature-dependent speed: algorithms based on Hamilton–Jacobi formulations, J. Comput. Phys. 79 (1988) 12.

[9] D.K. Purnell, Solution of the advective equation by upstream interpolation with a cubic spline, Mon. Weather Rev. 104 (1975) 42.

[10] R. Tanaka, T. Nakamura, T. Yabe, Constructing exactly conservative scheme in a non-conservative form, Comput. Phys. Comm. 126 (2000) 232.

[11] S.O. Unverdi, G.A. Tryggvasson, A front-tracking method for viscous, incompressible, multi-fluid flows, J. Comput. Phys. 100 (1992) 25.

[12] T. Utsumi, T. Kunugi, T. Aoki, Stability and accuracy of the cubic interpolated propagation scheme, Comput. Phys. Comm. 101 (1996) 9.

[13] T. Yabe, R. Tanaka, T. Nakamura, F. Xiao, An exactly conservative semi-Lagrangian scheme (CIP-CSL) in one dimension, Mon. Wea. Rev. 129 (2001) 332.

[14] T. Yabe, F. Xiao, T. Utsumi, Constrained interpolation profile method for multiphase analysis, J. Comput. Phys. 169 (2001) 556.

ELSEVIER Journal of Computational and Applied Mathematics 149 (2002) 279–296

JOURNAL OF COMPUTATIONAL AND APPLIED MATHEMATICS

www.elsevier.com/locate/cam

Advanced general-purpose computational mechanics system for large-scale analysis and design

Shinobu Yoshimura[a,*], Ryuji Shioya[b], Hirohisa Noguchi[c], Tomoshi Miyamura[d,1]

[a] *Institute of Environmental Studies, The University of Tokyo, 7-3-1 Hongo, Bunkyo, Tokyo 113-8656, Japan*
[b] *Department of Mechanical Engineering, Kyushu University, 6-10-1 Hakozaki, Higashi-ku, Fukuoka 812-8581, Japan*
[c] *Department of System Design Engineering, Keio University, 3-14-1 Hiyoshi, Kouhoku-ku, Yokohama 223-8522, Japan*
[d] *Department of Computer, Nihon University, 1 Nakagawara, Tokusada, Tamura-machi, Koriyama, Fukushima 963-8642, Japan*

Received 26 November 2001; received in revised form 13 March 2002

Abstract

We have been developing an advanced general-purpose computational mechanics system, named ADVENTURE, which is designed to be able to analyze a model of arbitrary shape with a 10–100 million degrees of freedom (DOFs) mesh, and additionally to enable parametric and nonparametric shape optimization. Domain-decomposition-based parallel algorithms are implemented in pre-processes (domain decomposition), main processes (system matrix assembling and solutions) and post-process (visualization), respectively. The hierarchical domain decomposition method with a pre-conditioned iterative solver (HDDM) is adopted in the main processes as one of the major solution techniques. Module-based architecture of the system with standardized I/O format and libraries are also developed and employed to attain flexibility, portability, extensibility and maintainability of the whole system. This paper describes some key technologies employed in the system, and shows some latest results including elastic stress analysis of a precise three-dimensional (3D) model of a nuclear reactor vessel with a 60 million DOF mesh on Hitachi SR2201 (1024 PEs), and nonparametric shape optimization of a support structure for an express way with a one million DOF mesh on a PC cluster (10 PEs).

Keywords: CAE system; Parallel processing; Hierarchical domain decomposition; Module-based architecture; I/O standardization; 100 million DOF mesh; Shape optimization

* Corresponding author.
E-mail addresses: yoshi@q.t.u-tokyo.ac.jp (S. Yoshimura), shioya@mech.kyushu-u.ac.jp (R. Shioya), noguchi@sd.keio.ac.jp (H. Noguchi), miyamura@cs.ce.nihon-u.ac.jp (T. Miyamura).
[1] Formerly Institute of Environmental Studies, The University of Tokyo.

0377-0427/02/$ - see front matter
PII: S0377-0427(02)00536-8

1. Introduction

Various general-purpose computational mechanics systems have been developed in the last three decades to quantitatively evaluate mechanical/physical phenomena such as deformation of solid, heat transfer, fluid flow and electromagnetics. Nowadays, such systems are regarded as infrastructural tools for the present industrialized society. The existing systems, however, cannot be used with massively parallel processors (MPPs) with the order of 100–10,000 processing elements (PEs), which are about to dominate the high-performance computing market in the 21st century, as they were developed for single-processor computers, which took leadership an age ago. Neither can the current systems be used in heterogeneous parallel and distributed computing environments, because they can deal with only medium-scale problems with millions DOFs at most.

The ADVENTURE project [3,25] is one of the research projects in the "Computational Science and Engineering" field selected for the "Research for the Future (RFTF)" Program sponsored by the Japan Society for the Promotion of Science (JSPS) [25,4]. In the project, we have been developing an advanced general-purpose computational mechanics system named ADVENTURE since August 1997. The system is designed to be able to analyze a three-dimensional (3D) finite element model of arbitrary shape with a 10–100 million DOF mesh, and additionally to enable parametric and nonparametric shape optimization.

In this paper, we first describe a design concept of the ADVENTURE system, and then explain some key technologies implemented in it, i.e. (a) hierarchical-domain-decomposition-based parallel solution algorithm, (b) parallel domain decomposer, (c) parallel visualization, (d) standardized I/O format and libraries, and (e) shape optimization. Finally, we show some latest analysis results including an elastic stress analysis of a precise 3D model of a nuclear reactor vessel with a 60 million DOF mesh on Hitachi SR2201 (1024 PEs), and shape optimization of a support structure for an express way with a one million DOF mesh.

2. Design concept of the system

2.1. Requirements for next-generation general-purpose computational mechanics systems

When we initiated the ADVENTURE project, we first summarized issues to be solved in next-generation general-purpose computational mechanics systems as follows:

(1) capability to perform a precise analysis of a whole structure using a 10^7–10^8 DOF mesh,
(2) high parallel performance in massively parallel processors (MPPs) with 10^2–10^4 processing elements (PEs),
(3) portability among various kinds of parallel and distributed environments ranging from PC cluster to MPPs,
(4) uni-phenomenon analyses, coupled analyses, design analyses in a parallel and distributed environment in an efficient and unified manner, and
(5) extensibility and maintainability.

To resolve these issues simultaneously, we have decided to develop in the project the advanced general-purpose computational mechanics system for large-scale analysis and design, named ADVENTURE.

2.2. Design and development strategies

As will be described in Section 3, the ADVENTURE system has employed a hierarchical domain-decomposition-based massively parallel algorithm as one of the major solution algorithms in order to efficiently handle a huge-scale finite element model with 10–100 million DOFs. In addition, the following strategies for design and development of the system have been employed:

(1) module-based architecture,
(2) standardized I/O format and libraries to collaborate modules in an efficient and unified manner,
(3) object-oriented design and programming for each module, and
(4) commodity technologies in hardware as well as software.

Conventional computational mechanics systems such as NASTRAN have increased their program size to the order of million lines, and become very complicated software through long histories of program development over 20 years. As a result, it has been a very hard task to even slightly modify their source programs in order to adjust them for version-up of operating system (OS) or to add a little amount of lines for new function. A tremendous amount of efforts has been put to pursue such tasks in daily development. This is one of the main reasons, why it is very difficult to develop efficient parallel versions of conventional computational mechanics systems [5].

Even in the case of the ADVENTURE system, it is easily imagined that its total system size would reach the order of sub-million lines, considering a variety of functions to be implemented in it. Furthermore, yearly version-up of the system would be indispensable, according to yearly version-up of parallel computing environments. Thus, keeping extensibility and maintainability of the system is one of the critical issues.

Considering the above aspects, we have decided to employ module-based architecture of the system together with standardized I/O format and libraries [21,12]. Each portion of the system requiring different background knowledge, theory, algorithm and programming style is packed into an independent module. Fig. 1 illustrates the module structure of the ADVENTURE system. The pre-process modules include surface patch generators which convert geometry model data into a collection of triangular or quadrilateral surface patch data, a tetrahedral mesh generator [24,27], a hexahedral mesh generator [28,22], an attachment tool of boundary conditions and material properties onto the mesh, and a domain decomposer of a finite element model. The main process modules include an implicit elastic–plastic analysis module [11,16], a thermal-fluid analysis module, a magnetic analysis module [6], an explicit impact analysis module, and a rigid plastic analysis module. The post-process module is for visualization of analysis results [18]. Each module is an independent software and its program size is limited to approximately thousands to 30,000 lines, so that each module could be kept manageable by one expert engineer.

There are two main reasons why we newly develop standardized I/O format and libraries. First, different main modules deal with different mechanics variables, while their relevant finite element models must be decomposed into a number of subdomains using a unique domain-decomposer module. To perform this process efficiently irrespective of mechanics, it is necessary to standardize I/O format and libraries. Second, the ADVENTURE system is basically required to deal with a large scale of finite element models with 10–100 million DOF mesh in various parallel and distributed computing environments. For this purpose, it is also indispensable to achieve efficient data I/O and

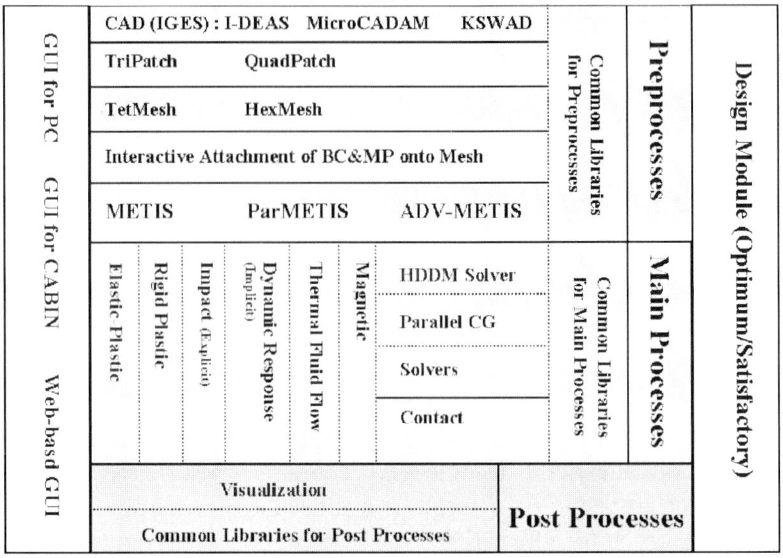

Fig. 1. Module-based architecture.

data transfer. The developed I/O system is named ADVENTURE_IO, which will be described in detail in the next section. Employing the ADVENTURE_IO also provides great benefit to coupled analyses [21,14] and design optimization calculations [26], in both of which efficient connection of several different modules is required. This feature will also be explained later.

Other design strategies of the project are to employ only commodity technologies in hardware as well as software, and not to employ any special tools or options. To make R& D processes in software projects efficient, it is usual to employ certain commercial tools, middleware and/or compiling options. However, considering the fact that application programs like CAE systems could have a longer life than those tools and environments, such a strategy of employing various tools and special environments might bring a kind of risk into the project. Sometimes licence of these tools limits the distribution of developed programs. To avoid such risk as much as possible, we have decided to employ only commodity technology in hardware as well as software, and not to employ any special tools or options. In the project, C and C++ are fundamental programming languages to build pre-, main and post-processing modules. MPI is utilized for communication among processors. Operating systems (OS) employed are Unix/Linux. OpenGL and Motif are adopted to deal with graphical data. In the project, we develop a minimum set of GUI and put most efforts to build kernel parts of the system. We expect that some private companies will build high-quality GUI for the ADVENTURE system, considering specific needs for potential applications and users. As described later, it is also essential that if required, all analyses can be performed fully automatically without any intervention by users. To do so, each module is designed to run by command-based operation as well. Although most modules have been developed by us from the start, only one exception is that graph partition tools named METIS and PartMETIS [7,8], which were developed in the University of Minnesota, have been employed as a kernel of the hierarchical domain decomposer of mesh. This is because these are efficient and high-quality software, and furthermore freely used when used for non-profit purpose.

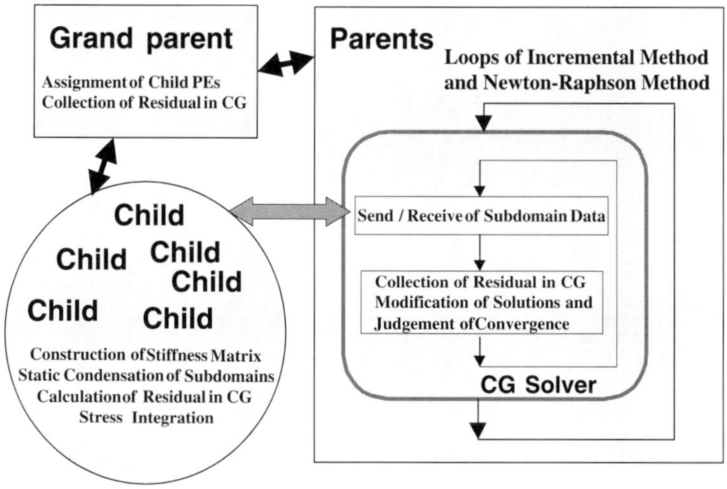

Fig. 2. Flow of HDDM for elastic–plastic analysis.

3. Key technologies

3.1. Parallel solver: hierarchical domain decomposition method (HDDM)

In domain decomposition methods (DDM), an analysis model, i.e., a finite element mesh with boundary conditions and material properties, is subdivided into a number of subdomains. Yagawa and Shioya proposed a hierarchical technique to implement the DDM on MPPs [23]. This technique was named the hierarchical domain decomposition method (HDDM). In this method, a group of processing elements (PEs) are subdivided into three sub-groups: one Grand Parent PE (Grand Parent), several Parent PEs (Parent/Parents), and many Child PEs (Child/Children). For implicitly solving the linear algebraic equations derived from the FEM, a simple substructure-based conjugate gradient (CG) method (the primal substructuring method) [2] is adopted. It was reported in Ref. [2] that the convergence property of this method is not so good compared with the dual substructuring methods such as the finite element tearing interconnection (FETI), especially when solving ill-conditioned problems such as large-scale thin structures with less constraints. Therefore, we are now implementing a pre-conditioner named balancing domain decomposition (BDD) [10] into the HDDM [16].

Fig. 2 shows the schematic data flow among the three groups of PEs, i.e., Grand Parent, Parents and Children [11]. An analysis model is also subdivided into several "parts" whose number is the same as the number of Parents, and then each part is subdivided into a number of subdomains. Fig. 3 shows an example of hierarchically decomposed mesh for a simplified pressure vessel model. Each Parent stores in its memory a set of the part data, i.e., a number of sets of subdomain data. Each subdomain data includes coordinates of nodes, connectivities of elements, material properties, and information on subdomain interfaces. In nonlinear analyses of solid, stresses and displacements are stored in the Parents as well. In a dynamic workload balancing mode as shown in Fig. 4 [23], an analysis is performed in the following. All the Children are dynamically allocated to the Parents, and the data associated with a subdomain stored in the Parents are sent to the Children. Each Child receives the data of one subdomain, conducts the computation associated with it, and sends its

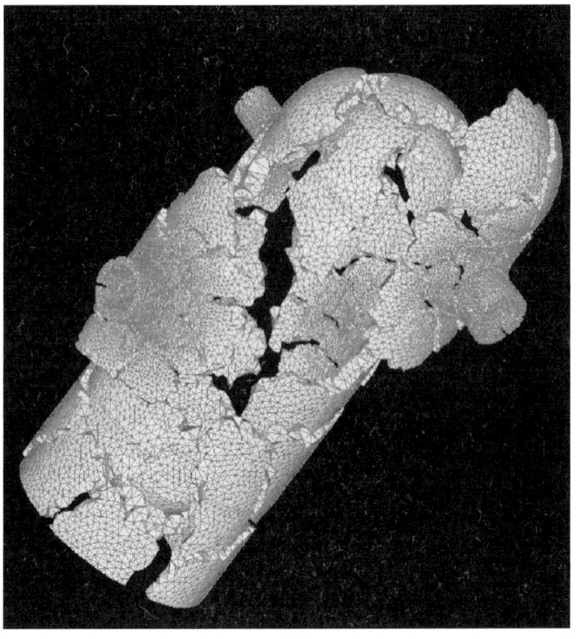

Fig. 3. Parts and subdomains of a simplified reactor vessel model.

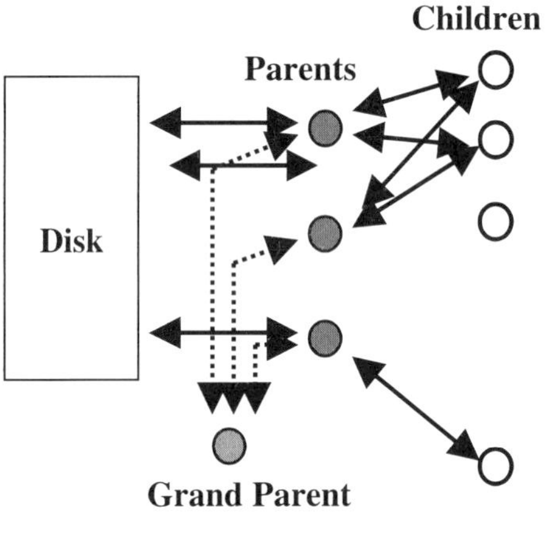

Fig. 4. Arrangement of PEs in HDDM and dynamic workload balancing process.

analysis results back to the Parents. The Grand Parent manages the following computation processes: check of the status of Parents and Children, and assignment of Children to Parents, and judgement of convergence of the iterative method.

The main features of the HDDM are summarized as follows:

(1) The number of Children is arbitrarily chosen when it is smaller than that of subdomains. As a result, it is possible for a user to change the number of PEs without changing that of subdomains.

(2) Dynamic workload balancing can be automatically achieved, if desired.
(3) Large-scale analysis data can be easily handled by increasing the number of Parents.
(4) A small number of Parents take care of disk I/O.
(5) Excellent parallel performance over 90% can be attained on MPPs.

We are now developing several kinds of main modules for elastic-plasticity, i.e., ADVENTURE_Solid [11], rigid plasticity, i.e., ADVENTURE_Forge, explicit impact, i.e., ADVENTURE_Impact [15], incompressible fluid flow, ADVENTURE_Fluid and magnetics, i.e., ADVENTURE_Magnetic [6]. Common functions related to finite elements are programmed as class libraries named libFEM.

3.2. Parallel domain decomposer

The requirements for domain decomposer are summarized as follows:

(1) *Performance*: There is a trade-off relation between processing speed and memory-use. In the project, we emphasize on the efficient use of memory rather than processing time reduction so as to deal with extra-large-scale meshes of millions to 100 million DOFs.

(2) *Minimizing the number of inner boundary conditions*: As described in the previous section, the present DDM employs an iterative solution algorithm such as a preconditioned CG method. It is well known that when analyzing the same FE problem using such an iterative DDM, the fewer the inner boundary conditions, the fewer the number of iterations to reach convergence. Therefore, minimizing the number of inner boundary conditions is very important to shorten a whole calculation time.

(3) *Balancing sizes of subdomains*: Although the present DDM has a feature of dynamic workload balancing during calculation [11,23], it is still desirable that the sizes of subdomains are well balanced to avoid unnecessary asynchronous overhead.

Several algorithms and software that can decompose an FE mesh into a number of subdomains have been developed so far in the literature. We have adopted the METIS code [7] and its parallel version ParMETIS [8], both of which are graph-partition packages that meet both the requirements (2) and (3). Although ParMETIS can partition graphs distributed over multiple processors, it has some limitation such that it can decompose graphs into only the same number of partitions as that of the processes working in the MPI environment. However, when solving a 100 million DOF problem, it is necessary to decompose an FE mesh into thousands of subdomains which are much larger than the number of available processes. Thus, using METIS and ParMETIS as kernel, we have developed a parallel domain decomposition module named ADVENTURE_Metis, which can create hierarchically decomposed model data, i.e., parts and subdomains (see Fig. 3). The present domain decomposition process consists of the following five phases as shown in Fig. 5. Each phase is fully parallelized.

(1) *Pre-filter phase*: An input FE mesh is converted into a graph, in which element is assigned to node of the graph and element connectivity is assigned to edge of the graph. Since the next partitioning-1 phase is performed with ParMETIS, the generated graph is distributed to multiple processors.

(2) *Partitioning-1 phase*: The graph obtained in the Pre-filter phase is partitioned into several sub-graphs named Parts by ParMETIS. This process is performed in parallel owing to an intrinsic feature of ParMETIS.

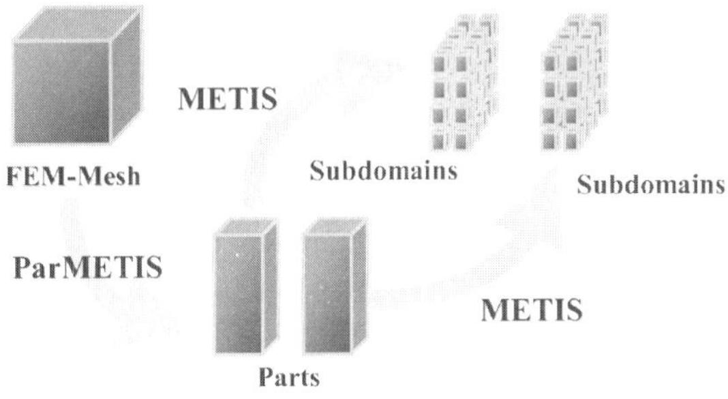

Fig. 5. Parallel domain decomposition.

(3) *Converter phase*: The data of the original FE mesh such as coordinates are localized in each partitioned part, and then inner boundary conditions between the parts are generated.

(4) *Partitioning-2 phase*: The localized graphs are partitioned with METIS. This process is performed in parallel as well.

(5) *Post-filter phase*: The graphs partitioned in the previous phases are regenerated into subdomains which correspond to the HDDM model. Inner boundary conditions among subdomains are generated and added to each subdomain. The final outputs are arranged into the same number of files as that of the processors.

3.3. Parallel visualization

Visualization for large-scale analyses of millions to 100 million DOFs requires an excessive amount of calculation and memory to be performed on a single processor system. For example, a problem with 10 million DOFs requires a memory of 3 GB to store the mesh itself. Thus parallel processing is also indispensable in the visualization process. In the ADVENTURE project, we have employed a client-server system for visualization [9], where a large number of analysis results are managed on a back-ended parallel environment, i.e., slaves, and both receiving user's requests and displaying image data are dealt with in a front-ended processor, i.e., master. As described in Section 3.2, the domain decomposer already generates the hierarchically decomposed data, i.e., several parts and numerous subdomains. The main solvers use them in their parallel calculations. The visualization module developed in the project also uses the same decomposed data, which are distributed on the back-ended parallel environment such as a PC cluster. Fig. 6 shows a typical computing environment for the present visualization module named ADVENTURE_Visual. In this module, a rendering process is dealt with in a single front-ended PE. If the amount of this process increases in accordance with further increase of model scale, it will be necessary to parallelize this process as well.

A front-ended PC receives a user's request for visualization via GUI or command-based operation, which is transferred into the back-ended PC cluster. The PC cluster performs most procedures starting from the calculation of physical quantities till the extraction of graphics primitives subdomain data by subdomain data in parallel. Then the extracted graphics primitives are transferred back to the front-ended PC. Rendering is finally performed there, i.e., the user operates the displayed object by zooming, rotating or shifting in a view space.

Another issue to be considered in the visualization of large-scale analyses is the trade-off between the number of calculations and the resolution of obtained image. Spatial resolution of a very

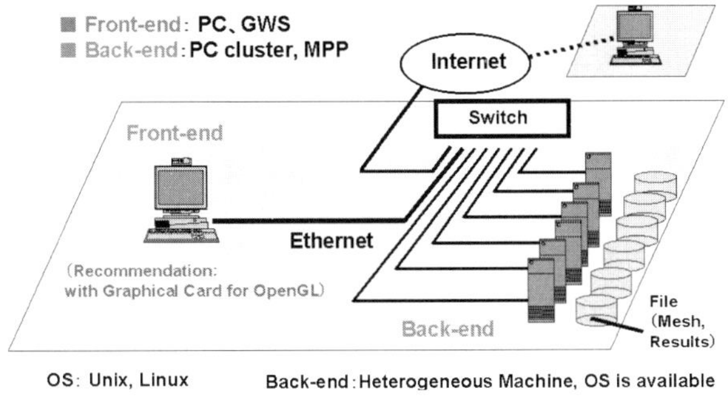

Fig. 6. A typical computing environment for ADVENTURE_Visual.

large-scale model often exceeds that of display monitor. In addition, even if the graphics primitives are quickly extracted due to parallel processing, there is some possibility that the rendering process on a front-ended processor is saturated due to a concentration of a huge amount of graphics data. For these reasons, the visualization should be performed, considering the resolution of monitor to be used. In the project, we have employed a method to simplify surface patches, based on a level-of-detail information. The current version of the ADVENTURE_Visual module can visualize surface contours, cross section and iso-surfaces [18].

3.4. Standardization of I/O

Characteristics of I/O data for large-scale computational mechanics analyses are summarized as follows:

(1) data size is very huge,
(2) the data must be dealt with in various types of parallel and distributed environments,
(3) the number of data files increases, and
(4) most of the data are of domain-decomposition type.

To efficiently handle such data with characteristics (1)–(3), we have developed "Media Independent Document" as a container of the data [12]. To efficiently deal with domain-decomposition type data for various kinds of mechanics analyses, i.e., characteristic (4), we have also developed a generalized I/O format [12].

The Media Independent Document (simply called the Document in this paper) can exist anywhere in a parallel and distributed environment, i.e., memory or files on hard disks of computers connected through a network. As shown in Fig. 7, the Document consists of the following two portions: the "Property" portion and the "Raw Data" portion. Stored in the Property portion are control data, the number of items stored in the Raw Data portion, labels and units of the data. The format of the Property portion is rather simple, i.e., key = value, where the value can be randomly accessed using the key. The Raw Data portion stores a huge amount of data such as nodal constraints, and connectivities of elements and analysis results in a binary format. These data can be accessed by several functions prepared in I/O libraries. The byte order of the binary data is unified to Little Endian. Different endian types among different hardware architectures are taken care of by the I/O libraries.

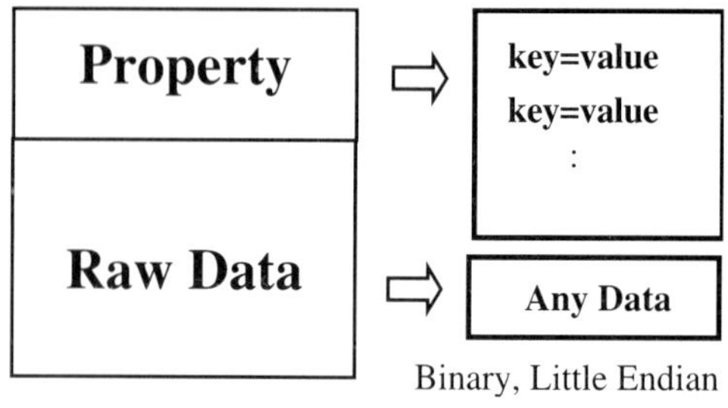

Fig. 7. Structure of media-independent document.

```
[Properties]
content_type=FEGenericAttribute
num_items=Number of Records
format=<value>'s format [((i|f)Number of Bytes)+]
fega_type=(All)?(Node|Element)(Variable|Constant)
label=Name with Physical Meaning (ex. Young, Poisson)
dd_option=Option for DomainDecomposer
index_byte=Number of Bytes of index (Usually 4(Bytes) because of int32))
(8 byte integer for ultra large scale problem)
[Data]
if fega_type=All(Node|Element)Variable
<value> x num_items
if fega_type=All(Node|Element)Constant
<value> x 1
if fega_type=(Node|Element)Variable
(<value>, <value>)  x num_items
if fega_type=All(Node|Element)Variable
<value>, <value> x num_items
```

Fig. 8. Media-independent document including FEGA.

Various kinds of input data are required to pass through the domain decomposer described in Section 3.2. To avoid complex specification for the domain decomposer, depending on different main modules, the model data other than the mesh are generalized in the following. The data used in a finite element analysis consist of a mesh, i.e., nodes and elements, and the data attached on to the mesh such as initial conditions, boundary conditions, material properties and analysis results. The latter data are attached to either nodes or elements. They can be abstracted in terms of the following eight general data types.

$$\text{fega\_type} = (\text{all})?(\text{node} \mid \text{element})(\text{variable} \mid \text{constant}).$$

For example, "NodeVariable" type means that some different sets of data in the same format are attached on to several nodes. The eight kinds of generalized data types are called FEGA (Finite Element Generic Attributes). The FEGA is similar to the UCD (Unstructured Cell Data) format of the commercial visualization software, AVS. The FEGA is stored in the Property portion of the FEGA document as shown in Fig. 8. Sets of data are stored in the Raw Data portion. The domain

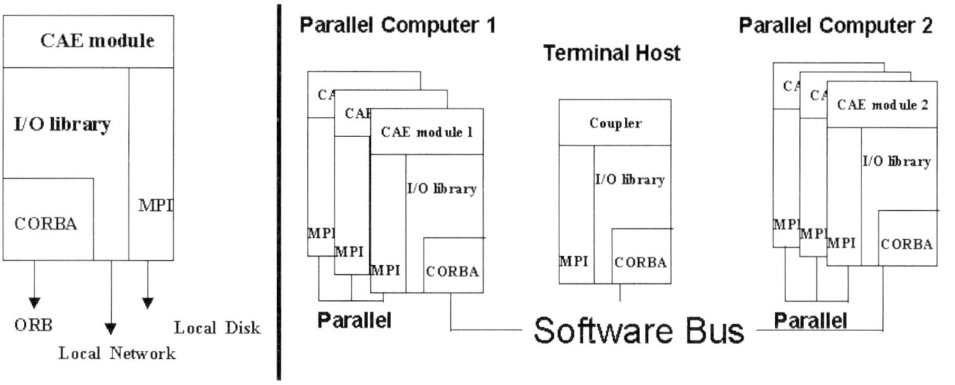

(a) CAE module with ADVENTURE_IO

(b) CAE modules collaboration via ADVENTURE_IO

Fig. 9.

decomposer of the ADVENTURE system can process any kinds of finite element model data once they are specified with the FEGA.

The ADVENTURE_IO [12], which is the name for the standardized I/O format and libraries, also possesses a generalized collaboration mechanism of multiple modules in a distributed computing environment, employing the CORBA technology. The CORBA is embedded in the ADVENTURE_IO libraries as illustrated in Fig. 9 [21]. By virtue of the mechanism, the modules can directly communicate among others in a parallel and distributed environment in a unified manner, which results in easy collaboration of multiple modules for coupled analyses [14] and optimization calculations [26].

3.5. Shape optimization

Owing to the standardization of I/O format and libraries, modules are easily integrated in several ways. Optimization calculations are easily performed by combining the ADVENTURE's pre-,main- and post-modules with Optimization modules. In the project, we are developing two kinds of design optimization modules. One is a module for parametric optimization, named the ADVENTURE_Opt, which is designed and programmed based on an object-oriented approach. The module implements various optimization algorithms including mathematical programming, gradient-based methods, simulated annealing, genetic/evolutionary algorithms.

The other is the nonparametric shape optimization module based on the traction method [1], which assures the smoothness of boundaries during shape varying processes. With this module, we can solve volume minimization problems of linear elastic continua under constraint on mean compliances with respect to given external forces. Reshaping is accomplished by the traction method in which boundary value problems of partial differential equations are defined, the shape gradient on the boundary is then evaluated and the boundary is varied by acting the traction in proportion to the negative value of the shape gradient.

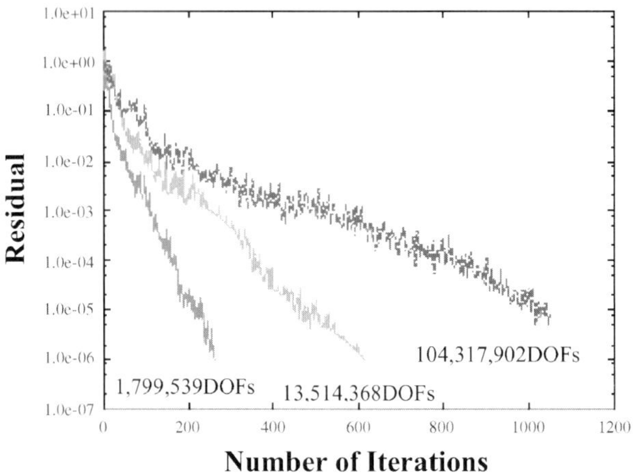

Fig. 10. History of CG residuals via iterations.

4. Numerical examples

As the first example, we performed a static stress analysis of a simplified model of a nuclear pressure vessel with a 100 million DOF mesh [17]. Fig. 10 shows a history of CG residual plotted against CG iterations. Using a Hitachi SR2201 with 1024 PEs, it took 2.1 h for 421 CG iterations to attain a solution with a residual of 10^{-3}. A preliminary study shows that the CG residual of 10^{-3} was sufficient to attain converged solutions of displacements in a practical sense [23]. Parallel performance was 82.3%.

The second example is a static stress analysis of a 60 million DOF mesh for a precise model of an advanced boiling water reactor (ABWR) vessel using the Hitachi SR2201 with 1024 PEs. Fig. 11(a) shows a CAD model created with MicroCADAM/Helix, and Fig. 11(b) shows a mesh of second-order tetrahedral elements generated by ADVENTURE_TriPatch and ADVENTURE_TetMesh. The former module is for surface triangle patch generation, while the latter is for tetrahedral mesh generation. Fig. 11(c) shows 40 parts of the HDDM model decomposed by ADVENTURE_Metis. Fig. 11(d) shows stress distributions on the surface of the vessel. As a model analysis, uniform displacement-controlled compressible load was applied to the edge of one of the nozzles. It took 19.3 h for 2043 CG iterations to attain a solution with a residual of 10^{-3}. An ordinary PC or a PC cluster was used for pre- and post-processes. A whole pre-process of this model excluding a CAD process took one day, while the visualization of surface stress distribution took about a few minutes.

The third example is an elasto-dynamic analysis of an ancient Roman dome Pantheon (see Fig. 12) subjected to El Centro NS earthquake loading. This is a hypothetical problem. A 2.44 million DOF mesh is used. Time increment is 0.01 s. Fig. 13 shows 14 parts of the mesh. Time history of displacement at a middle point of the cylinder portion is shown in Fig. 14. Fig. 15 shows the deformation after 2 s. Fig. 16 show a stress distribution on a vertical cross-section. Calculation time is 4 min/step using a cluster of 24 Alpha 21164 (533 MHz, 1 GB). A whole calculation time for 200 time steps is 13.5 h. Although concrete behaviors are nonlinear, material properties are assumed to be isotropic for the purpose of demonstration [13].

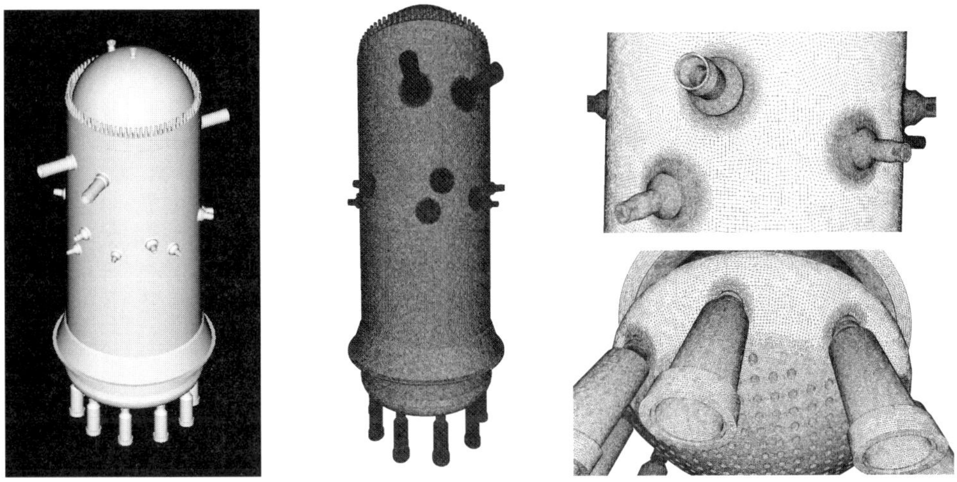

(a) CAD model of ABWR vessel (b) 60 million mesh of ABWR vessel

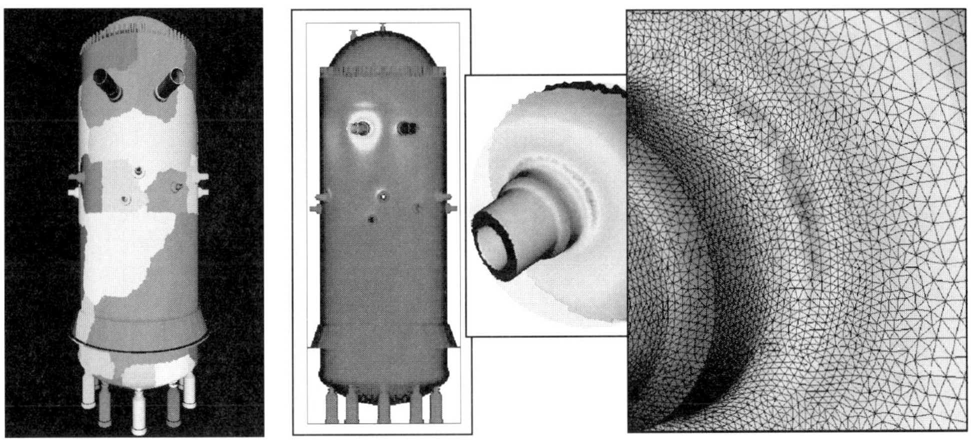

(c) 40 parts of HDDM (d) Stress distribution on the surface of ABWR model

Fig. 11.

The fourth example is a nonparametric shape optimization of support structure of express way as shown in Fig. 17 [12]. Fig. 18 shows a mesh with one million DOFs. The bottom surface is completely fixed, its right-hand surface was fixed in one direction, and both edges of the I-shape beam were fixed in one direction. Uniform pressure was applied to the top surface of the I-shape beam. In this analysis, a PC cluster consisting of 10 PCs (Alpha 21264 (667 MHz, 1 GB)) was used. Fig. 19 shows the obtained optimized shape. For one iteration of the traction force method, the main portion of the traction force method took only 197 s, while the portion of the parallel finite element analyses took 77,058 s. To get the optimized shape, three iterations for the traction force method were needed. It means that 15 times of the parallel finite element analysis with a one million DOF mesh were needed.

(a) Exterior (b) Interior

Fig. 12. An ancient Roman dome, Pantheon.

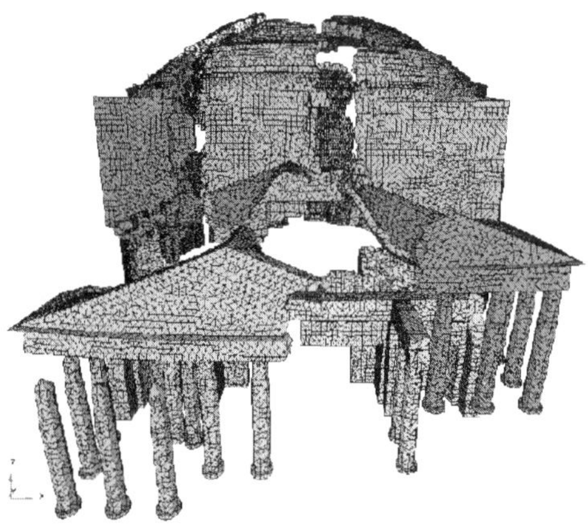

Fig. 13. 14 parts of a 2.44 million mesh of Pantheon.

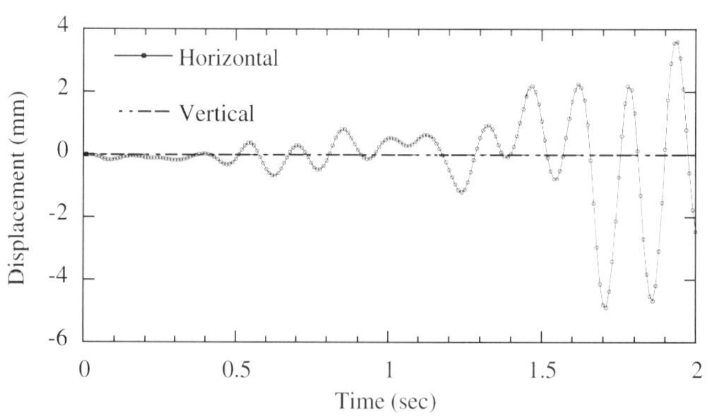

Fig. 14. Displacements vs. time.

Fig. 15. Deformation of Pantheon after 2 s.

Fig. 16. Stress distribution along vertical cross section.

Fig. 17. Support structure of express way.

Fig. 18. One million mesh of support structure.

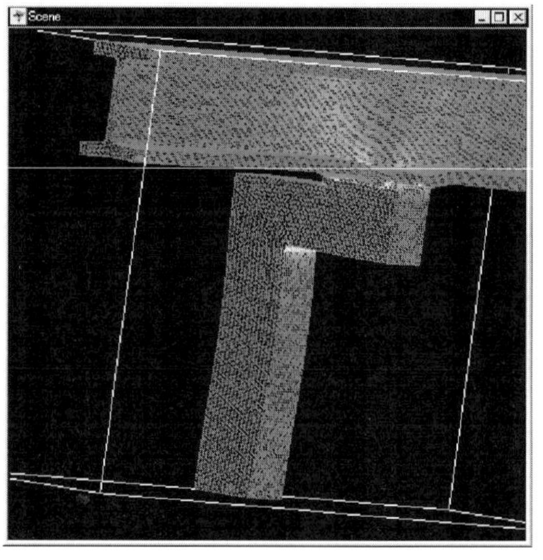

Fig. 19. Optimum shape.

5. Conclusions

The present paper describes the module-based parallel computational mechanics system for large-scale analysis and design named ADVENTURE. After describing the design concept of the system, some key technologies employed to develop the system are presented, i.e., a parallel solver based on the hierarchical domain decomposition, standardization of I/O format and libraries, the parallel domain decomposer and the parallel visualizer. Practical performances of the ADVENTURE system are demonstrated through some static and dynamic analyses, and shape optimization calculation of 3D precise models with one million to 100 million DOF meshes.

The β version of the ADVENTURE system has been released from the project website as free source programs since December 1, 2000 [3]. About 520 registered users in academia and industries are now testing the programs, while one private company has developed and released its commercial version named ADVentureCluster in May 2001 [19,20].

Acknowledgements

This work was performed as a part of Research for the Future Program sponsored by the Japan Society for the Promotion of Science. The authors also wish to thank Professor Yagawa, University of Tokyo and all the members of the ADVENTURE project.

References

[1] H. Azegami, Z.C. Wu, Domain optimization analysis in linear elastic problems (Approach using traction method) JSME Internat. J. Ser. A 39 (1996) 272–278.
[2] C. Farhat, F.X. Roux, Implicit parallel processing in structural mechanics, Comput. Mech. Adv. 2 (1994) 1–124.
[3] http://adventure.q.t.u-tokyo.ac.jp/.
[4] http://proton.is.s.u-tokyo.ac.jp/cse/index-e.html/.
[5] http://www.gmd.de/SCAI/europort.
[6] H. Kanayama, R. Shioya, D. Tagami, M. Saito, Numerical analysis of 3D eddy current problems by the hierarchical domain decomposition method, Trans. Japan Soc. Comput. Eng. Sci. 3 (2001) 151–156 (in Japanese).
[7] G. Karypis, V. Kumar, Multilevel k-way partitioning scheme for irregular graphs, Technical Report TR 95-064, Department of Computer Science, University of Minnesota, 1995.
[8] G. Karypis, V. Kumar, Parallel multilevel k-way partitioning scheme for irregular graphs, Technical Report TR 96-036, Department of Computer Science, University of Minnesota, 1996.
[9] P.W. Liu, L.S. Chen, S.C. Chen, J.P. Chen, F.Y. Lin, S.S. Hwang, Distributed computing: New power for scientific visualization IEEE Comput. Graphics Appl. 16 (1996) 42–51.
[10] J. Mandel, Balancing domain decomposition, Comm. Numer. Methods Eng. 9 (1993) 233–241.
[11] T. Miyamura, H. Noguchi, R. Shioya, S. Yoshimura, G. Yagawa, Elastic-plastic analysis of nuclear structures with millions of DOFs using the hierarchical domain decomposition method, Nuclear Eng. Design 212 (2002) 335–355.
[12] T. Miyamura, S. Tanaka, H. Takubo, S. Yoshimura, G. Yagawa, Standardization of Input/Output data in large scale parallel computational mechanics system, Trans. Japan Soc. Comput. Eng. Sci. 2 (2000) 219–226 (in Japanese).
[13] T. Miyamura, S. Yoshimura, Parallel stress analyses of ancient architecture Pantheon on PC cluster, Trans. Architect. Inst. Japan 550 (2001) 95–102. (in Japanese).
[14] Y. Nakabayashi, S. Yoshimura, G. Yagawa, Fluid-structure weak coupled analysis in parallel and distributed environment, Comput. Methods Appl. Mech. Eng., in press.
[15] A. Oishi, S. Yoshimura, G. Yagawa, A parallel explicit finite element analysis based on hierarchical domain decomposition, Trans. Japan Soc. Comput. Eng. Sci. 3 (2000) 131–226 (in Japanese).
[16] R. Shioya, H. Kanayama, D. Tagami, M. Ogino, 3D large scale structural analysis using a balancing domain decomposition method, Trans. Japan Soc. Comput. Eng. Sci. 2 (2000) 139–144 (in Japanese).
[17] R. Shioya, G. Yagawa, Parallel finite element analysis of 100 million DOFs based on the hierarchical domain decomposition method, Trans. Japan Soc. Comput. Eng. Sci. 3 (2001) 201–206 (in Japanese).
[18] S. Shoui, S. Yoshimura, H. Akiba, T. Ohyama, G. Yagawa, Parallel visualization of finite element solutions with ten million DOFs using PC cluster, Proceedings of European Congress on Computational Methods in Science and Engineering (ECCOMAS2000), Balcelona, CD-ROM, 2000.
[19] M. Suzuki, H. Akiba, S. Yoshimura, G. Yagawa, Analysis of stress intensity factor of piping using large scale analysis code ADVentureCluster, Transactions of 16th International Conference on Structural Mechanics in Reactor Technology, G05/5, Washington, DC, CD-ROM, 2001.

[20] M. Suzuki, T. Ohyama, H. Akiba, H. Noguchi, S. Yoshimura, Development of fast and robust parallel Coarse-grid based CG solver for large scale finite element analyses, Trans. Japan Soc. Mech. Eng. 68A–671 (2002) (in Japanese).
[21] S. Tanaka, Y. Wada, S. Yoshimura, G. Yagawa, A framework for integration of CAE modules in a distributed environment, Internet Transactions of Japan Society of Computational Engineering and Science, Paper No. 19990026, 1999 (in Japanese).
[22] Y. Wada, S. Yoshimura, Hexahedral mesh generation of nuclear structures using intelligent local approach, Nuclear Eng. Design 212 (2002) 321–333.
[23] G. Yagawa, R. Shioya, Parallel finite elements on a massively parallel computer with domain decomposition, Comput. Systems Eng. 4 (1993) 495–503.
[24] G. Yagawa, S. Yoshimura, K. Nakao, Automatic mesh generation of complex geometries based on fuzzy knowledge processing and computational geometry, Integrated Comput. Aided Eng. 2 (1995) 265–280.
[25] S. Yoshimura, Development of computational mechanics system for large scale analysis and design, Report on Computational Science and Engineering, JSPS-RFTF Program 2000–2001, 2001, pp. 13–24.
[26] S. Yoshimura, N. Oshio, H. Kawai, H. Azegami, K. Takeuchi, Large-scale parallel shape optimization using hierarchical domain decomposition with traction method, J. Japan Soc. Simul. Technol. 20-4 (2001) 32–38. (in Japanese).
[27] S. Yoshimura, H. Nitta, G. Yagawa, H. Akiba, Parallel automatic mesh generation of nuclear structures with ten-million nodes, Transactions of 15th International Conference on Structural Mechanics in Reactor Technology, Seoul, Vol. II, 1999, pp. 21–28.
[28] S. Yoshimura, Y. Wada, G. Yagawa, Automatic mesh generation of quadrilateral elements using intelligent local approach, Comput. Methods Appl. Mech. Eng. 179 (1999) 125–138.

ELSEVIER

Journal of Computational and Applied Mathematics 149 (2002) 297–305

JOURNAL OF COMPUTATIONAL AND APPLIED MATHEMATICS

www.elsevier.com/locate/cam

A class of product-type Krylov-subspace methods for solving nonsymmetric linear systems

Shao-Liang Zhang

Department of Applied Physics, University of Tokyo, 7-3-1 Hongo, Bunkyo, Tokyo 113-8656, Japan

Received 31 October 2001; received in revised form 6 December 2001

Abstract

In the present paper, a class of product-type Krylov-subspace methods for solving nonsymmetric linear systems is discussed. A characteristic of this class is the relationship $\boldsymbol{r}_n = H_n(A)\boldsymbol{r}_n^{\mathrm{BCG}}$ where $\boldsymbol{r}_n$ is the residual vector corresponding to the nth iterate $\boldsymbol{x}_n$, and $\boldsymbol{r}_n^{\mathrm{BCG}}$ is the nth residual generated in bi-conjugate gradient method. The polynomial H_n is chosen to speed up and stabilize convergence, while satisfying standard three-term recurrence relations. These product-type methods can be regarded as unifications and generalizations of CGS and Bi-CGSTAB.

1. Introduction

There is a strong need for the fast solution of large, nonsymmetric linear systems $A\boldsymbol{x} = \boldsymbol{b}$ which arise frequently in the field of scientific computing as subproblems or as intermediate steps. For solving these systems, direct methods, such as LU factorization, are often prohibitive both in terms of computation time and memory storage. For this reason, research on Krylov-subspace methods as an alternative has been the dominant paradigm in the field of numerical algebra since the publication of a paper by Lanczos [4] in the 1950s. In his paper, Lanczos proposed the bi-conjugate gradient method (Bi-CG hereafter) which was revived in [2].

In Bi-CG computation of the transpose matrix–vector multiplications must be performed because another Krylov subspace generated from the transpose matrix is used. Many efforts have been devoted to deriving more efficient methods from restructuring Bi-CG in past two decades to avoid calculating the transpose matrix–vector multiplications and to improve the convergence rate in Bi-CG.

E-mail address: zhang@zzz.t.u-tokyo.ac.jp (S.-L. Zhang).

0377-0427/02/$ - see front matter © 2002 Elsevier Science B.V. All rights reserved.
PII: S0377-0427(02)00537-X

A common technique is to define new residual polynomial by product of two polynomial factors: the Lanczos polynomial in Bi-CG and an undetermined polynomial of degree n that will be referred to hereafter as the accelerating polynomial. CGS [9], Bi-CGSTAB [10], Bi-CGSTAB2 [3], Bi-CGSTAB(L) [8] and GPBi-CG [11] were derived from Bi-CG by this common technique. Sonneveld defined the accelerating polynomial in CGS by the same Lanczos polynomial, and CGS was recognized as a powerful variant of Bi-CG by numerical experiments [9]. In Bi-CGSTAB, Van der Vorst defined the accelerating polynomial by using two-term recurrence relations to design the residual polynomial of Bi-CGSTAB. Gutknecht replaced the two-term recurrent polynomial in Bi-CGSTAB with a three-term recurrent polynomial only at even iterations and proposed Bi-CGSTAB2. Bi-CGSTAB2 is a generalization of Bi-CGSTAB, i.e., the combination Bi-CG with GMRES(2) [7], in the sense that Bi-CGSTAB is the combination of Bi-CG with GMRES(1). In [8], Sleijpen and Fokkema generated Bi-CGSTAB by combining Bi-CG with GMRES(L) and implemented a new algorithm, Bi-CGSTAB(L) for modest L. In GPBi-CG [11], Zhang proposed a different generalization of Bi-CGSTAB that uses a three-term recurrent polynomial at all iterations. He constructed the polynomial from a pair of coupled two-term recurrent polynomials. Recently, Röllin and Gutknecht [6] proposed a variant of GPBi-CG, named BICG $\times$ MR2_2 $\times$ 2. Although GPBi-CG and BICG $\times$ MR2_2 $\times$ 2 are mathematically equivalent, there are some differences between the two methods.

The primary aim of this paper is to highlight the most important development of product-type Krylov-subspace methods in past two decades. This paper is organized as follows: In the next section, we characterize the product-type Krylov-subspace methods based on Bi-CG and derive a set of recurrence formulas among the related iterates. In Section 3, several implementations of algorithms of the product-type methods are considered, and some well-known variants are recalled. In Section 4, we report results from numerical experiments that indicate that GPBi-CG may be more attractive than Bi-CGSTAB. We conclude with a summary of our findings in Section 5. Throughout this paper, the superscript BCG is used to distinguish iterates generated in the algorithm of Bi-CG.

2. Product-type Krylov-subspace methods

The Bi-CG algorithm in [2] for solving linear system $A\boldsymbol{x} = \boldsymbol{b}$ where A be an $N \times N$ large and sparse nonsymmetric matrix is as follows:

Algorithm 1. Bi-CG

$\boldsymbol{x}_0^{\mathrm{BCG}}$ *is an initial guess*, set $\boldsymbol{p}_0^{\mathrm{BCG}*} = \boldsymbol{r}_0^{\mathrm{BCG}*} = \boldsymbol{p}_0^{\mathrm{BCG}} = \boldsymbol{r}_0^{\mathrm{BCG}} = \boldsymbol{b} - A\boldsymbol{x}_0^{\mathrm{BCG}}$;

for $n = 0, 1, \ldots$ *until* $\| r_n^{\mathrm{BCG}} \| \leqslant \varepsilon \|\boldsymbol{b}\|$ do:

$$\alpha_n = (\boldsymbol{r}_n^{\mathrm{BCG}*}, \boldsymbol{r}_n^{\mathrm{BCG}})/(\boldsymbol{p}_n^{\mathrm{BCG}*}, A\boldsymbol{p}_n^{\mathrm{BCG}}),$$

$$\boldsymbol{x}_{n+1}^{\mathrm{BCG}} = \boldsymbol{x}_n^{\mathrm{BCG}} + \alpha_n \boldsymbol{p}_n^{\mathrm{BCG}},$$

$$\boldsymbol{r}_{n+1}^{\mathrm{BCG}} = \boldsymbol{r}_n^{\mathrm{BCG}} - \alpha_n A\boldsymbol{p}_n^{\mathrm{BCG}},$$

$$\boldsymbol{r}_{n+1}^{\mathrm{BCG}*} = \boldsymbol{r}_n^{\mathrm{BCG}*} - \alpha_n A^T \boldsymbol{p}_n^{\mathrm{BCG}*}, \tag{1}$$

$$\beta_n = (\boldsymbol{r}_{n+1}^{\mathrm{BCG}*}, \boldsymbol{r}_{n+1}^{\mathrm{BCG}})/(\boldsymbol{r}_n^{\mathrm{BCG}*}, \boldsymbol{r}_n^{\mathrm{BCG}}),$$

$$\boldsymbol{p}_{n+1}^{\mathrm{BCG}} = \boldsymbol{r}_{n+1}^{\mathrm{BCG}} + \beta_n \boldsymbol{p}_n^{\mathrm{BCG}},$$

$$\boldsymbol{p}_{n+1}^{\mathrm{BCG}*} = \boldsymbol{r}_{n+1}^{\mathrm{BCG}*} + \beta_n \boldsymbol{p}_n^{\mathrm{BCG}*}. \tag{2}$$

In the Bi-CG algorithm, two Krylov subspaces $K_n(A;\boldsymbol{r}_0) := \mathrm{span}\{\boldsymbol{r}_0, A\boldsymbol{r}_0, \ldots, A^{n-1}\boldsymbol{r}_0\}$ and $K_n(A^{\mathrm{T}};\boldsymbol{r}_0^*) := \mathrm{span}\{\boldsymbol{r}_0^*, A^{\mathrm{T}}\boldsymbol{r}_0^*, \ldots, (A^{\mathrm{T}})^{n-1}\boldsymbol{r}_0^*\}$ are generated, and the approximate solution $\boldsymbol{x}_n^{\mathrm{BCG}}$ has an associated residual $\boldsymbol{r}_n^{\mathrm{BCG}} (= \boldsymbol{b} - A\boldsymbol{x}_n^{\mathrm{BCG}}$ theoretically) that is orthogonal with respect to $K_n(A^{\mathrm{T}};\boldsymbol{r}_0^*)$. Therefore, we have the following orthogonalities of Bi-CG [2]:

$$\boldsymbol{r}_n^{\mathrm{BCG}} \perp K_n(A^{\mathrm{T}};\boldsymbol{r}_0^*) \quad \text{and} \quad A\boldsymbol{p}_n^{\mathrm{BCG}} \perp K_n(A^{\mathrm{T}};\boldsymbol{r}_0^*). \tag{3}$$

Note that $\boldsymbol{r}_n^{\mathrm{BCG}*}$ and $\boldsymbol{p}_n^{\mathrm{BCG}*}$ can be written as $\boldsymbol{r}_n^{\mathrm{BCG}*} = (-1)^n \prod_{i=0}^{n-1} \alpha_i (A^{\mathrm{T}})^n \boldsymbol{r}_0^* + \boldsymbol{g}_1$ and $\boldsymbol{p}_n^{\mathrm{BCG}*} = (-1)^n \prod_{i=0}^{n-1} \alpha_i (A^{\mathrm{T}})^n \boldsymbol{r}_0^* + \boldsymbol{g}_2$, where $\boldsymbol{g}_1$ and $\boldsymbol{g}_2 \in K_n(A^{\mathrm{T}}, \boldsymbol{r}_0^*)$. Using orthogonalities (3), auxiliary formulas for computing α_n and β_n can be recovered:

$$\alpha_n = \frac{((A^{\mathrm{T}})^n \boldsymbol{r}_0^*, \boldsymbol{r}_n^{\mathrm{BCG}})}{((A^{\mathrm{T}})^n \boldsymbol{r}_0^*, A\boldsymbol{p}_n^{\mathrm{BCG}})} \quad \text{and} \quad \beta_n = -\alpha_n \frac{((A^{\mathrm{T}})^{n+1} \boldsymbol{r}_0^*, \boldsymbol{r}_{n+1}^{\mathrm{BCG}})}{((A^{\mathrm{T}})^n \boldsymbol{r}_0^*, \boldsymbol{r}_n^{\mathrm{BCG}})}. \tag{4}$$

Here, we attempt to use an n-degree polynomial $H_n(A)$ to accelerate r_n^{BCG}, i.e., set $H_n(A)\boldsymbol{r}_n^{\mathrm{BCG}}$ to be the new residual to accelerate the convergence toward zero. By doing so, we can derive a class of methods whose characteristic is the relationship

$$\boldsymbol{r}_n = H_n(A)\boldsymbol{r}_n^{\mathrm{BCG}},$$

where $\boldsymbol{r}_n$ is the residual vector corresponding to the nth iterate $\boldsymbol{x}_n$. The polynomial $H_n(A)$ is chosen to speed up and stabilize convergence.

In practice, the polynomial H_n is designed to have the following desirable properties:

(1) it satisfies short-term recurrence relations so that little computational work and storage costs are required per iteration;
(2) its parameters are chosen to lead to fast and stable convergence behavior.

Next, we will describe the basic idea to establish a standard polynomial H_n that leads to a class of product-type methods based on Bi-CG.

We introduce two undetermined parameters ζ_n and η_n and define the polynomial H_n as follows:

$$H_0(\lambda) := 1, \quad H_1(\lambda) := (1 - \zeta_0\lambda)H_0(\lambda), \tag{5}$$

$$H_{n+1}(\lambda) := (1 + \eta_n - \zeta_n\lambda)H_n(\lambda) - \eta_n H_{n-1}(\lambda). \tag{6}$$

Note that $H_n(0) = 1$ holds for any n, so we have $H_{n+1}(0) - H_n(0) = 0$. Thus, we can find an auxiliary polynomial $G_n(\lambda)$ with degree n and obtain pairs of polynomials $H_n(\lambda)$ and $G_n(\lambda)$ that are mutually interlocked by the recurrence relations:

$$H_{n+1}(\lambda) = H_n(\lambda) - \lambda G_n(\lambda) \quad \text{and} \quad G_{n+1}(\lambda) = \zeta_{n+1}H_{n+1}(\lambda) + \eta_{n+1}G_n(\lambda). \tag{7}$$

Now, let us derive the product-type methods with residual

$$\boldsymbol{r}_n (= \boldsymbol{b} - A\boldsymbol{x}_n) = \boldsymbol{r}_n^{H_n} := H_n(A)\boldsymbol{r}_n^{\mathrm{BCG}} \tag{8}$$

that can be obtained using the iterates $\boldsymbol{r}_{n+1}^{H_n}:=H_n(A)\boldsymbol{r}_{n+1}^{\mathrm{BCG}}$, $\boldsymbol{p}_n^{H_n}:=H_n(A)\boldsymbol{p}_n^{\mathrm{BCG}}$, $\boldsymbol{r}_{n+1}^{AG_{n-1}}:=AG_{n-1}(A)\boldsymbol{r}_{n+1}^{\mathrm{BCG}}$, $\boldsymbol{p}_{n+1}^{AH_n}:=AH_n(A)\boldsymbol{p}_{n+1}^{\mathrm{BCG}}$, $\boldsymbol{p}_n^{AG_n}:=AG_n(A)\boldsymbol{p}_n^{\mathrm{BCG}}$ and $\boldsymbol{r}_{n+1}^{G_n}:=G_n(A)\boldsymbol{r}_{n+1}^{\mathrm{BCG}}$.

According to recurrence relations (1)–(2) and (5)–(7), we have a set of recurrence formulas among the sequences of the iterates $\boldsymbol{r}_n$, $\boldsymbol{p}_n^{H_n}$, $\boldsymbol{r}_{n+1}^{H_n}$, $\boldsymbol{p}_n^{AG_n}$, $\boldsymbol{p}_{n+1}^{AH_n}$, $\boldsymbol{r}_{n+1}^{AG_{n-1}}$ and $\boldsymbol{r}_{n+1}^{G_n}$:

$$\boldsymbol{r}_{n+1}=\boldsymbol{r}_{n+1}^{H_n}-\eta_n\boldsymbol{r}_{n+1}^{AG_{n-1}}-\zeta_nA\boldsymbol{r}_{n+1}^{H_n} \tag{9}$$

$$=\boldsymbol{r}_n-\alpha_nA\boldsymbol{p}_n^{H_n}-A\boldsymbol{r}_{n+1}^{G_n}, \tag{10}$$

$$\boldsymbol{r}_{n+1}^{H_n}=\boldsymbol{r}_n-\alpha_nA\boldsymbol{p}_n^{H_n}, \tag{11}$$

$$\boldsymbol{r}_{n+2}^{AG_n}=\boldsymbol{r}_{n+1}^{H_n}-\boldsymbol{r}_{n+1}-\alpha_{n+1}\boldsymbol{p}_{n+1}^{AH_n}+\alpha_{n+1}A\boldsymbol{p}_{n+1}^{H_{n+1}}, \tag{12}$$

$$\boldsymbol{p}_{n+1}^{H_{n+1}}=\boldsymbol{r}_{n+1}+\beta_n(\boldsymbol{p}_n^{H_n}-\boldsymbol{p}_n^{AG_n}), \tag{13}$$

$$\boldsymbol{p}_{n+1}^{AH_n}=A\boldsymbol{r}_{n+1}^{H_n}+\beta_nA\boldsymbol{p}_n^{H_n}, \tag{14}$$

$$\boldsymbol{p}_n^{AG_n}=\zeta_nA\boldsymbol{p}_n^{H_n}+\eta_n(\boldsymbol{r}_n^{H_{n-1}}-\boldsymbol{r}_n+\beta_{n-1}\boldsymbol{p}_{n-1}^{AG_{n-1}}), \tag{15}$$

$$\boldsymbol{r}_{n+1}^{G_n}=\zeta_n\boldsymbol{r}_n+\eta_n\boldsymbol{r}_n^{G_{n-1}}-\alpha_n\boldsymbol{p}_n^{AG_n}. \tag{16}$$

From (8) and (10), we have the formula to update the approximating solution: $\boldsymbol{x}_{n+1}=\boldsymbol{x}_n+\alpha_n\boldsymbol{p}_n^{H_n}+\boldsymbol{r}_{n+1}^{G_n}$.

Since the coefficient of the highest order term of H_n is $(-1)^n\prod_{i=0}^{n-1}\zeta_i$, we have

$$(\boldsymbol{r}_0^*,\boldsymbol{r}_n)=(-1)^n\prod_{i=0}^{n-1}\zeta_i((A^{\mathrm{T}})^n\boldsymbol{r}_0^*,\boldsymbol{r}_n^{\mathrm{BCG}}),\quad (\boldsymbol{r}_0^*,A\boldsymbol{p}_n^{H_n})=(-1)^n\prod_{i=0}^{n-1}\zeta_i((A^{\mathrm{T}})^n\boldsymbol{r}_0^*,A\boldsymbol{p}_n^{\mathrm{BCG}}).$$

Therefore, from formula (4), α_n and β_n can be recovered from the iterates $\boldsymbol{r}_{n+1}$, $\boldsymbol{r}_n$ and $\boldsymbol{p}_n^{H_n}$ as follows:

$$\alpha_n=(\boldsymbol{r}_0^*,\boldsymbol{r}_n)/(\boldsymbol{r}_0^*,A\boldsymbol{p}_n^{H_n})\quad\text{and}\quad\beta_n=(\alpha_n/\zeta_n)\cdot[(\boldsymbol{r}_0^*,\boldsymbol{r}_{n+1})/(\boldsymbol{r}_0^*,\boldsymbol{r}_n)]. \tag{17}$$

3. Implementation details

Following requirement (2) described in Section 2, we summarize several possibilities for selecting ζ_n and η_n for implementing product-type methods based on Bi-CG. As well-known variants, CGS, Bi-CGSTAB and GPBi-CG will be reviewed in terms of special choices.

3.1. The choice for GPBi-CG

It is convenient to determine parameters ζ_n and η_n for minimizing the residual 2-norm as a function of ζ and η: $\|\boldsymbol{r}_{n+1}^{H_n} - \eta \boldsymbol{r}_{n+1}^{AG_{n-1}} - \zeta A \boldsymbol{r}_{n+1}^{H_n}\|$.

Thus, we have a variant of the product-type methods, GPBi-CG [11]:

Algorithm 2. GPBi-CG

$\boldsymbol{x}_0$ is an initial guess, $\boldsymbol{r}_0 = b - A\boldsymbol{x}_0$; set $\boldsymbol{r}_0^* = \boldsymbol{r}_0$, $\boldsymbol{r}_0^{H_{-1}} = \boldsymbol{p}_0^{AH_{-1}} = 0$, $\beta_{-1} = 0$;
for $n = 0, 1, \ldots$ until $\|r_n\| \leqslant \varepsilon \|b\|$ do:

$$\boldsymbol{p}_n^{H_n} = \boldsymbol{r}_n + \beta_{n-1}(\boldsymbol{p}_{n-1}^{H_{n-1}} - \boldsymbol{p}_{n-1}^{AG_{n-1}}),$$

$$\alpha_n = (\boldsymbol{r}_0^*, \boldsymbol{r}_n)/(\boldsymbol{r}_0^*, A\boldsymbol{p}_n^{H_n}),$$

$$\boldsymbol{r}_{n+1}^{AG_{n-1}} = \boldsymbol{r}_n^{H_{n-1}} - \boldsymbol{r}_n - \alpha_n \boldsymbol{p}_n^{AH_{n-1}} + \alpha_n A\boldsymbol{p}_n^{H_n},$$

$$\boldsymbol{r}_{n+1}^{H_n} = \boldsymbol{r}_n - \alpha_n A\boldsymbol{p}_n^{H_n},$$

$$\zeta_n = \frac{(\boldsymbol{r}_{n+1}^{AG_{n-1}}, \boldsymbol{r}_{n+1}^{AG_{n-1}})(A\boldsymbol{r}_{n+1}^{H_n}, \boldsymbol{r}_{n+1}^{H_n}) - (\boldsymbol{r}_{n+1}^{AG_{n-1}}, \boldsymbol{r}_{n+1}^{H_n})(A\boldsymbol{r}_{n+1}^{H_n}, \boldsymbol{r}_{n+1}^{AG_{n-1}})}{(A\boldsymbol{r}_{n+1}^{H_n}, A\boldsymbol{r}_{n+1}^{H_n})(\boldsymbol{r}_{n+1}^{AG_{n-1}}, \boldsymbol{r}_{n+1}^{AG_{n-1}}) - (\boldsymbol{r}_{n+1}^{AG_{n-1}}, A\boldsymbol{r}_{n+1}^{H_n})(A\boldsymbol{r}_{n+1}^{H_n}, \boldsymbol{r}_{n+1}^{AG_{n-1}})},$$

$$\eta_n = \frac{(A\boldsymbol{r}_{n+1}^{H_n}, A\boldsymbol{r}_{n+1}^{H_n})(\boldsymbol{r}_{n+1}^{AG_{n-1}}, \boldsymbol{r}_{n+1}^{H_n}) - (\boldsymbol{r}_{n+1}^{AG_{n-1}}, A\boldsymbol{r}_{n+1}^{H_n})(A\boldsymbol{r}_{n+1}^{H_n}, \boldsymbol{r}_{n+1}^{H_n})}{(A\boldsymbol{r}_{n+1}^{H_n}, A\boldsymbol{r}_{n+1}^{H_n})(\boldsymbol{r}_{n+1}^{AG_{n-1}}, \boldsymbol{r}_{n+1}^{AG_{n-1}}) - (\boldsymbol{r}_{n+1}^{AG_{n-1}}, A\boldsymbol{r}_{n+1}^{H_n})(A\boldsymbol{r}_{n+1}^{H_n}, \boldsymbol{r}_{n+1}^{AG_{n-1}})},$$

(if $n = 0$, then $\zeta_n = (A\boldsymbol{r}_{n+1}^{H_n}, \boldsymbol{r}_{n+1}^{H_n})/(A\boldsymbol{r}_{n+1}^{H_n}, A\boldsymbol{r}_{n+1}^{H_n})$, $\eta_n = 0$)

$$\boldsymbol{p}_n^{AG_n} = \zeta_n A\boldsymbol{p}_n^{H_n} + \eta_n(\boldsymbol{r}_n^{H_{n-1}} - \boldsymbol{r}_n + \beta_{n-1}\boldsymbol{p}_{n-1}^{AG_{n-1}}),$$

$$\boldsymbol{r}_{n+1}^{G_n} = \zeta_n \boldsymbol{r}_n + \eta_n \boldsymbol{r}_n^{G_{n-1}} - \alpha_n \boldsymbol{p}_n^{AG_n},$$

$$\boldsymbol{x}_{n+1} = \boldsymbol{x}_n + \alpha_n \boldsymbol{p}_n^{H_n} + \boldsymbol{r}_{n+1}^{G_n},$$

$$\boldsymbol{r}_{n+1} - \boldsymbol{r}_{n+1}^{H_n} - \eta_n \boldsymbol{r}_{n+1}^{AG_{n-1}} - \zeta_n A\boldsymbol{r}_{n+1}^{H_n},$$

$$\beta_n = (\alpha_n/\zeta_n) \cdot [(\boldsymbol{r}_0^*, \boldsymbol{r}_{n+1})/(\boldsymbol{r}_0^*, \boldsymbol{r}_n)],$$

$$\boldsymbol{p}_{n+1}^{AH_n} = A\boldsymbol{r}_{n+1}^{H_n} + \beta_n A\boldsymbol{p}_n^{H_n}.$$

3.2. The choice for CGS

Suppose that $\zeta_n = \alpha_n$ and $\eta_n = (\beta_{n-1}/\alpha_{n-1})\alpha_n$ in recurrence relations (5) and (6). We obtain a significant variant of the product-type methods that only depends on information of Bi-CG. It is easy to see that this variant is mathematically equivalent to CGS.

Note that $\boldsymbol{r}_n^{H_{n-1}} - \boldsymbol{r}_n = \boldsymbol{r}_n^{AG_{n-1}}$. In this case, we have $H_n = R_n$ and $G_n = P_n$, and we use the recurrence formula (10) to update $\boldsymbol{r}_{n+1}$. This fact leads to the relation $\boldsymbol{p}_n^{H_n} - \boldsymbol{p}_n^{AG_n} = \boldsymbol{r}_{n+1}^{G_n/\alpha_n}$ for any n, then the iterates $\boldsymbol{r}_{n+1}^{H_n}$, $\boldsymbol{r}_{n+1}^{AG_{n-1}}$ and $\boldsymbol{p}_{n+1}^{AH_n}$ can be omitted in the recurrence formulas (9)–(16). And we obtain the CGS algorithm [9].

Algorithm 3. CGS

$\boldsymbol{x}_0$ is an initial guess, $\boldsymbol{r}_0 = b - A\boldsymbol{x}_0$; *set* $\boldsymbol{r}_0^* = \boldsymbol{r}_0$, $\beta_{-1} = 0$;

for $n = 0, 1, \ldots$ *until* $\|\boldsymbol{r}_n\| \leqslant \varepsilon\|b\|$ do:

$$\boldsymbol{p}_n^{H_n} = \boldsymbol{r}_n + \beta_{n-1}\boldsymbol{r}_n^{G_{n-1}/\alpha_{n-1}},$$

$$\boldsymbol{p}_n^{G_n/\alpha_n} = \boldsymbol{p}_n^{H_n} + \beta_{n-1}(\boldsymbol{r}_n^{G_{n-1}/\alpha_{n-1}} + \beta_{n-1}\boldsymbol{p}_{n-1}^{G_{n-1}/\alpha_{n-1}}),$$

$$\alpha_n = (\boldsymbol{r}_0^*, \boldsymbol{r}_n)/(\boldsymbol{r}_0^*, A\boldsymbol{p}_n^{G_n}),$$

$$\boldsymbol{r}_{n+1}^{G_n/\alpha_n} = \boldsymbol{p}_n^{H_n} - \alpha_n A\boldsymbol{p}_n^{G_n/\alpha_n},$$

$$\boldsymbol{x}_{n+1} = \boldsymbol{x}_n + \alpha_n(\boldsymbol{p}_n^{H_n} + \boldsymbol{r}_{n+1}^{G_n/\alpha_n}),$$

$$\boldsymbol{r}_{n+1} = \boldsymbol{r}_n - \alpha_n A(\boldsymbol{p}_n^{H_n} + \boldsymbol{r}_{n+1}^{G_n/\alpha_n}),$$

$$\beta_n = (\boldsymbol{r}_0^*, \boldsymbol{r}_{n+1})/(\boldsymbol{r}_0^*, \boldsymbol{r}_n);$$

3.3. The choice for Bi-CGSTAB

If we attempt to get a variant of the product-type methods with little computational work, we would define η_n by a quantity ω called the relaxation factor in advance. Suppose that $\eta_n = 0$ for any n, and ζ_n is selected to minimize the residual 2-norm as a function of ζ: $\|\boldsymbol{r}_{n+1}^{H_n} - \zeta A\boldsymbol{r}_{n+1}^{H_n}\|$. In this case, $\boldsymbol{p}_n^{AG_n} = \zeta_n A\boldsymbol{p}_n^{H_n}$ and $\boldsymbol{r}_{n+1}^{G_n} = \zeta_n\boldsymbol{r}_{n+1}^{H_n}$. Note that the iterates $\boldsymbol{r}_{n+1}^{AG_{n-1}}$, $\boldsymbol{p}_n^{AG_n}$, $\boldsymbol{r}_{n+1}^{G_n}$ and $\boldsymbol{p}_{n+1}^{AH_n}$ become worthless. In this manner, we obtain an important and economical variant called Bi-CGSTAB [10]:

Algorithm 4. Bi-CGSTAB

$\boldsymbol{x}_0$ is an initial guess, $\boldsymbol{r}_0 = b - A\boldsymbol{x}_0$; *set* $\boldsymbol{r}_0^* = \boldsymbol{r}_0$, $\beta_{-1} = 0$;

for $n = 0, 1, \ldots$ *until* $\|\boldsymbol{r}_n\| \leqslant \varepsilon\|b\|$ do:

$$\boldsymbol{p}_n^{H_n} = \boldsymbol{r}_n + \beta_{n-1}(\boldsymbol{p}_{n-1}^{H_{n-1}} - \zeta_{n-1}A\boldsymbol{p}_{n-1}^{H_{n-1}}),$$

$$\alpha_n = (\boldsymbol{r}_0^*, \boldsymbol{r}_n)/(\boldsymbol{r}_0^*, A\boldsymbol{p}_n^{H_n}),$$

$$\boldsymbol{r}_{n+1}^{H_n} = \boldsymbol{r}_n - \alpha_n A\boldsymbol{p}_n^{H_n},$$

$$\zeta_n = (A\boldsymbol{r}_{n+1}^{H_n}, \boldsymbol{r}_{n+1}^{H_n})/(A\boldsymbol{r}_{n+1}^{H_n}, A\boldsymbol{r}_{n+1}^{H_n}),$$

$$\boldsymbol{x}_{n+1} = \boldsymbol{x}_n + \alpha_n\boldsymbol{p}_n^{H_n} + \zeta_n\boldsymbol{r}_{n+1}^{H_n},$$

$$\boldsymbol{r}_{n+1} = \boldsymbol{r}_{n+1}^{H_n} - \zeta_n A\boldsymbol{r}_{n+1}^{H_n},$$

$$\beta_n = (\alpha_n/\zeta_n) \cdot [(\boldsymbol{r}_0^*, \boldsymbol{r}_{n+1})/(\boldsymbol{r}_0^*, \boldsymbol{r}_n)].$$

4. Numerical experiments

For solving actual problems, all variants are combined with efficient preconditioning techniques, such as the incomplete LU factorization [5].

Let K be a suitable preconditioning matrix, i.e., $K \approx A$. We write $K = K_1K_2$, and apply CGS, Bi-CGSTAB and GPBi-CG to the explicitly preconditioned system $\tilde{A}\tilde{\boldsymbol{x}}=\tilde{\boldsymbol{b}}$ with $\tilde{A}=K_1^{-1}AK_2^{-1}, \tilde{\boldsymbol{x}}=K_2\boldsymbol{x}$, and $\tilde{\boldsymbol{b}} = K_1^{-1}\boldsymbol{b}$.

Preconditioned GPBi-CG requires the evaluation of two matrix–vector products with A, two solvers for K, $32N$ flops for vector updates, and seven inner products. Preconditioned CGS requires evaluation of two matrix–vector products with A, two solvers for K, $13N$ flops for vector updates, and two inner products. Preconditioned Bi-CGSTAB requires evaluation of two matrix–vector products with A, two solvers for K, $12N$ flops for vector updates, and four inner products. In practical situations, a few vector updates and inner products lead to only a small increase in computational work per iteration step. On vector and parallel computers vector updates and inner products can be computed faster than matrix vector products with A and solvers for K.

In this section, we report some numerical experiments in which CGS, Bi-CGSTAB and GPBi-CG are applied to the explicitly preconditioned system $L^{-1}AU^{-1}(Ux)=L^{-1}b$ in double precision floating point arithmetic on a SUN SPARCstation IPX computer. In all experiments the iteration started with $\boldsymbol{x}_0 = 0$. Our convergence plots show the relative residual 2-norms $\|\boldsymbol{r}_n\|/\|\boldsymbol{r}_0\|$ (on the vertical axis) vs. the iteration number n (on the horizontal axis). The convergence plots of CGS were cut because of irregular and oscillatory convergence behavior.

4.1. Example 1

Consider a linear system that comes from discretization of the Poisson equation $-(a(x, y)u_x)_x - (a(x, y)u_y)_y = f(x, y)$ over the unit square with Dirichlet boundary conditions: $u=1$, for $y=0$, $x=0$ and 1, and $u = 0$ for $y = 1$. The function $a(x, y)$ is defined as shown in Fig. 1; $f(x, y) = 0$

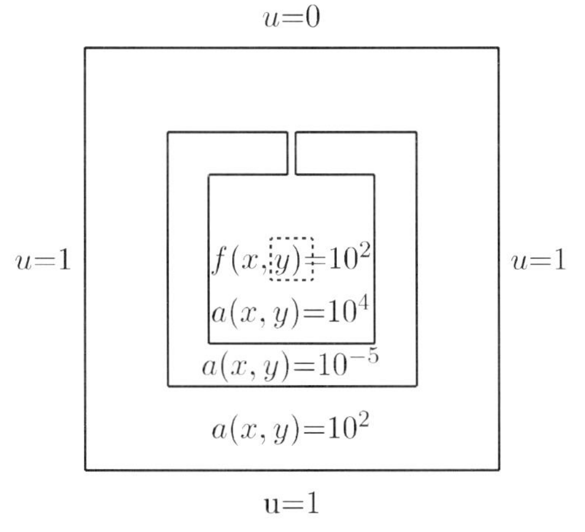

Fig. 1. Coefficients in the Poisson equation.

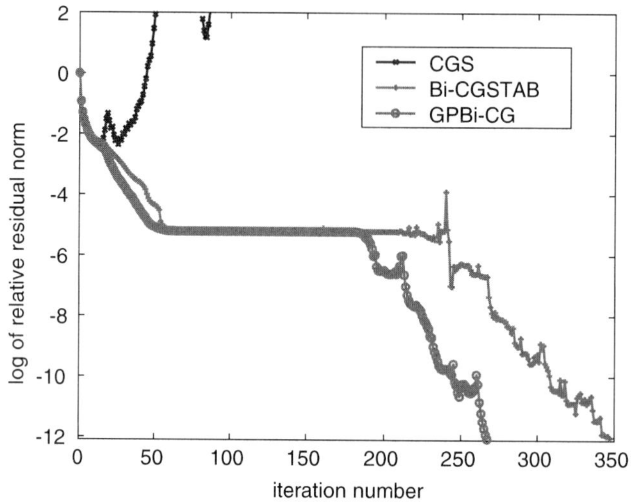

Fig. 2. The residual norm history $((\log(\|\boldsymbol{r}_n\|/\|\boldsymbol{r}_0\|))$ vs. $n)$.

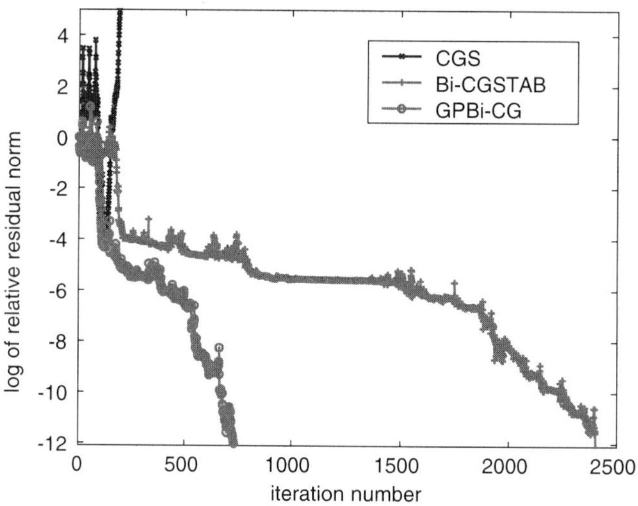

Fig. 3. The residual norm history $((\log(\|\boldsymbol{r}_n\|/\|\boldsymbol{r}_0\|))$ vs. $n)$.

everywhere except the smallest square in the center in which $f(x, y) = 100$. We discretized the equation over a rectangular grid using central differences and stepsize $\frac{1}{201}$ in each direction, which led to a system with 200^2 unknowns. The linear system was preconditioned using the incomplete LU factorization. Fig. 2 shows that GPBi-CG converges slightly faster than Bi-CGSTAB.

4.2. Example 2

Consider a linear system with complex spectrum that comes from the 200×200 central difference discretization of the Helmholtz equation $u_{xx} + u_{yy} + k^2 u = 0$ over $[0,\pi] \times [0,\pi]$ described in [1] with Dirichlet condition $u = 0$ along $y = \pi$, Neumann conditions $u_x = i\sqrt{k^2 - \frac{1}{4}}\cos(y/2)$ along $x = 0$ and $u_y = 0$ along $y = 0$, and radiation condition $u_x - i\sqrt{k^2 - \frac{1}{4}}u = 0$ along $x = \pi$. This leads to a system with 201×200 unknowns. We only consider the case $k = 2.27$. The linear system was preconditioned using the incomplete LU factorization.

Although GPBi-CG is more expensive with respect to the number of inner products and vector updates than Bi-CGSTAB, Fig. 3 shows that GPBi-CG only required 734 iteration steps to get the residual 2-norm below 10^{-12} while Bi-CGSTAB required 2404 iteration steps to achieve the same level of accuracy. More specifically, GPBi-CG only required 41% of the CPU-time of Bi-CGSTAB.

5. Concluding remarks

In view of more stable convergence behavior and reduced CPU-time and storage requirements, the polynomial H_n generated by the three-term recurrence relation (6) is better than others which come from truncated iterative method such as GMRES(k) ($k > 2$) [7] for requirement (1) described in Section 2. Our experiments indicate that GPBi-CG may be an attractive method.

References

[1] A. Bayliss, C.I. Glodstein, E. Turkel, An iterative method for the Helmholtz equation, J. Comput. Phys. 49 (1983) 443–457.

[2] R. Fletcher, Conjugate Gradient Methods for Indefinite Systems, Lecture Notes in Mathematics, Vol. 506, Springer, Berlin, 1976, pp. 73–89

[3] M.H. Gutknecht, Variants of Bi-CGSTAB for matrices with complex spectrum, SIAM J. Sci. Comput. 14 (1993) 1020–1033.

[4] C. Lanczos, Solution of systems of linear equations by minimized iterations, J. Res. Nat. Bur. Standards 49 (1952) 33–53.

[5] J.A. Meijerink, H.A. Van der Vorst, An iterative solution method for linear systems of which the coefficient matrix is a symmetric M-matrix, Math. Comput. 31 (1977) 148–162.

[6] S. Röllin, M.H. Gutknecht, Variations of Zhang's Lanczos-type product method, Appl. Numer. Math., to appear.

[7] Y. Saad, M.H. Schultz, GMRES: a generalized minimal residual algorithm for solving nonsymmetric linear systems, SIAM J. Sci. Statist. Comput. 7 (1986) 856–869.

[8] G.L.G. Sleijpen, D.R. Fokkema, BICGSTAB(L) for linear equations involving unsymmetric matrices with complex spectrum, Electron. Trans. Numer. Anal. 1 (1993) 11–32.

[9] P. Sonneveld, CGS, A fast Lanczos-type solver for nonsymmetric linear systems, SIAM J. Sci. Statist. Comput. 10 (1989) 36–52.

[10] H.A. Van der Vorst, Bi-CGSTAB: a fast and smoothly converging variant of Bi-CG for the solution of nonsymmetric linear systems, SIAM J. Sci. Statist. Comput. 13 (1992) 631–644.

[11] S.-L. Zhang, GPBi-CG: generalized product-type methods based on Bi-CG for solving nonsymmetric linear systems, SIAM J. Sci. Comput. 18 (1997) 537–551.

ELSEVIER

Journal of Computational and Applied Mathematics 149 (2002) 307–324

JOURNAL OF COMPUTATIONAL AND APPLIED MATHEMATICS

www.elsevier.com/locate/cam

Coupled, macro–micro modeling for hot deformation and sintering

Tatstuhiko Aizawa[a,*], Yunan Prawoto[a], Fujio Tsumori[b]

[a] *High Performance Materials, Research Center for Advanced Science and Technology, University of Tokyo, 4-6-1 Komaba, Meguro-ku, Tokyo 153-8904, Japan*

[b] *Department of Mechanical Engineering, Kyoto University, Yoshida-Honcho, Sakyou-ku, Kyoto 606-8501, Japan*

Received 27 November 2001; received in revised form 14 March 2002

Abstract

Coupled, macro–micro modeling is proposed to describe hot deformation and sintering behavior of materials at the elevated temperature. Micro-model for description of microstructure evolution is coupled in the hierarchical structure with the macro-model for deformation of structural members or specimens. Owing to the homogenization theory and the selected unit-cell model, hot pressing as well as hipping processes can be quantitatively described by the present method. Direct coupling among the elasto-creep deformation, the thermal transient and the diffusion process enables us to construct the theoretical frame for quantitative description of various material behavior at the elevated temperature. Numerical examples are shown to demonstrate the validity and effectiveness of the present methodology.

Keywords: Multi-level modeling; Thermo-mechanical coupling; Homogenization theory; Unit-cell model; Elasto-creep behavior

1. Introduction

Metallic alloys and monolithic ceramics have been developed to have homogeneous microstructure and to improve their intrinsic strength and ductility. In nature, however, most of the plants and the sea-shells are composed of porous-structured or cellular materials [7]. Bamboo has a graded structure in density, resulting in its flexibility in bending. Since fine porous structure leads to high thermal insurance, porous polymers or rubbers have been designed to increase the porosity as much as possible. Hence, most of matters and materials daily used are thought to have their own

* Corresponding author. Tel.: +81-3-5452-5086; fax: +81-3-5452-5116.
E-mail address: aizawa@odin.hpm.rcast.u-tokyo.ac.jp (T. Aizawa).

PII: S0377-0427(02)00538-1

heterogeneous structure as demanded from functional requirement. Their mechanical response can never be precisely described without a rational model of their microstructure. In other words, their mechanical response cannot be described until their microstructure is adequately modeled. Through research and development of optimization theories for geometric design, the homogenization method has become the most reliable tool for this purpose [5].

During materials processing and manufacturing, the starting materials are shaped or often near-net shaped to a designed geometric configuration. To be noticed, their microstructure is also modified or improved to satisfy the design demand. Typical examples can be seen in hot deformation and sintering of materials. Large nonlinear deformation is positively used for shaping of materials at the elevated temperature. During this hot working, original microstructure can be improved to be free from defects through densification, diffusion and joining. During sintering, the discrete connection of particles in the powder compact can be changed to dense continuum. In order to describe quantitatively the above geometric change in shaping as well as the microstructure evolution, macro–micro modeling is also indispensable [14,13,1,3].

Macro–micro modeling is based on the homogenization theory and the finite element method. In the conventional approaches, no coupling was considered between macro- and micro-models. Although nonlinear finite element analysis is effective to describe large deformation of materials in hot working, little data can be provided for microstructure evolution [6]. In addition, very complex constitutive equations are necessary to deal with varying state of materials [9]. While, the Monte-Carlo (MC) methods [11] have been actively used to simulate the mass transportation and grain growth during the holding time in sintering. Although various microstructure evolution can be qualitatively simulated without use of the constitutive equations, no process conditions such as loading and temperature setup can be taken into account.

In the present methodology, a macro-model for deformation and stress analysis of specimens and structural members is coupled in the hierarchical manner with a micro-model for physical description of microstructure evolution [12]. In particular, degradation of materials takes place by generation of voids and defects at the elevated temperature. Mass transportation is activated by diffusion and grain growth. Hence, the above macro–micro modeling must be used to consider the physical coupling behavior even in each level of models, that is, elasto-creep or elasto-plastic model for description of mechanical response, thermal conduction model for description of temperature transients, and diffusion model for description of material evolution. Each model has mutual interaction in the hierarchical structure of the macro–micro modeling. As illustrated in Fig. 1, elasto-creep model is strongly coupled with thermal conduction model, while diffusion model is weakly coupled with the elasto-creep model for multi-scaled modeling in hot deformation and sintering. In the present paper, the above coupled, macro–micro modeling for hot deformation and sintering of materials is theoretically explained with some comments on the basic formulation and the coupling terms. Numerical examples are used to demonstrate the validity and effectiveness of the present methodology. Further advancement in the related research is also discussed.

2. Coupled, macro–micro modeling

Sintering is a process to recover the continuity of matters from the discrete powder compact through the mass diffusion or transportation mechanism. As shown in Fig. 2, since sintering and hot

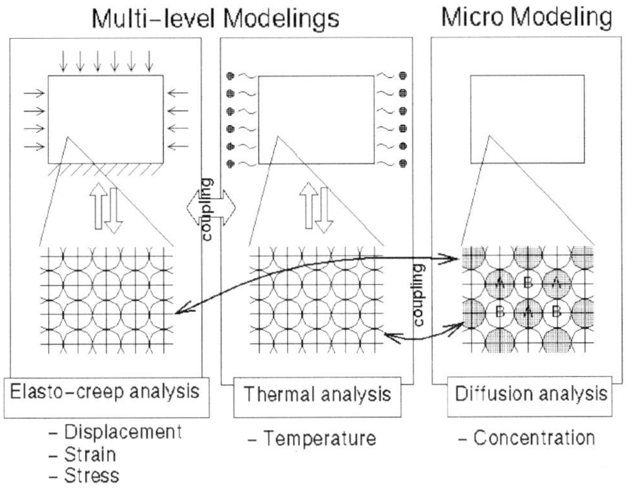

Fig. 1. Coupled, hierarchical modeling for hot deformation and sintering analysis.

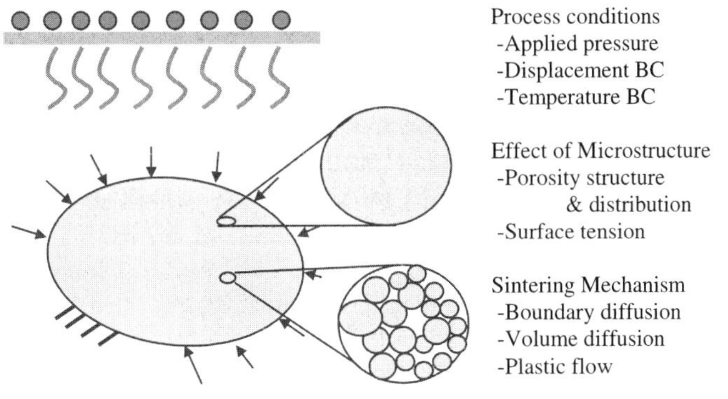

Fig. 2. Heterogeneous advancement of microstructure evolution with densification.

deformation advance heterogeneously with time, local microstructure of materials grows with great difference. This difference reflects on the macroscopic deformation and stress distributions; hence, hierarchical modeling must be constructed with consideration of processing conditions.

The first item required for the coupled macro–micro modeling, to analyze the sintering behavior, is a measure to describe quantitatively the recovery of continuity from powder compact to full dense continuum. Variation of the homogenized stiffness D^{H}_{ijkl} with time and its distribution in the macro-model is effective to build up this kind of metrics [3]. Since the powder metallurgy (P/M) is a near-net shaping process, the geometric variation of products during P/M process must be accurately analyzed for the final designed shape. Hence, macroscopic change in geometry due to shrinkage must be traced with time. Diffusion process drives nonlinear mass transportation and related mechanical behavior at the elevated temperature. Then, microstructure evolution in sintering is indispensable for a micro-model.

A prototype system is constructed for the coupled, macro–micro modeling in the present paper. To be discussed later, the macro-FEM (finite element method) models for deformation and thermal transients are formulated by using the averaged variational equation or weak form. In the similar manner, the micro-FEM models for deformation, thermal transients and diffusion process, are formulated by using the physical variational equation or weak form. Simultaneous solution of these variational equations for each physical system provides total response of models. Time integration of coupled variational equation systems provides physically coupling effects among deformation, thermal transient and diffusion process.

3. Theory of coupled, macro–micro modeling

Homogenization theory on the basis of the unit cell modeling is used to formulate the above coupled, macro–micro modeling in the frame of the finite element method.

3.1. General frame on the basis of the homogenization theory

Essential contributions in the homogenization theory are briefly summarized in the following. In this theory, the local constitutive structure is thought to have a locally specified periodic unit cell as shown in Fig. 3(a). Hence, every physical variable field on this material support can be expressed by the combination of the averaged part with the local disturbance as depicted in Fig. 3(b). Assuming that the periodicity, ε, should be sufficiently small, every physical field $u(x,y)$ in total can be represented by the following asymptotic expansion in ε:

$$u(x,y)=u^0(x)+\varepsilon u^1(x,y)+\tfrac{1}{2}\varepsilon^2u^2(x,y)+\cdots \quad \text{for } y\equiv x/\varepsilon \tag{1}$$

or

$$u(x,y)\approx u^0(x)+\varepsilon u^1(x,y), \tag{2}$$

where $\{u^0,u^1\}$ are functions for the averaged part and disturbance in the coordinates $\{x,y\}$, respectively.

Owing to the periodicity of function u in the y-coordinates, the following two equations are power tool for formulation:

$$\frac{\partial}{\partial x_i}u(x,y)=\frac{\partial u}{\partial x_i}+\frac{1}{\varepsilon}\frac{\partial u}{\partial y_i}, \tag{3}$$

$$\lim_{\varepsilon\to 0}\int_V u(x,y)\,\mathrm{d}V=\int_V\frac{1}{|Y|}\left(\int_Y u(x,y)\,\mathrm{d}Y\right)\mathrm{d}V, \tag{4}$$

where Y denotes for a volume of unit cell. Substituting Eq. (1) or (2) to any convex-type variational equation or weak form, $A(u,v)=F(v)$ for $v\in H(V)$, and rearranging it in each term of ε, the following

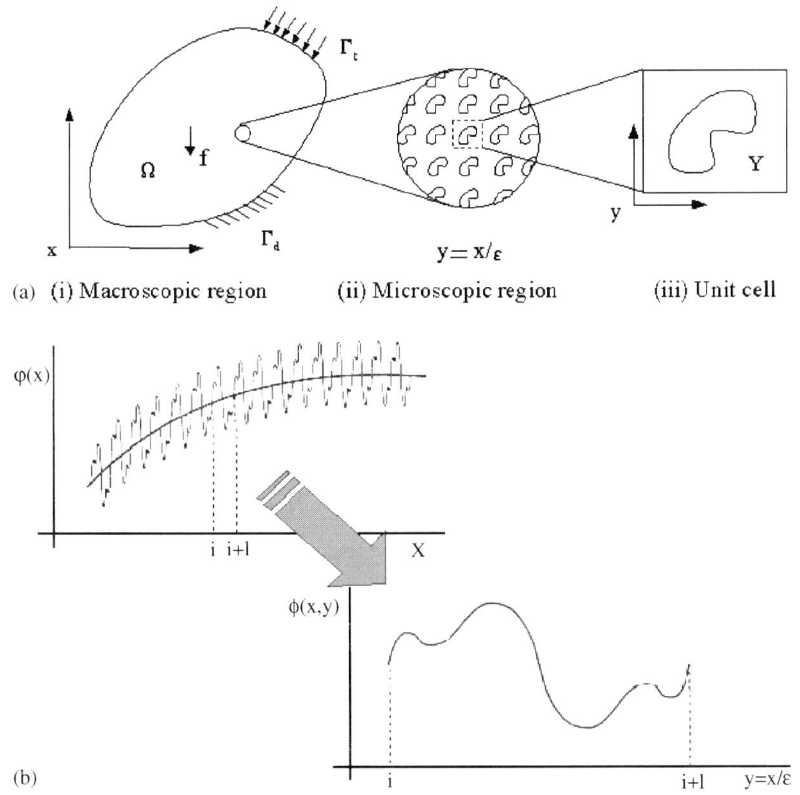

Fig. 3. (a) A frame of multi-scaling in materials using the homogenization theory. (b) Local deviation in the field variable to be embedded into the rationally smoothed distribution.

system of variational equations can be obtained, respectively, for macro- and micro-models:

$$A'(u^0, v) = F'(v) \quad \text{for } v \in H(V) \tag{5}$$

and

$$A''(u^1, v) = F''(v) \quad \text{for } v \in H(Y). \tag{6}$$

In the above, A' and A'' are deduced from $A(u, v)$, and F' and F'' from $F(v)$ through mathematical manipulation, respectively.

In the coupled macro–micro modeling, various field variables and their related governing equations must be formulated into the weak form and solved simultaneously. Table 1 lists typical variables and their governing equations in the coupled analysis for the elasto-creep deformation, the thermal transients and the diffusion process. From the equivalent weak form to the governing equations and the natural boundary conditions in Table 1, the homogenized variational equation (5) for a macro-model and the physical variational equation (6) for a micro-model can be deduced in nearly the same manner. In the following, the finite element formulation for elasto-creep deformation is used as a common frame for variational formulation by the homogenization theory.

Table 1
A list of field variables and related governing equations

	Deformation	Thermal conduction	Diffusion process
Field variable	Displacement (u_i) Strain (ε_{ij}) Stress (σ_{ij})	Temperature (T)	Mass concentration (C)
Governing equations	$\sigma_{ij,j}=0$ $\varepsilon_{ij}=\frac{1}{2}(u_{i,j}+u_{j,i})$ $\varepsilon_{ij}^{e}=\varepsilon_{ij}-\varepsilon_{ij}^{c}$ $\varepsilon_{ij}^{c}=\dfrac{3}{2}\dfrac{\bar{\varepsilon}^{c}}{\dot{\bar{\sigma}}}S_{ij}$ in V	$\rho C_p\dfrac{\partial T}{\partial t}=\kappa\dfrac{\partial^2 T}{\partial x_i^2}$ in V	$\dfrac{\partial C}{\partial t}=D\dfrac{\partial^2 C}{\partial x_i^2}$ in V
Boundary conditions	$u_i=\bar{u}_i$ on S_u $T_i\equiv\sigma_{ij}n_j=\bar{T}_i$ on S_T	$T=\bar{T}$ on S_T $-\kappa\dfrac{\partial T}{\partial x_n}=\bar{q}$ on S_q	$C=\bar{C}$ on S_C $-D\dfrac{\partial C}{\partial x_n}=\bar{q}$ on S_q
Initial conditions	$\sigma_{ij}=\varepsilon_{ij}=u_i=0$ at $t=0$	$T=T_0(x,y)$ at $t=0$	$T=C_0(x,y)$ at $t=0$

Notes: (1) Superscripts "e" and "c" denote the elastic and creep contributions, respectively. (2) S is the surface of V, which is divided into two parts $S_u\cup S_q$, $S_T\cup S_q$ or $S_c\cup S_q$.

3.2. Elasto-creep modeling as a common frame of formulation

Visco-elastic deformation or time-dependent material nonlinearity must be taken into account in order to describe the hot deformation and sintering. Time marching scheme is used to convert the rate form in the governing equation into the incremental formulation. Then, three governing equations for the elasto-creep model in Table 1, are represented by the following:

$$\left.\begin{aligned}&\Delta\sigma_{ij,j}^{e}=0,\\&\Delta\varepsilon_{ij}=\tfrac{1}{2}(\Delta u_{i,j}+\Delta u_{j,i})=B_{ijk}\Delta u_k,\\&\Delta\sigma_{ij}^{e}=\Delta\sigma_{ij}-\Delta\sigma_{ij}^{c}=D_{ijkl}(\Delta\varepsilon_{kl}-\Delta\varepsilon_{kl}^{c}),\end{aligned}\right\}\tag{7}$$

where $\Delta\sigma_{ij}$, $\Delta\varepsilon_{ij}$ and Δu_i are incremental stress, strain and deformation, respectively, during the specified time increment, Δt. The superscripts e and c denote the elastic and creep parts, respectively. In the above, the incremental creep strain $\Delta\varepsilon_{ij}^{c}$ is prescribed by the multi-axial constitutive equation of elasto-creep model

$$\Delta\varepsilon_{ij}^{c}=\frac{3}{2}\frac{\Delta\dot{\bar{\varepsilon}}^{c}}{\bar{\sigma}}S_{ij},\tag{8}$$

where $\Delta\varepsilon^{c}$ is defined by $\Delta\dot{\bar{\varepsilon}}^{c}\Delta t$ for the uniaxial creep rate ($\Delta\dot{\bar{\varepsilon}}^{c}$), the equivalent stress ($\bar{\sigma}$) and the deviatoric stress (S_{ij}). Only $\Delta\dot{\bar{\varepsilon}}^{c}$ is provided by the uniaxial creep testing for the constitutive matrix material.

The variational equation equivalent to both the governing equations (7) and the natural boundary conditions on the surface S_σ, is given by

$$\int_V v_j B_{lmj} D_{lmpq} B_{pqi} \Delta u_i \, \mathrm{d}V = \int_V v_j B_{pqj} \Delta\sigma^{\mathrm{c}}_{pq} \, \mathrm{d}V + \int_{S_\sigma} v_j \Delta \bar{T}_j \, \mathrm{d}S \quad \text{for } v_j \in H(V) \tag{9a}$$

or

$$A(\Delta u_i, v_j) = F(v_j) \quad \text{for } v_j \in H(V). \tag{9b}$$

Substituting Eq. (2) into (9) and arranging the equations on the basis of the homogenization theory, Eq. (9) provides a simultaneous system of variational equations for macro- and micro-models. That is, the variational equation in the macro-model is given by

$$\int_V v_j B^x_{lmj} D^{\mathrm{H}}_{lmpq} B^x_{pqi} \Delta u^0_i \, \mathrm{d}V = \int_V v_j B^x_{lmj} \Delta\sigma^{\mathrm{H}}_{lm} \, \mathrm{d}V + \int_{S_\sigma} v_j \Delta \bar{T}_j \, \mathrm{d}S \quad \text{for } v_j \in H(V), \tag{10}$$

where B^x_{ijk} denotes B-operator in the x-coordinates, and, the homogenized stiffness D^{H}_{ijkl} and incremental creep stress $\Delta\sigma^{\mathrm{H}}_{ij}$ are, respectively, given by

$$\left(D^{\mathrm{H}}_{ijkl} = \frac{1}{|Y|} \int_Y D_{ijkl} - D_{ijpq} \frac{\partial \chi^{kl}_p}{\partial y_q} \right) \mathrm{d}Y, \tag{11}$$

$$\Delta\sigma^{\mathrm{H}}_{ij} = \frac{1}{|Y|} \int_Y (\Delta\sigma^{\mathrm{c}}_{ij} - D_{ijkl} B^y_{klj} \psi_j) \, \mathrm{d}Y. \tag{12}$$

While, since a solution Δu^1 for the micro-model can be expressed by $\Delta u^1_i = -\chi^{(kl)}_i \Delta\varepsilon^x_{kl}(u^0) + \phi_i(x, y)$, its variational equations are given by

$$\int_Y v_j B^y_{lmj} D_{lmpq} B^y_{pqi} \chi_i \, \mathrm{d}Y = \int_Y v_j B^y_{lmj} D_{lmpq} \bar{\varepsilon}_{pq} \, \mathrm{d}Y \quad \text{for } v_j \in H(Y), \tag{13}$$

where $\bar{\varepsilon}_{pq}$ is a prescribed uniform strain in the unit cell, and, B^y_{ijk} the B-operator in the y-coordinates, and

$$\int_Y v_j B^y_{lmj} D_{lmpq} B^y_{pqi} \psi_i \, \mathrm{d}Y = \int_Y v_j B^y_{lmj} \Delta\sigma^{\mathrm{c}}_{lm} \, \mathrm{d}Y + \int_{S_\sigma} v_j \Delta\gamma n_j \, \mathrm{d}S \quad \text{for } v_j \in H(Y), \tag{14}$$

where $\Delta\gamma$ is the prescribed surface tension increment on the surface of porosity with the normal directional vector n_i. Simultaneous solution of Eqs. (10), (13) and (14) determines the total deformation, the accumulated strain and current stress state at the time (t) together with the estimate of microstructure by D^{H}_{ijkl}. To be noted, the macroscopic, geometric configuration (V) and the unit cell configuration (Y) can change themselves with strong dependency on the process transients and shrinkage of local porosity.

3.3. Physical coupling among three field variables

Each modeling field can be designed to have coupling among related fields at the specified level in the hierarchical structure. In the present macro–micro modeling, three types of coupling are considered: elasto-creep model vs. thermal conduction model, thermal effect on the diffusion process and diffusion effect on the elastic response. In the first coupling, the creep strain rate is determined as a function of temperature, while temperature distribution is considered in the deformed material domain. This coupling is considered in the micro-models. In the second coupling, the thermal dependency of diffusion coefficient is directly considered in the micro-model. In the third coupling, the elastic properties in the micro-model is assumed to be dependent on the mass fraction of diffusing elements.

4. Numerical examples

Validity and effectiveness of the present method is demonstrated through the following numerical examples.

4.1. Numerical validation of the present method

A two-dimensional elastic problem was employed here to utilize the analytical solution by the micromechanics [10] for description of the convergence behavior of the present solution. Hassani et al. [8] solved analytically the equivalent stiffness for a two-dimensional cell with a square cavity as depicted in Fig. 4. The full-dense material is assumed to be in isotropic elasticity: $D_{1111} = D_{2222} = 30$, $D_{1122} = 10$ and $D_{1212} = 10$. At the presence of the square cavity located at the center in the unit-cell, the equivalent stiffness D^{H}_{ijkl} becomes anisotropic; due to the local stress distribution in the uniformly strained unit cell, each modulus in D^{H}_{ijkl} is further reduced from the original value in D_{ijkl}. In the present method, the unit cell in Fig. 4 was subdivided by six-node triangular elements. For each prescribed uniform strain state, the local stress distribution can be determined by solving Eqs. (13) and (14). Then, every component of D^{H}_{ijkl} is deduced by using Eq. (11) from the calculated deformations and stresses.

Table 2 compared the calculated components in D^{H}_{ijkl} with the Hassani's analytical solutions [8] and the Kikuchi's data [4]. In the present computation, the number of elements is 1946 and the number of nodes, 4092. Under this fine mesh model, the present solutions of D_{1111}, D_{2222}, D_{1122} and D_{1212} all converge to the analytical ones with sufficient accuracy. As partially discussed in [8], the Kikuchi's approach provided the upper bound of the analytical solution, while the present

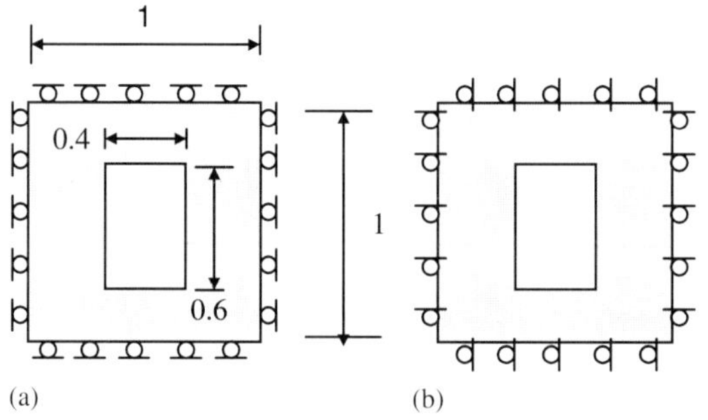

Fig. 4. Unit cell in the two-dimensional elasticity with the square cavity.

Table 2
Comparison of the present solution with the analytical solution and the Kikuchi's data

Solution	D_{1111}	D_{1122}	D_{2222}	D_{1212}
Kikuchi	13.015	3.241	17.552	2.78
Hassani	12.839	3.124	17.422	2.648
This method	12.826	3.12445	17.4083	2.6348

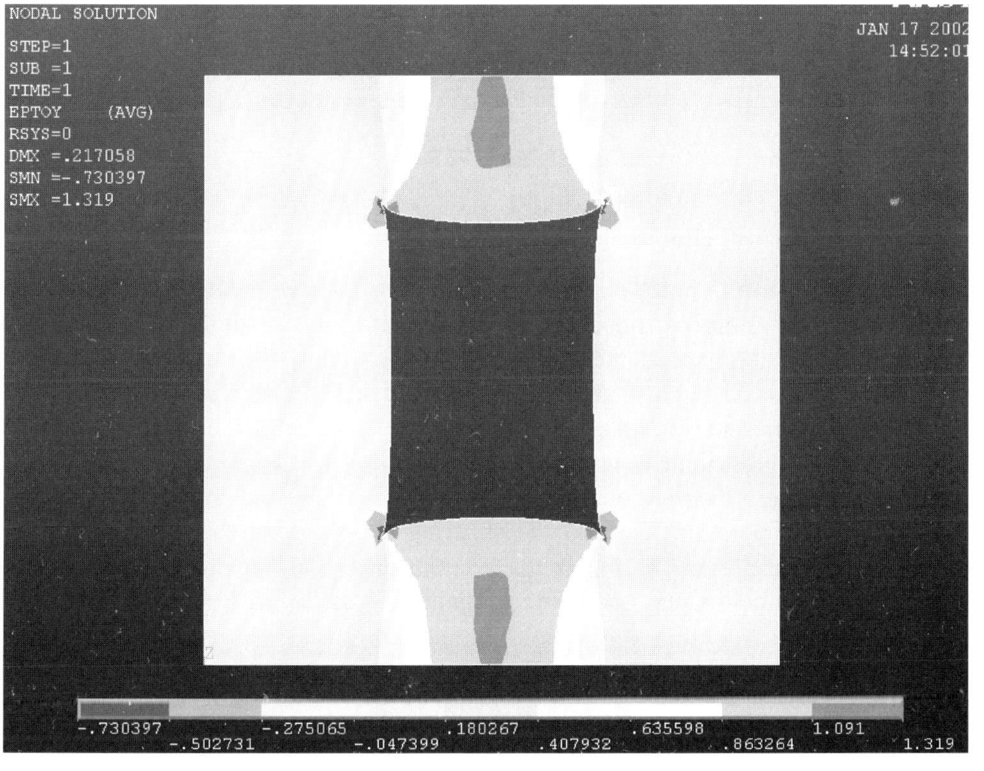

Fig. 5. Normal stress distribution in the unit cell when the normal strain is uniformly prescribed.

solution converges to the analytical one from the lower bound. As had already been noticed in [13], this might be partially because of the stress concentration around the cavity in the unit cell. As shown in Fig. 5, the normal stress state significantly changes at the vicinity around each corner of a square cavity. Hence, without adequate selection of finite element subdivision and scheme, sufficient convergence might never be attained in computation of the equivalent stiffness.

4.2. Hot deformation in sintering without pressure

In the conventional sintering, the isolated pores in the powder compact are closed by the surface tension at the elevated temperature. In the case when the pore size distributes in the compact,

Table 3
Fundamental data for hot deformation and sintering analysis

Uniaxial creep strain rate[a]	$\varepsilon^c = 16\alpha\beta\Omega D \exp(-U/RT)\sigma/(\pi l^2 kT)$
Parameters	Material constants: $\alpha = 1$, $\beta = 1$ Diffusion coefficient: $D = 1.45 \times 10^{-13}$ m^2/s Atomic volume: $\Omega = 1.0 \times 10^{-29}$ m^3 Edge length of a grain: $l = 1$ μm k: Boltzmann coefficient U: Activation energy T: Temperature
Elastic properties	Young's modulus: E = 130 GPa Poisson's ratio: $\nu = 0.343$
Surface traction	$\gamma = 1.72$ J/m^2

[a]Copper powders are employed as a reference material. This functional form stands on the body-diffusion mechanism in sintering.

macroscopic shrinkage in geometry never takes place uniformly, resulting in the distortion of a sintered product. Table 3 summarized the computational conditions for sintering analysis. Although those material properties listed in this table were cited for the copper, various kinds of creep laws can be installed into the present system.

In Fig. 6, the pore size distribution was modeled by using the different unit cells, and the variation of geometry and stress distribution was simulated by the present method. Local shrinkage is strongly dependent on the pore size, so that the sintering strain and stress distribute in the sintered product. In the present case, since more porous sections were easy to be sintered, a product was totally bent to the upper direction and geometric irregularity was frozen in the final shape as reported in experiment [3].

4.3. Sintering in hot pressing

Hot pressing is a typical P/M processing to fabricate the dense sintered compact. Through various experiments, temperature and pressure schedules in processing are important to make high densification. In this example, porosity size distribution is also considered by using the different unit cells as illustrated in Fig. 7. Unit-cell models from "A" to "C" in Fig. 7 have the same average porosity density in area, but local size distribution is assumed to be different. Coupling between the thermal distribution and the hot deformation is also considered to discuss the effect of pressurizing in hot pressing on the sintering behavior.

In the absence of isostatic pressure, since fine porosities are easy to be closed by surface tension, the product was forced to be bent upward even in the uniform temperature condition. This might be because local shrinkage is exaggerated in section "A" where finer porosity distributes. In fact, as shown in Fig. 8(a), porous materials with finer porosity undergo larger sintering strains. In case of hot pressing, the externally applied stress effect is superposed on the above sintering stress state. Figs. 8(b) and (c) show the macroscopic deformation and stress state, and, the microscopic shrinkage

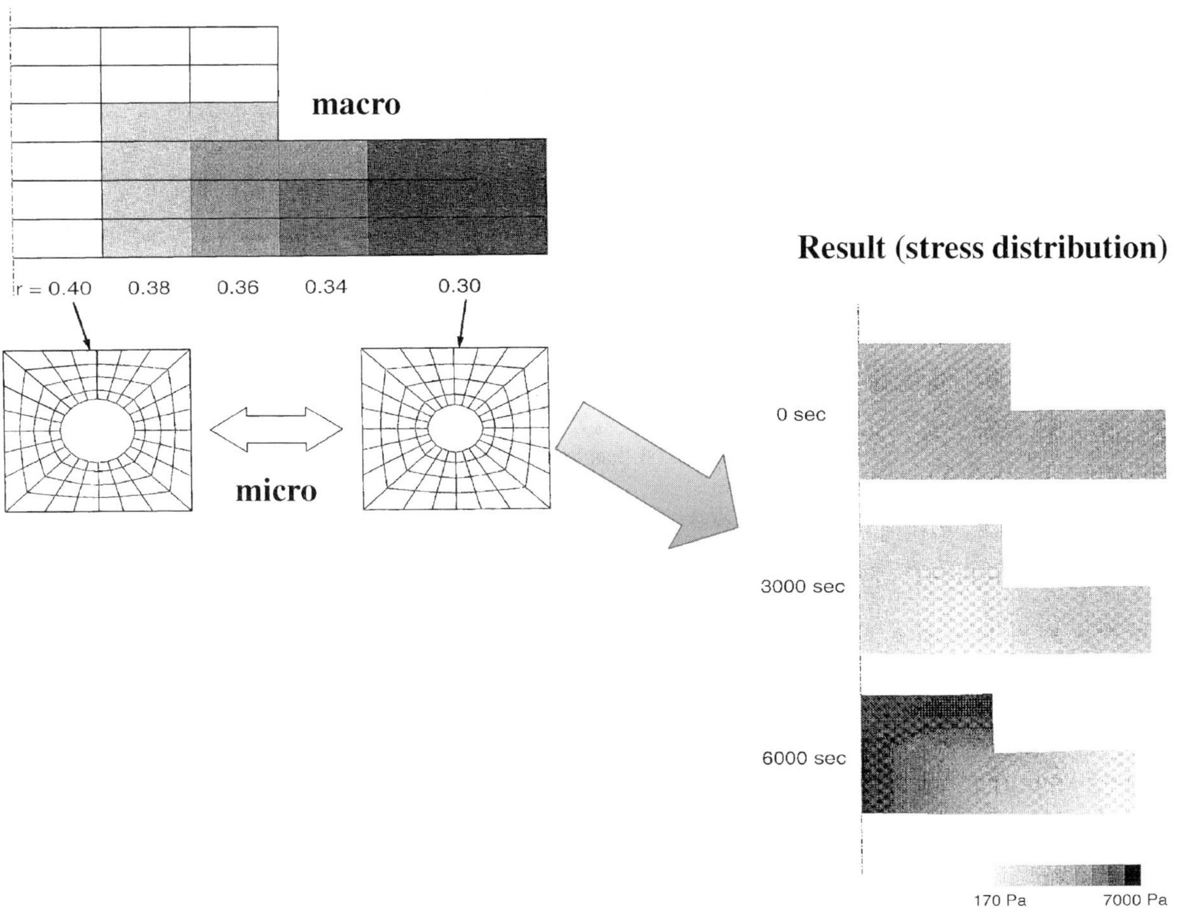

Fig. 6. Hot deformation of powder compact with the porosity distribution during sintering.

during sintering, respectively. No geometric distortion was generated since the externally applied stress has more dominant influence on the activation of sintering. To be noticed in comparison between Figs. 8(a) and (c), local shrinkage was enhanced by this application of external pressure.

4.4. Sintering during hipping process

Hot isostatic pressing or hipping process is frequently utilized for sintering ceramic materials to full density. In this process, the starting porous materials are thought to be gradually sintered to the state, where isolated pores are only left in dense materials, and these pores are absolutely closed in high dense compact. Fig. 9(a) illustrates the hipping experimental condition of austenitic stainless steel (SUS316L) powders with the average size of 50 μm. Table 4 lists the experimentally measured material properties and constitutive equations for SUS316L bulk materials. Starting from the initial geometry with uniform porosity distribution, both hipping experiment and simulation trace the prescribed thermal and pressurizing schedules in Fig. 9(a). Fig. 9(b) compares the volumetric strain

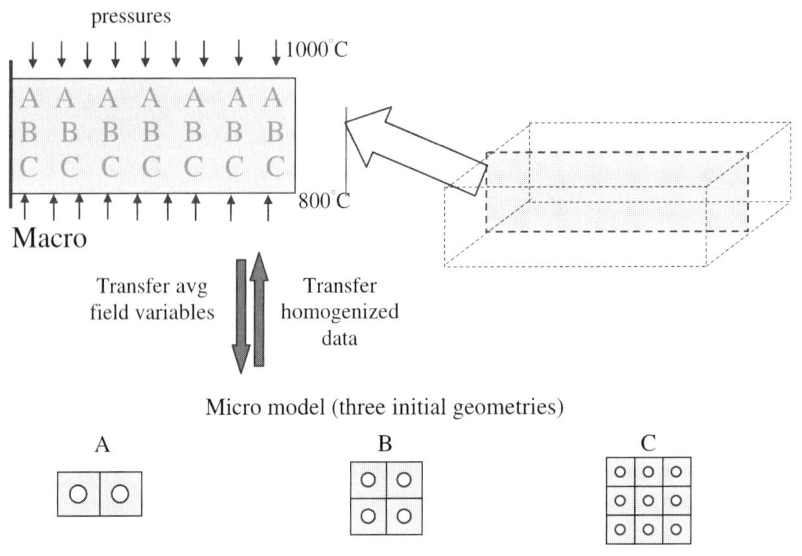

Fig. 7. Unit-cell modeling to consider the local porosity distribution with the constant average porosity.

Table 4
Computational conditions for hipping process simulation

Materials property	Functional form
Young's modulus[a] (MPa)	19400–81.4T for $T < 973$ K
Poisson's ratio	0.3
Thermal conductivity (W/m K)	$13.651 + 0.01434T$
Specific heat (J/kg K)	$465.45 + 0.21045T + 8.6641 \times 10^{-7}T^2$

[a]Above 973 K, Young's modulus is supposed to decrease monotonically to zero with T up to the melting temperature (1648 K).

rate and the average relative density between experimental and simulated results, respectively. The present modeling can predict the time history of volumetric strain rate with fairly good accuracy. This might be because geometric sintering strain rate experimentally measured is approximately governed by local shrinkage rate of a closed pore. On the other hand, gradual densification observed in experiment revealed that heterogeneous shrinkage took place in the macroscopic sintering process.

4.5. Coupling of deformation model with diffusion process

Mass transportation in the diffusion process must be taken into account in the microstructure evolution. Coupling between deformation and diffusion models is considered to investigate the effect of constitutive element redistribution during diffusion on the variation of equivalent elastic properties. Mixture of pure nickel and copper was considered as a starting material, and, mutual diffusion was

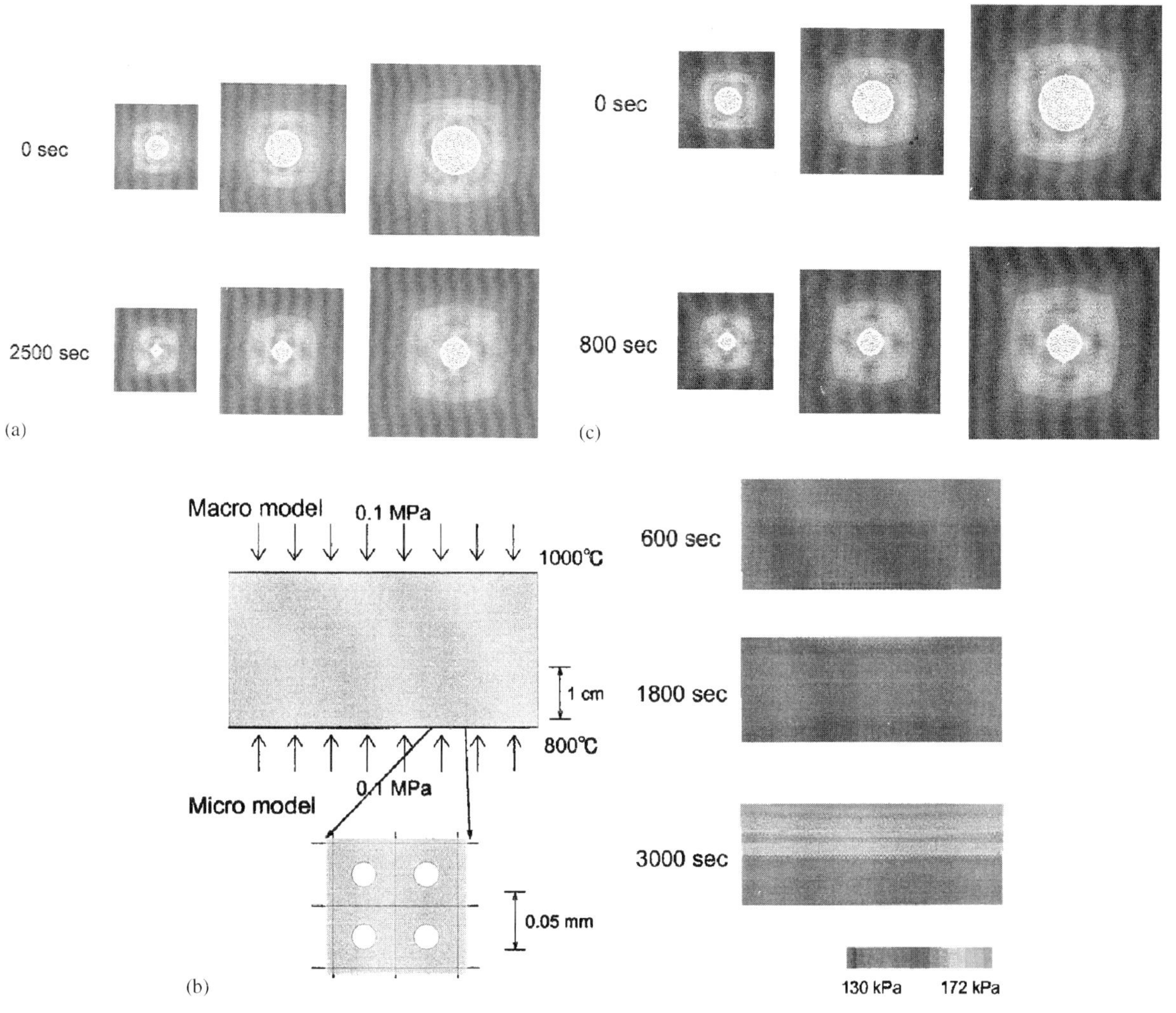

Fig. 8. (a) Local shrinkage behavior observed in the unit cells with different initial porosity size in the absence of the externally applied pressure. (b) Macroscopic geometry change during hot pressing. (c) Local shrinkage behavior observed in the unit cells with different initial porosity size in the presence of the externally applied pressure.

assumed to take place between both elements. As shown in Fig. 10(a), both elements are mixed with time through the mutual diffusion process. In parallel with that, the homogenized stiffness D^{H}_{ijkl} also varies with the diffusion time. Fig. 10(b) depicts a typical increase of D^{H}_{1111} component in D^{H}_{ijkl} with diffusion time.

5. Discussion

In the conventional nonlinear analysis, construction of constitutive equations to represent actual materials nonlinearity is one thing but theoretical modeling is another. Hence, when using the

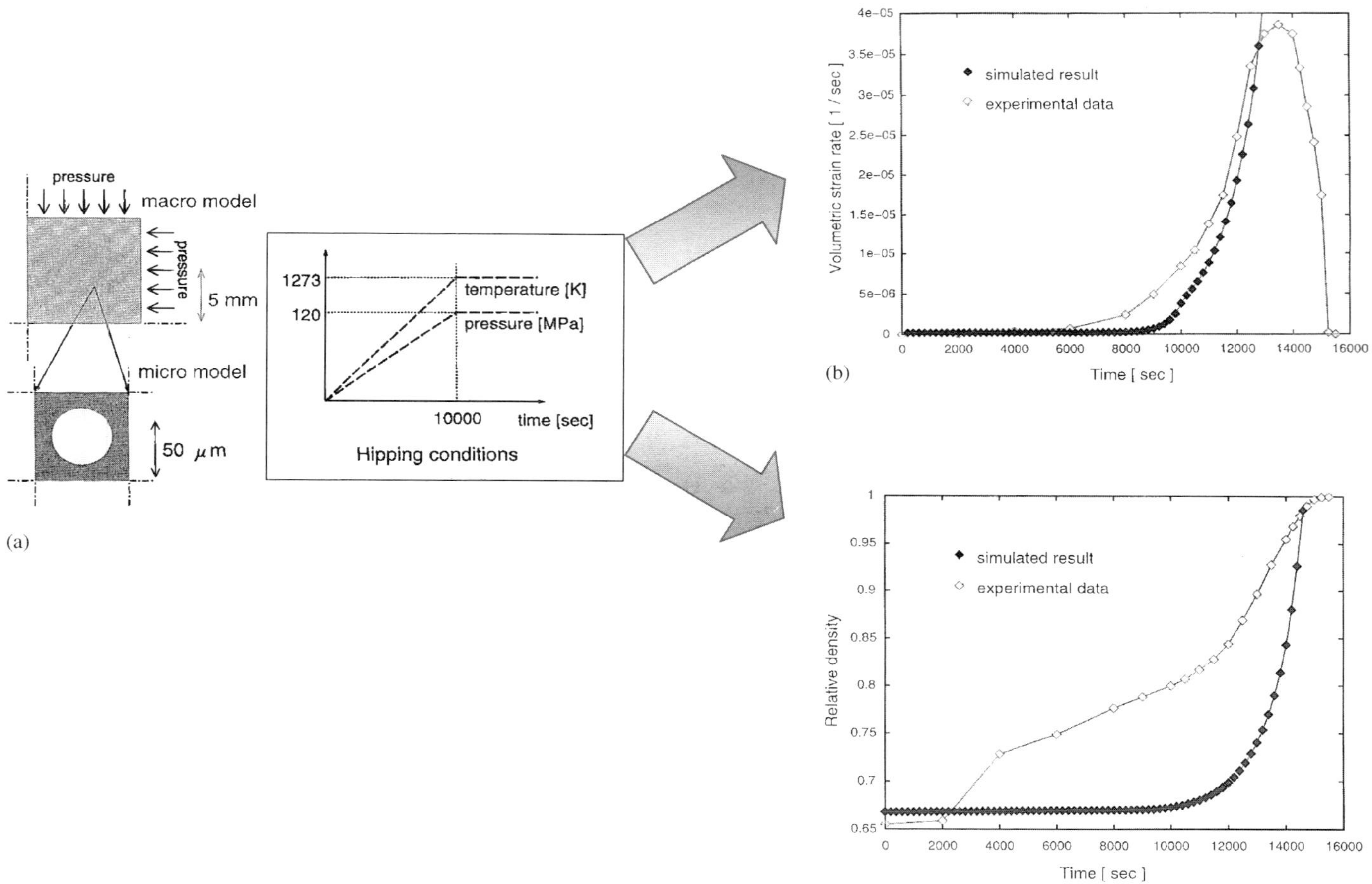

Fig. 9. (a) Specification of unit cell and thermal/loading transients to be used in simulation. (b) Comparison of experimental and calculated time histories of volumetric strain rate and relative density.

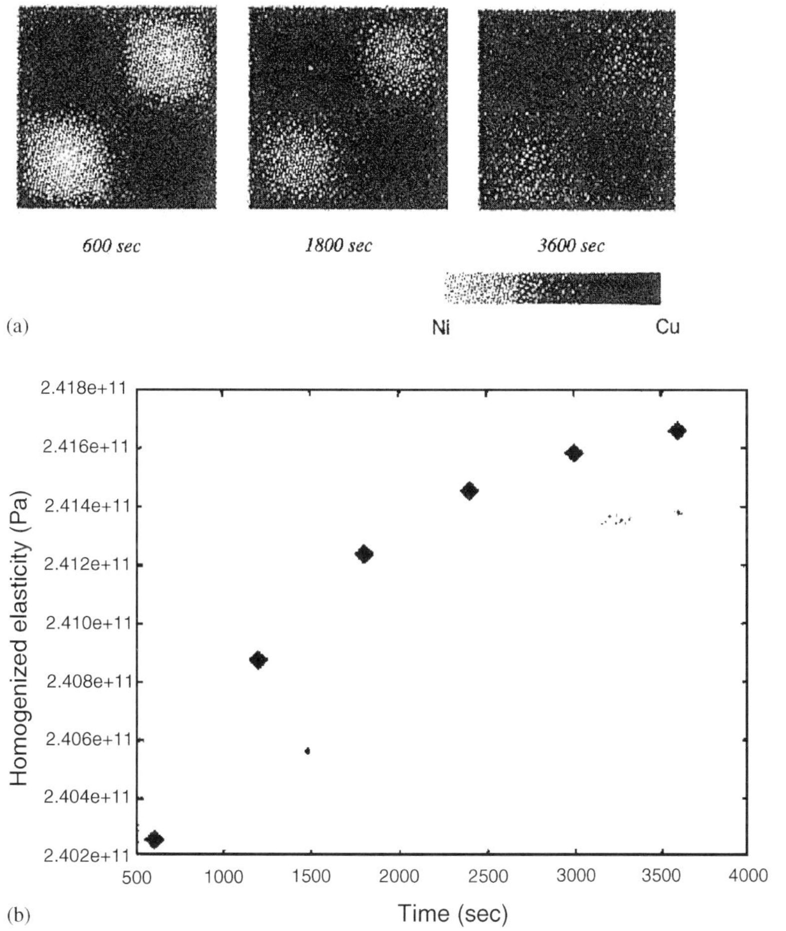

(a) (b)

Fig. 10. (a) Redistribution of constitutive elements with the diffusion time. (b) Variation of D^{H}_{1111} component in the homogenized stiffness D^{H}_{ijkl} with the diffusion time.

nonlinear finite element method for simulations, complex constitutive equations must be obtained to explain the targeting nonlinear phenomena physically. While, the microstructure information, which had been obtained through precise observation, is difficult to utilize in the above nonlinear simulation. Aiming for more accurate nonlinear analysis, only macroscopic data by experiments is insufficient as an essential key to determine the quality of simulation. Promising, quantitative simulation must be equipped with prediction of microstructure change at the elevated temperature. The present coupled, multi-level modeling is the first prototype for this kind of simulation.

As discussed in the above, the homogenized stiffness becomes a measure to mechanically evaluate the microstructure evolution. During sintering the powder compact, the equivalent elastic stiffness increases itself with gradual reduction of porosities with time. Fig. 11 showed the comparison of Young's modulus between the experimental and calculated results. In experiment, Young's modulus was evaluated from the measured lateral and shear wave velocities of each sintered SUS304 powder

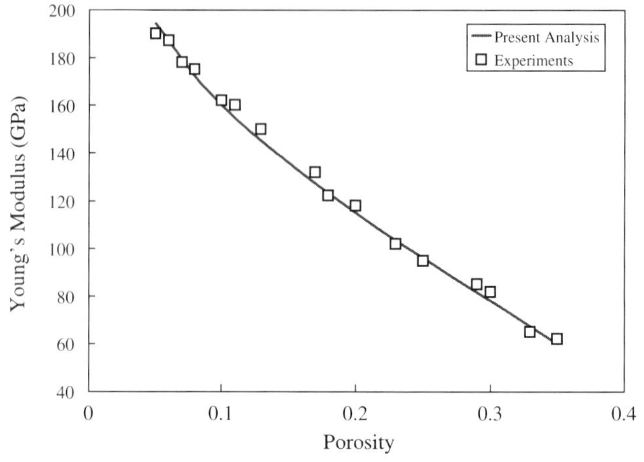

Fig. 11. Comparison of the evolution of Young's modulus during sintering between the experimental and calculated results.

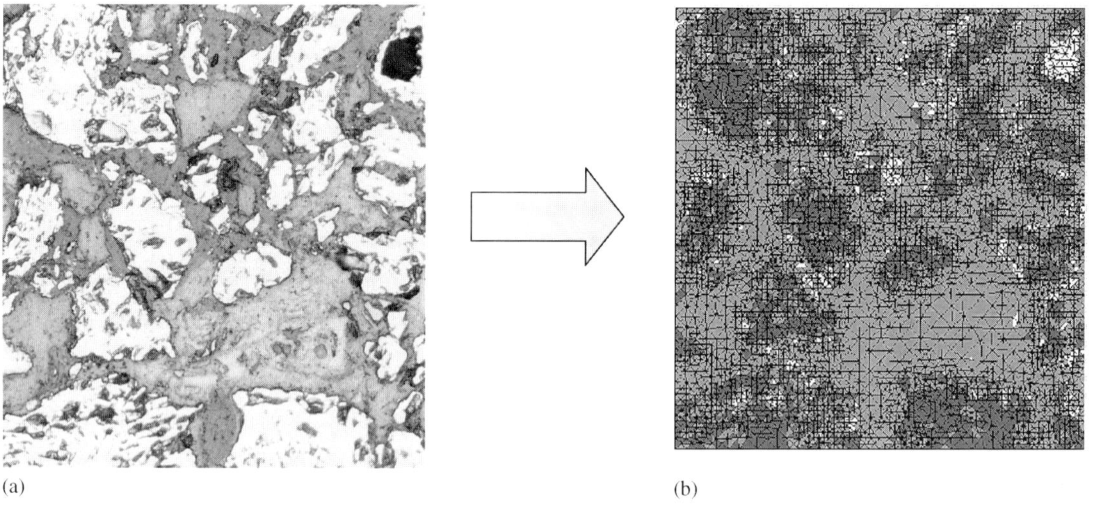

Fig. 12. Conversion of a microstructure image obtained by optical microscope to a mathematical model for the finite element analysis: (a) mixture of iron ore and coal with porosities, and (b) finite element mesh model with each material property allocated to element.

compact with different porosity by the ultrasonic method. In the present computation, the unit cell was set up to the initial powder compact with the porosity of 40%. Young's modulus was deduced from the calculated stiffness matrix D^{H}_{ijkl} by the time marching simulation in the present method. Good agreement between two in Fig. 11 reveals that the homogenized stiffness should be equivalent to actual elastic properties of sintered materials.

In the present approach, the microstructure observation and analysis reflects on the coupling model at the microscopic level and on the unit cell to be employed. In the above numerical examples, the unit-cell models were selected to correspond to the powder particle configuration or the initial particle assembly. Through recent our work, however, actual microstructure observed by optical microscope or scanning electron microscope (SEM) can be image processed and converted to a mathematical model with a unit cell. Typical example was shown in Fig. 12. Using the computer-aided scanner and utilizing the image processor combined with automatic mesh generator, the microstructure information for a mixture of iron ore and coal with porosities can be converted to a mathematical model for finite element analysis [2]. The unit cell size can be also determined by computing the homogenized stiffness; an optimum cell size is just corresponding to the convergent value of the calculated equivalent stiffness.

6. Conclusion

Hot deformation during the sintering process was selected in the present method as a target for coupled, macro–micro modeling. In this theoretical frame, grain growth or microstructure evolution, which has been described by the discrete modeling like the MC method, can be also formulated by using the nonlinear diffusion model [11]. Effective diffusion coefficients needed for the above formulation must be determined by physical conversion of the simulated time history of mass transportation by MC. The most favorite feature to this approach lies in the capability to consider the chemical reaction as a nonlinear term. Various microstructure changes with chemical reaction taking place in the reactive sintering or the solid-state synthesis, can be dealt with by this method.

Coupling among the stress/deformation model, the thermal transient model and the transportation of chemical potential, is also necessary as an extension of the present coupled macro–micro modeling. Phase transformation often accompanies with significant volumetric change, resulting in large strains or fracture in hot deformation. Time history of temperature is directly affected by phase transformation. Through this new coupled, multi-level modeling, qualitative understanding of mechano-chemical interactions can be changed into quantitative description by theoretical simulation.

Acknowledgements

This study is financially supported in part by collaboration with Japan Nuclear Cycle Development Institute (JNC).

References

[1] T. Aizawa, Superplastic forging analysis by the multi-level modeling, Mater. Sci. Forum 306 (1999) 621–630.

[2] T. Aizawa, Stiffness and strength analysis of porous media by using the unit cell modeling, Proceedings of the 141st Annual Meeting of ISIJ, 2001.

[3] T. Aizawa, F. Tsumori, Coupled-multilevel modeling for hot deformation and sintering, Key Eng. Mater. 180 (2000) 661–666.

[4] J.M. Guedes, N. Kikuchi, Preprocessing and postprocessing for materials based on the homogenization method with adaptive finite-element method, Comput. Methods Appl. Mech. Eng. 83 (1990) 143–198.

[5] M.P. Bendsoe, N. Kikuchi, Generating optimum topologies in structural design using a homogenization method, Comput. Methods Appl. Mech. Eng. 71 (1988) 197–224.

[6] D. Besdo, E. Stein (Eds.), Finite Inelastic Deformation Theory and Applications, Springer, Berlin, 1991.

[7] L.J. Gibson, M.F. Ashby, Cellular Solids, Cambridge University Press, Cambridge, 1997.
[8] B. Hassani, E. Hinton, A review of homogenization and topology optimization, Comput. Struct. 69 (1998) 719–737.
[9] JNC: Creep Data Handbook, Manual for FINAS (Finite Element Nonlinear Structural Analysis Systems). Version 13.0, 1995.
[10] T. Mura, Micromechanics of Defects in Solids, Martinus Nijhoff Publishers, The Hague, 1982.
[11] Y. Saito, Microstructure Formation and Diffusion Equation, Korona-sha, 2000.
[12] F. Tsumori, T. Aizawa, Macro–micro modeling for coupled elasto-creep and theral analyses under sintering, Mater. Sci. Eng. A 285 (2000) 357–362.
[13] F. Tsumori, T. Aizawa, J. Kihara, Elasto-creep sintering analysis by macro–micro modeling, Sintering Technol. (1996) 29–36.
[14] F. Tsumori, T. Aizawa, J. Kihara, Evaluation of pressurizing effects on the sintering process by coupled macro–micro modeling, Jpn. J. Powder Metall. 43 (1996) 1165–1170.

JOURNAL OF COMPUTATIONAL AND APPLIED MATHEMATICS

ELSEVIER Journal of Computational and Applied Mathematics 149 (2002) 325–333

www.elsevier.com/locate/cam

The high precise/efficient pre/postprocessor

Ichiro Hagiwara *, Junichi Shinoda

Department of Mechanical Engineering and Science, Tokyo Institute of Technology, 2-12-1, Ookayama, Meguro-ku, Tokyo, Japan

Received 31 October 2001; received in revised form 30 March 2002

Abstract

According to the advancement of computers, communication technologies and other devices, digitization of data such as design information and so on has advanced rapidly. Hence, the development of unified systems for CAD/CAM/CAE is extremely important. In this article, we shall state four techniques as technical tools useful for digital processing, which are to be comprised in our unified system for CAD/CAM/CAE.

Keywords: Subdivision method; Characteristic topology method; Dual cycle elimination method; Morphing; Compactly supported radial-based functions; Lifting subdivision method

1. Introduction

As represented by CAD/CAM/CAE and computer graphics, digitization of informations has been widespread rapidly. However, there are many cases which need a person's help. For example, in mesh generation necessary for the finite element method, it takes too much time to modify, for example, irregular shaped patches. Hence, it cannot be said that the digital design in the true sense is realized. The performance of computers has made rapid progress and it enables us to visualize the results of analysis which need a large amount of data such as animation. On the contrary, the difference of data structures is a bottleneck to connect smoothly CAD, CAM and CAE. So it is necessary to construct a unified system for CAD/CAM/CAE. In this article we introduce technical tools for digital processing which are to compose our future system and are to be of benefit to the readers.

As a methodology of our research, we have two ways, one for preprocessing and one for post-processing. This paper is comprised as follows. In the preprocessing part where mainly the mesh

* Corresponding author.
E-mail address: hagiwara@mech.titech.ac.jp (I. Hagiwara).

0377-0427/02/$ - see front matter © 2002 Elsevier Science B.V. All rights reserved.
PII: S0377-0427(02)00542-3

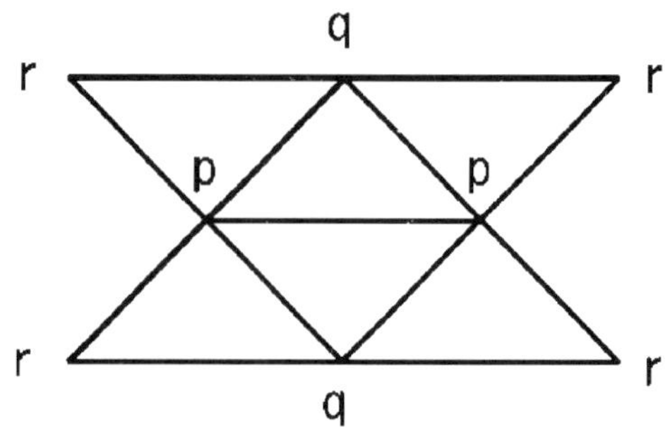

Fig. 1. Mask of modified butterfly scheme for regular vertices.

generation is discussed, there are two research themes, one is the subdivision method (Section 2), and another is the characteristic topology method (Section 3). In the postprocessing part there are also two research themes. We treat the morphing in Section 4, which is applicable to animation, and in Section 5 we discuss the lifting subdivision method efficient to reduce the rendering cost.

2. Subdivision method

In the 1990s, the subdivision method was noticed in the field of computer graphics and has been progressing with vast speed. This method is that, for a given triangular or quadrilateral mesh, by dividing each triangle or quadrilateral into four pieces in an appropriate way and repeating this procedure, we finally can get a smooth surface in the limit. Subdivision schemes are classified by the types of mesh, i.e., triangular or quadrilateral, and by the refinement rules, i.e., interpolation or approximation, etc. For details of the subdivision method, we refer to [22] and the literature therein.

Characteristics of the subdivision method are the following: (1) applicable to arbitrary topology and unnecessary to make an effort to preserve the compatibility of patches as in the parametric method such as B-spline and NURBS; (2) low computational cost because of simplicity of definition. As an example, we show the mask of one of the subdivision schemes (modified butterfly scheme for regular vertices) in Fig. 1. Here $p = 1/2$, $q = 1/8$, $r = -1/16$.

Although the subdivision method is used for construction of smooth surfaces, we think that this is available in the construction of master models of products. Firstly, we scan the coordinates of points on the physical object by the scanning machine from the engineering point of view. Secondly, we make an initial mesh so as to satisfy the users' aspect ratio demand as a first stage for the subdivision method. Next, we apply the subdivision method to the initial mesh obtained by the previous step; here we use the modified butterfly scheme which is interpolatory and applicable to triangular meshes. The reasons for applying this scheme are that the interpolation scheme does not change the points obtained in the previous steps, in particular, the initial points. After the repeated application of the subdivision scheme several times, we get the refined triangular mesh as a polygonal master model. And finally we can utilize this master model for FEM analysis, and apply the results further to CAM and rapid prototyping. The flow of this process is described in Fig. 2.

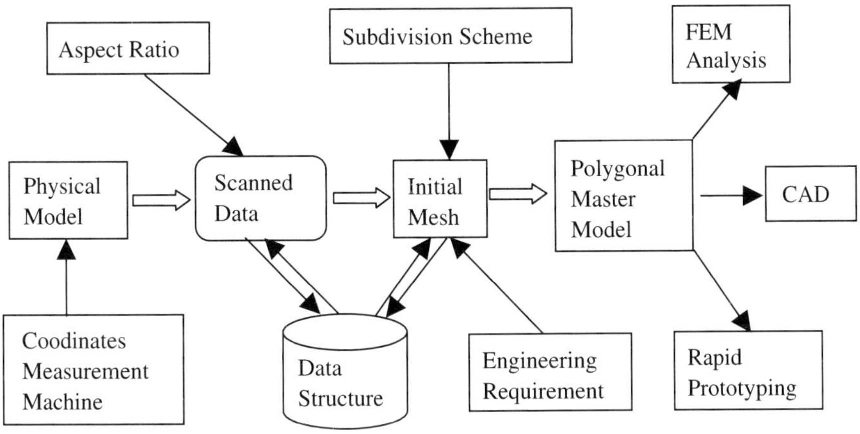

Fig. 2. General scheme for constructing a master model.

Moreover, subdivision schemes can control the local mesh density, see for example [11]. And a new type of subdivision scheme ($\sqrt{3}$-subdivision) discussed in [9] has the possibility of the control of mesh size. Since the subdivision method was originally introduced to generate smooth surfaces, if we apply the subdivision scheme to the resulting mesh more and more, we can obtain a CAD model of a physical model. Hence, it is also possible to construct a unified CAD/CAM/CAE system using the subdivision method. We expect that such a system will replace the existing system using the parametric method as a new technology.

At the end of this section, we mention initial meshes briefly. As stated above, it is necessary to construct an initial mesh for applying a subdivision scheme. And to keep the quality of the resulting mesh, we must generate a good initial mesh. Our algorithm for generating an initial mesh is for triangular meshes. This is accomplished by looking for the point to a fixed segment connecting two initial points in a reasonable way in order that these three points constitute a good triangle, see [16] for details.

3. Characteristic topology method

In the development of products in industry, it is an important subject to reduce the cost and the period for development. For this aim, it is efficient to exploit the analysis by computers with FEM and so on, in the initial stage of design. On the contrary, it takes too much time to generate a mesh necessary for FEM analysis with large structure. As for the accuracy of FEM analysis for solids, it is known that a hexahedral mesh is superior to a tetrahedral one. But there does not exist the system automatically generating hexahedral meshes with good quality applied to products and parts appearing in practical affairs. So the appearance of such a system is eagerly wished.

On the other hand, several methods were proposed to generate hexahedral meshes. Among them, here, we quote the plastering method in [2,1] and the whisker weaving method in [21,5]. Both of them start from a surface quadrilateral mesh. Recently, Müller-Hannemann proposed the dual cycle

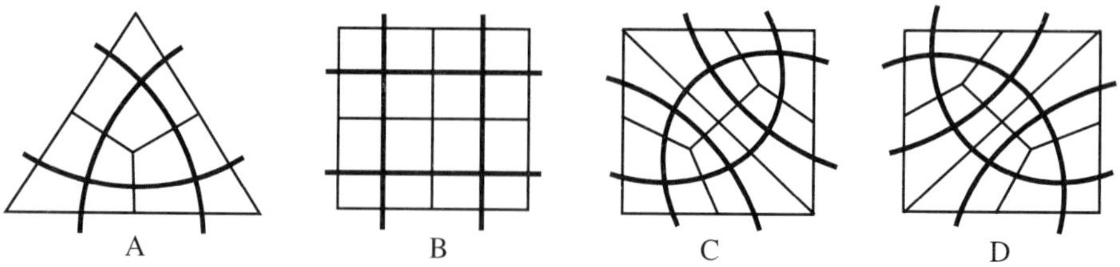

Fig. 3. Templates.

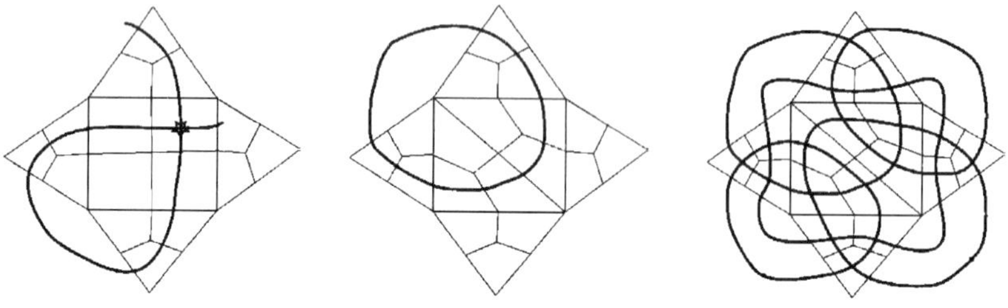

Fig. 4. Example of insertion of templates.

elimination method to this end in [13]. His method also starts from a mesh on the surface of an object, but it needs the condition that every dual cycle has no self-intersection. It is divided into two parts. Firstly, by eliminating dual cycles in an appropriate order, we reduce the original object to a single hexahedron. Secondly, we generate a hexahedral mesh by tracing back in the reverse way and supplementing hexahedrons inside of the object layer by layer. But he could not solve the problem of how to construct the quadrilateral mesh without self-intersection of dual cycles on the surface at that time.

Now we propose a simple method to solve the self-intersection problem, see also [19]. Our problem is formulated as follows. Given a polygonal surface homeomorphic to the sphere and nodes on the edges, i.e., the boundaries on the polygonal faces, construct the quadrilateral mesh which has no self-intersection such that every given node is that of an element of the resulting mesh.

Note that the configuration of nodes is arbitrary but we permit a little change of the number of nodes.

We state the outline of our algorithm. Firstly, we divide each face of the given polygon into triangles and quadrilaterals. Secondly, we prepare four templates; one is a triangle and the others are quadrilaterals (Fig. 3), and insert these templates into the divided polyhedron in an appropriate manner. For example, in Fig. 4, templates A and B are inserted into the pyramid-shaped object, but if we start from the fixed edge and go along the path shown in Fig. 3, we meet an intersection point. Then, we replace template B with template C and consequently we get a loop without a self-intersection which we call a characteristic dual cycle. In a similar way, we get a set of loops describing the structure of the divided polygon. Next we insert these characteristic dual cycles as many as possible so as to go between each neighboring two nodes. In this case, by the arbitrariness of the configuration of nodes, it is impossible to place the characteristic dual cycles so as to go between every neighboring two nodes. To do this, we need supplementary loops which we call quasi-characteristic dual cycles. By inserting these quasi-characteristic dual cycles in an appropriate way, the topological structure of the quadrilateral mesh is determined and hence we can get a mesh satisfying the demands.

Finally we remark that the way of inserting loops is only two; this leads to the simplification of the order in the elimination of dual cycles when applying Hannemann's method. Further, we note that very recently Müller-Hannemann himself also proposed a method to generate the surface quadrilateral meshes without self-intersection of dual cycles which is discussed in [14].

4. Morphing

We carry out our research in the framework of volume modeling, which can be regarded as a sub-domain of computer graphics. One of the main features of volume modeling is a functional representation (F-rep) of three-dimensional objects (solids), that is representation in the implicit form $f(x,y,z) \geqslant 0$.

Suppose we have a set of scattered points on the surface of some undefined solid. To reconstruct the surface of the solid (in other words, to find the F-rep of the solid) from the set of points, we use a new method, namely we reconstruct the surface with special kinds of volume splines—compactly supported radial-based functions (CSRBF).

The previous approach using radial-based functions (RBF) is seen in [18], and its application in [12,3]. In our approach using CSRBF, we use the following spline function of the form as an implicit function,

$$f(x,y,z) = 1/2\sum_{i=1}^{N}\lambda_i\phi(|P-P_i|) + \lambda_{N+1} + \lambda_{N+2}x + \lambda_{N+3}y + \lambda_{N+4}z - f_c(x,y,z),$$

where $\{P_i\}_{i=1}^{N}$ is a set of given points, (x,y,z) is the coordinates of point P, f_c is a defining function and, for some parameter $R > 0$,

$$\phi(r) = \begin{cases} \left(1 - \dfrac{r}{R}\right)^2, & 0 \leqslant r \leqslant R, \\ 0, & r > R. \end{cases}$$

Further we adopt the following steps to determine the unknown coefficients $\{\lambda_i\}_{i=1}^{N+4}$:

1. Octal tree creation.
2. Searching neighbors of each point, using tree.
3. Sorting data using the octal tree.
4. Solving a system $Ax = b$ of linear algebraic equations with a band-diagonal matrix A.

The method realizes the efficient reconstruction. In fact, from 1487 points, we can reconstruct "head" shown in Fig. 5 in almost 3 s by PC (Pentium III 700 MHz, 320 MB RAM), see [10].

One of the applications of our novel and robust technique is morphing. Morphing is an operation on two geometric objects resulting in a new object with an intermediate shape. Suppose we have an initial object represented by $f(x,y,z) \geqslant 0$, and a final object represented by $g(x,y,z) \geqslant 0$. Then,

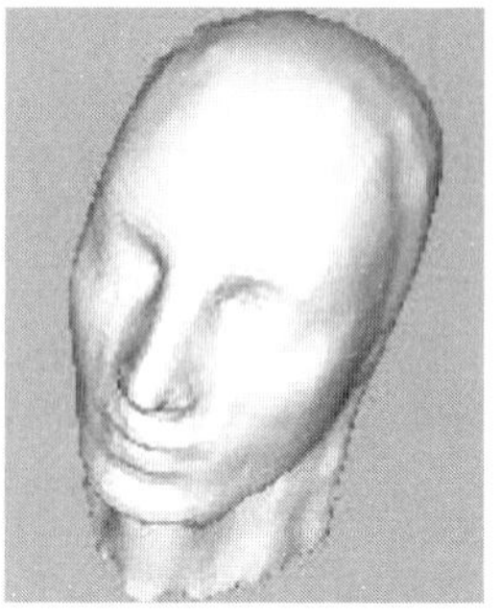

Fig. 5. Head reconstruction by CSRBF.

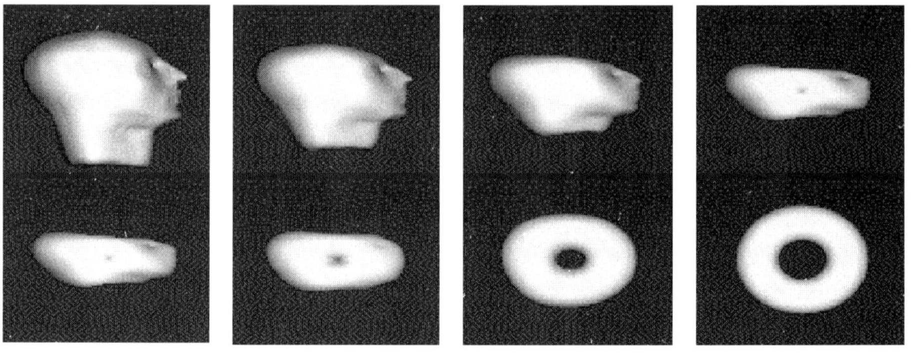

Fig. 6. Example of morphing.

using our new method we can effectively and quickly construct an intermediate object $h(x, y, z, t) \geqslant 0$, where $h(x, y, z, t) = (1 - t)f(x, y, z) + tg(x, y, z)$ and t is time from [0,1]. Details are described in [10]. Here we show the example of morphing with constraints in Fig. 6.

The initial volumetric head ($t = 0$) is gradually transforming into torus ($t = 1$). The goal of the choice of torus is to show that our method can easily transform an object to the other one with a different topology. The pictures above are results of our C++ programs implementation.

Our technique for surface reconstruction is applicable to other directions. One of them is two-dimensional inpainting problem. Some examples are seen in [17]. Also, we connect our future research with various applications of our technique, for example, to animation, collision detection, restoration design and so on.

5. Lifting subdivision

In order to represent three-dimensional geometric models in detail, a large number of vertices and meshes are used and a vast amount of computational costs for its rendering is required. Hence, mesh simplification for reducing rendering costs is necessary and the multiresolutional representation method is used to resolve this problem. The most typical method is Hoppe's progressive mesh

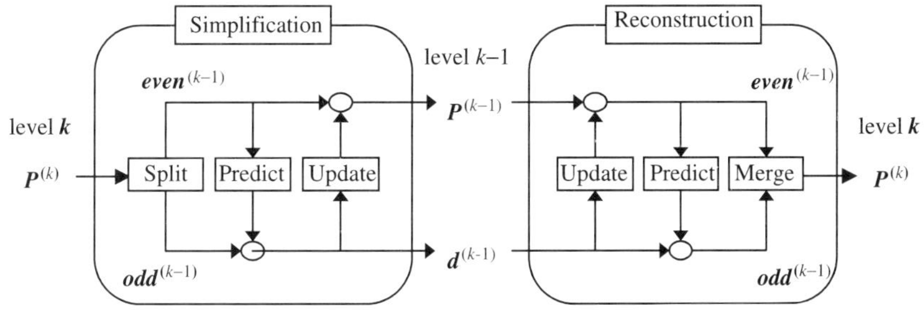

Fig. 7. The lifting scheme.

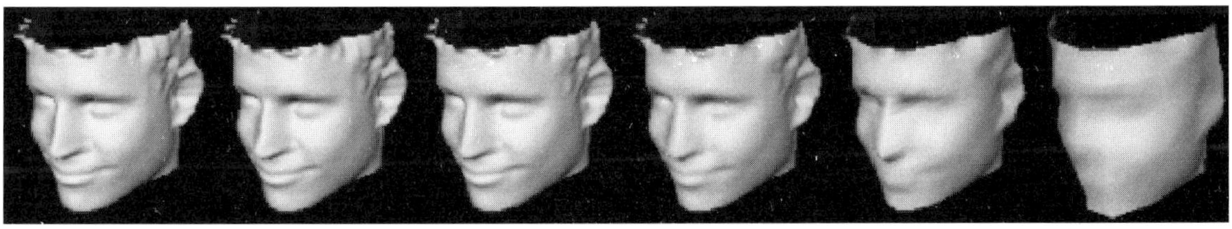

Fig. 8. Multiresolutional representation of quadrilateral mesh.

referring to [6,7]. Although this method can reconstruct the original mesh from simplified meshes completely and achieve the reverse transformation with high speed, it needs a great number of operations in simplification. In this section we restrict ourselves to indexed quadrilateral meshes, which especially appear in medical imaging, image reconstruction using computed tomography and so on. And we propose a new mesh simplification method to reduce the rendering costs of models with detailed quadrilateral meshes. This mesh simplification method leads a sequence of level of detail (LOD) without having to allocate new memory space. We call this method Lifting Subdivision since it is based on lifting scheme [20] and subdivision method [4].

The lifting scheme is one of wavelet transform methods, which can design the wavelet and is faster than the fast wavelet transform under some conditions. The scheme for the one-dimensional case is shown in Fig. 7. The sequence of the one-dimensional sequence $P^{(k)}$ is translated into the simplified sequence $P^{(k-1)}$ and the sequence of wavelet coefficients $d^{(k-1)}$ in the simplification stage. The original signal $P^{(k)}$ is completely obtained from $P^{(k-1)}$ and $d^{(k-1)}$ in the reconstruction stage. There are four operators: 'Split' divides the sequence $P^{(k)}$ into even and odd indexed signals; 'Predict' estimates the odd indexed signals from evens; 'Update' determines the order of vanishing moments of the wavelet; Merge.

On the other hand, the subdivision, four-point scheme, is a method to generate interpolating surfaces in the limit from indexed quadrilateral meshes. We apply this four-point scheme to the predictor of the lifting subdivision. This predictor leads an interpolating scaling function and we regard the coefficients as the simplified coordinates of the original model. For the precise description of the lifting subdivision method, we refer to [15].

An example of multiresolutional representation of quadrilateral meshes is shown in Fig. 8. Level k of multiresolutional representation is set to $0, -1, \ldots, -5$ from the left to the right. The number of

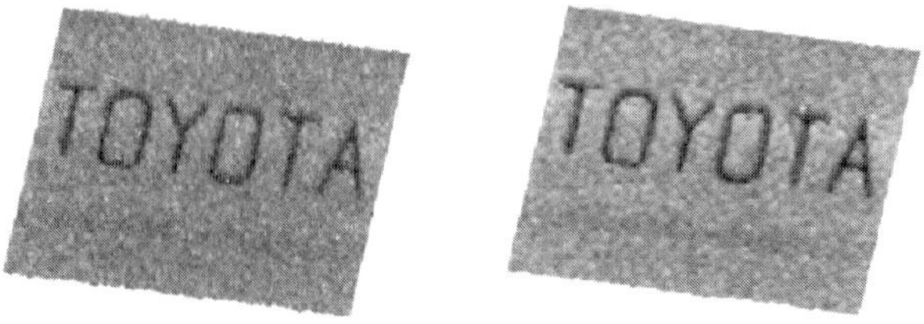

Fig. 9. Example of noise reduction (left: before reduction; right: after reduction).

vertices is 400×400 at level 0 and 13×13 at level -5. The CPU time to obtain the mesh at level -1 (200×200 vertices) from that of level 0 (400×400 vertices) is 0.18 s (Celeron 600 MHz).

The lifting subdivision method is also applicable to get rid of noise. Too much noise is generated when the observation is carried out by the recently developed hydrogen microscope. We give an example of noise reduction (Fig. 9), which is described precisely in [8].

References

[1] T.D. Blaker, R.J. Meyers, Seams and wedges in plastering: 3D hexahedral mesh generation algorithm, Eng. Comput. 9 (1993) 83–93.

[2] S.A. Canann, Plastering: A New Approach to Automated 3D Hexahedral Mesh Generation, American Institute of Aeronautics and Astronautics, Reston, VA, 1992.

[3] J.C. Carr, et al., Surface interpolation with radial basis functions for medical imaging, IEEE Trans. Med. Imaging 16 (1997) 96–107.

[4] N. Dyn, J. Gregory, D. Levin, A four-point interpolatory subdivision scheme for curve design, Comput. Aided Geom. Design 4 (1987) 257–268.

[5] N.T. Folwell, S.A. Mitchell, Reliable whisker weaving via curve contraction, Eng. Comput. 15 (1999) 292–302.

[6] H. Hoppe, Progressive meshes, computer graphics, Proceedings of SIGGRAPH'96, 1996, pp. 99–108.

[7] H. Hoppe, New quadric metric for simplifying meshes with appearance attributes, IEEE Visual (1999) 59–66.

[8] K. Iwashige, An application of image analysis to surface science, B4 Thesis, Tokyo Institute of Technology, 2002 (in Japanese).

[9] L. Kobbelt, $\sqrt{3}$-subdivision, preprint.

[10] N. Kojekine, V. Savchenko, D. Berzin, I. Hagiwara, Compactly-supported radial basis splines for functional representation of shapes, in: Proceedings of the International Association of Science Technology for Development, 2001.

[11] L. Labsik, et al., Progressive transmission of subdivision surfaces, Comput. Geom. 15 (2000) 25–39.

[12] P. Litwinovicz, L. Williams, Animating images with drawing, computer graphics, Proceedings of SIGGRAPH'94, 1994, pp. 409–412.

[13] M. Müller-Hannemann, Improving the surface cycle structure for hexahedral mesh generation, in: Proceedings of the 16th Symposium on Computational Geometry, Assoc. Comput. Machinery, New York, 2000, pp. 19–28.

[14] M. Müller-Hannemann, Quadrilateral surface meshes without self-intersecting dual cycles for hexahedral mesh generation, Comput. Geom., to appear.

[15] S. Nakata, I. Hagiwara, Multiresolution representation of three-dimensional shape model using lifting scheme, J. Inform. Process. Soc. Japan, submitted for publication (in Japanese).

[16] B.-Y. Ren, I. Hagiwara, Q.-X. Meng, Construction of master model from 3D scanned data using recursive subdivision scheme, Initiatives of Precision Engineering at the Beginning of a Millennium, 2001, pp. 902–906.

[17] V. Savchenko, http://www.savchenko.com/products/csrbf/gallery/.

[18] V. Savchenko, V. Vishnjakov, The use of the "specialization" approach in the design of parallel programs exemplified by problems in applied celestial mechanics, performance evaluation of parallel systems, Proceedings of PEPS'93, University of Warwick, Coventry, UK, 1993, pp. 126–133.
[19] J. Shinoda, et al., Characteristic topology method for quadrilateral mesh without self-intersection of dual cycles, Proceedings of the 22nd Computers and Information in Engineering Conference, submitted for publication.
[20] W. Sweldens, The lifting scheme: a custom-design construction of biorthogonal wavelets, J. Appl. Comput. Harmonic Anal. 3 (1996) 186–199.
[21] T.J. Tautges, et al., The whisker weaving algorithm: a connectivity-based method for constructing all-hexahedral finite element meshes, Int. J. Numer. Methods Eng. 39 (1996) 3327–3349.
[22] D. Zorin, P. Schröder, Subdivision for modeling and animation, Proceedings of SIGGRAPH 2000 Course Notes.

ELSEVIER

Journal of Computational and Applied Mathematics 149 (2002) 335–339

JOURNAL OF COMPUTATIONAL AND APPLIED MATHEMATICS

www.elsevier.com/locate/cam

Comparison of criteria on the direction of crack extension

Kohji Ohtsuka

Department of Computer Science, Hiroshima Kokusai Gakuin University, 6-20-1 Nakano, Aki-ku, Hiroshima 739-0321, Japan

Received 1 November 2001; received in revised form 13 March 2002

Abstract

There are three famous criteria for the prediction on the direction of crack extension: maximum energy release rate criterion, local symmetric criterion and maximum stress criterion. It has already been proved that these criteria are different. In this paper, we show how to measure the difference between maximum energy release rate criterion and maximum stress criterion. The results in this paper and special solutions in Amestoy and Leblond (Internat. J. Solids Struct. 29 (1992) 465) indicate that it is enough to use the maximum stress criterion for the numerical prediction of crack path, if the stress intensity factor of the shearing mode is not too big.

Keywords: Fracture mechanics; Functional analysis; Crack path prediction

To simplify the situation, we study the prediction of the simple elastic plate with straight crack defined on Ω_Σ, $\Omega_\Sigma = \Omega\backslash\Sigma$, $\Omega = \{(x_1,x_2);\ -b < x_1 < b, -b < x_2 < b\}$, $\Sigma = \{(x_1,x_2);\ -a \leqslant x_1 \leqslant a, x_2 = 0\}$ $(a \ll b)$. The side $\Gamma_D = \{(-b,x_2);\ -b < x_2 < b\}$ is fixed. The force $\boldsymbol{g} = (g_1,g_2) \in L^2(\Gamma_N)^2$ is given on $\Gamma_N = \partial\Omega - \overline{\Gamma_D}$. Also body force $\boldsymbol{f} \in L^2(\Omega)^2$ is in Ω. We use the notation $H^s(O)$ for Sobolev space of order s defined on a domain O with the norm $\|\cdot\|_{s,O}$ and $H^0(O) = L^2(O)$. The symbol ":=" means the left-hand side is defined by the right-hand side. The restriction of a function φ on a set S is denoted by $\varphi|_S$.

We now consider the virtual crack extension (see Fig. 1) where

$$\Sigma^\alpha(t) := \Sigma \cup \delta\Sigma^\alpha(t), \quad \gamma_0 := (-a,0), \quad \gamma := (a,0), \quad \gamma(t) := (a + t\cos\alpha, t\sin\alpha),$$

$$\delta\Sigma^\alpha(t) := \{(x_1,x_2);\ x_1 = a + s\cos\alpha,\ x_2 = s\sin\alpha,\ 0 \leqslant s \leqslant t\},\ -\pi < \alpha < \pi.$$

E-mail address: ohtsuka@barnard.cs.hkg.ac.jp (K. Ohtsuka).

0377-0427/02/$ - see front matter
PII: S0377-0427(02)00539-3

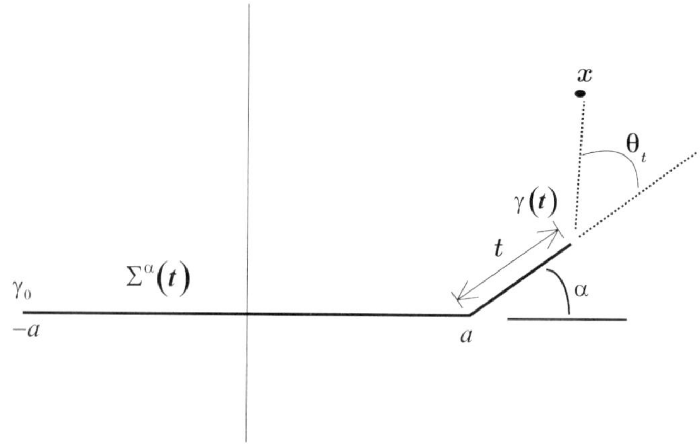

Fig. 1. L-shape crack extension and local polar coordinate system.

The question is the following: Find the angle α^* which is close to the real one? The answer to this question is very important in numerical calculation of fracture phenomena because the simulation of the crack extension has been approximated by polygonal lines, and these polygonal lines are created in a step-by-step manner as follows: (1) Find the direction. (2) Extend the crack with suitable length in this direction and back to (1). Quasi-static (stable) crack extension paths are obtained numerically as described here. However, there are many criteria in finding the direction based on different theoretical results. Unfortunately, we cannot select the true criterion by experiment because it is very difficult to measure the exact angle. The difference among criteria is discussed here in this very simple case, but we can derive similar results for a very general situation including the three-dimensional case.

For the virtual crack extension in the α-direction, let $\boldsymbol{u}^\alpha(t)=(u_i^\alpha(t))$, $\boldsymbol{\varepsilon}^\alpha(t)=(\varepsilon_{ij}^\alpha(t))$ and $\boldsymbol{\sigma}^\alpha(t)=(\sigma_{ij}^\alpha(t))$ denote the displacement vector, the strain tensor and the stress tensor, respectively, that is,

$$\varepsilon_{ij}^\alpha(t)=\varepsilon_{ij}(\boldsymbol{u}^\alpha(t)):=(u_{i,j}^\alpha(t)+u_{j,i}^\alpha(t))/2,\quad u_{i,j}^\alpha(t)=\partial u_i^\alpha(t)/\partial x_j,$$

$$\sigma_{ij}^\alpha(t)=\sigma_{ij}(\boldsymbol{u}^\alpha(t)):=c_{ijkl}\varepsilon_{kl}(\boldsymbol{u}^\alpha(t)),$$

where Hooke's tensor c_{ijkl} $(i,j,k,l=1,2)$ is in C^2 class in $\mathbb{R}^2$ and $u_i^\alpha(t)$ are the components of the displacement $\boldsymbol{u}^\alpha(t)$. For each load $L=(\boldsymbol{f},\boldsymbol{g})\in L^2(\Omega)^2\times L^2(\Gamma_N)^2$, the displacement $\boldsymbol{u}^\alpha(t)$ is obtained as the element of $V(\Omega_{\Sigma^\alpha(t)})$, $\Omega_{\Sigma^\alpha(t)}=\Omega\backslash\Sigma^\alpha(t)$, minimizing the potential energy functional

$$E(\boldsymbol{v};L,\Omega_{\Sigma^\alpha(t)})=\int_{\Omega_{\Sigma^\alpha(t)}}\{w(x,\boldsymbol{v})-\boldsymbol{f}\cdot\boldsymbol{v}\}\,\mathrm{d}x-\int_{\Gamma_N}\boldsymbol{g}\cdot\boldsymbol{v}\,\mathrm{d}\ell \tag{1}$$

over $\boldsymbol{v}\in V(\Omega_{\Sigma^\alpha(t)})$. Here $w(x,\boldsymbol{v})=c_{ijkl}(x)\varepsilon_{ij}(\boldsymbol{v})\varepsilon_{kl}(\boldsymbol{v})/2$,

$$V(\Omega_{\Sigma^\alpha(t)})=\{\boldsymbol{v}\in H^1(\Omega_{\Sigma^\alpha(t)})^2;\ \boldsymbol{v}=0 \text{ on } \Gamma_D\}.$$

To measure crack extension force, we now introduce the two-dimensional version of the GJ-integral proposed in [4] originally constructed for three-dimensional fracture. For each solution $\boldsymbol{u}(t)=\boldsymbol{u}^\alpha(t)$, the GJ-integral is defined as a functional depending on the domain D and a vector field $\boldsymbol{X}\in C^2(\mathbb{R}^2)^2$

as follows:

$$J_D(\boldsymbol{u}(t),\boldsymbol{X}) := P_D(\boldsymbol{u}(t),\boldsymbol{X}) + R_D(\boldsymbol{u}(t),\boldsymbol{X}),$$

$$P_D(\boldsymbol{u}(t),\boldsymbol{X}) := \int_{\partial(D\cap\Omega_{\Sigma^\alpha(t)})} \{w(t)(\boldsymbol{X}\cdot\boldsymbol{n}) - \sigma_{ij}(\boldsymbol{u}(t))n_j(\boldsymbol{X}\cdot\nabla u_i(t))\}\,\mathrm{d}\ell,$$

$$R_D(\boldsymbol{u}(t),\boldsymbol{X}) := \int_{D\cap\Omega_{\Sigma^\alpha(t)}} \{\sigma_{ij}(\boldsymbol{u}(t))\partial_j X_k \partial_k u_i(t) - \boldsymbol{f}(\boldsymbol{X}\cdot\nabla u(t))$$

$$- \boldsymbol{X}\cdot\nabla_x w(t) - w(t)\,\mathrm{div}\,\boldsymbol{X}\}\,\mathrm{d}x. \tag{2}$$

Here $w(t)=w(x,\boldsymbol{u}(t))$, $\boldsymbol{n}=(n_1,n_2)$ the outward unit normal on $\partial(D\cap\Omega_{\Sigma^\alpha(t)})$ and $\mathrm{d}\ell$ the line element of $\partial(\omega\cap\Omega_{\Sigma^\alpha(t)}))$. The integral $R_D(\boldsymbol{u},\boldsymbol{X})$ is well defined for all weak solutions $\boldsymbol{u}(t)$, but $P_D(\boldsymbol{u}(t),\boldsymbol{X})$ needs the regularity of $\boldsymbol{u}(t)$. We notice that $P_D(\boldsymbol{u}(t),\boldsymbol{X})$ contains no integral over $D\cap\Sigma^\alpha(t)$, if $\boldsymbol{X}|_D$ is a tangent to $\Sigma^\alpha(t)$.

Let $\omega(t)$ be a disk of radius $t/2$ with origin $\gamma(t)$. Then the following expansion by the local polar coordinate (r_t,θ_t) is proved in the isotropic case (see references in [5]),

$$u_i^\alpha(x,t) = \sum_{l=1}^{2} \frac{K_{l,\alpha}(\gamma(t))}{2\mu}\sqrt{\frac{r_t}{2\pi}}\,S_{i,l}^C(\theta_t) + u_{i,R}^\alpha(x,t) \quad \text{on } \omega(t), \tag{3}$$

where $S_{i,l}^C \in C^\infty((-\pi,\pi))$, $r_t = |x-\gamma(t)|$, $\theta_t = \pm\pi$ express the plus/minus side of $\Sigma^\alpha(t)$ (see Fig. 1) and $u_{i,R}^\alpha(t)\in H^2(\omega(t)\cap\Omega_{\Sigma^\alpha(t)})$. $K_{1,\alpha}(\gamma(t))$ is called *stress intensity factor* of opening mode at $\gamma(t)$ and $K_{2,\alpha}(\gamma(t))$ shearing mode. The most important parameter in fracture mechanics is the *energy release rate*

$$G(L;\Omega_{\Sigma^\alpha(t)}) := \lim_{\delta t\downarrow 0} \frac{E(\boldsymbol{u}^\alpha(t);L,\Omega_{\Sigma^\alpha(t)}) - E(\boldsymbol{u}^\alpha(t+\delta t);L,\Omega_{\Sigma^\alpha(t+\delta t)})}{|\Sigma^\alpha(t+\delta t)-\Sigma^\alpha(t)|}, \tag{4}$$

where $|\Sigma^\alpha(t+\delta t)-\Sigma^\alpha(t)|$ denotes the length of $\Sigma^\alpha(t+\delta t)-\Sigma^\alpha(t)$. The energy release rate at $t>0$ in our case is expressed as (see [3])

$$G(L;\Omega_{\Sigma^\alpha(t)}) = J_{\omega(t)}(\boldsymbol{u}^\alpha(t),\cos\alpha\boldsymbol{e}_1 + \sin\alpha\boldsymbol{e}_2) - \frac{1}{E}(K_{1,\alpha}(\gamma(t))^2 + K_{2,\alpha}(\gamma(t))^2), \tag{5}$$

because the crack extension $\delta t \mapsto \Sigma^\alpha(t+\delta t)$; $\delta t > -t/2$ of $\Sigma(t)$ is smooth. When the crack extension $t\mapsto\Sigma(t)$ is not smooth, we can get the expression of the energy release rate at $t=0$ by the mean value theorem,

$$G(L;\Omega_{\Sigma^\alpha(+0)}) = \frac{1}{E}(K_{1,\alpha}^2 + K_{2,\alpha}^2) \quad \text{with } K_{l,\alpha} = \lim_{t\downarrow 0} K_{l,\alpha}(\gamma(t)). \tag{6}$$

Three famous criteria are defined as follows: *maximum energy release rate criterion* is to find the maximum value of $\alpha\mapsto G(L;\Omega_{\Sigma^\alpha(\cdot)})$ on $-\pi<\alpha<\pi$. *Local symmetry criterion* is to find the angle $\alpha^\#$ that satisfies the condition $K_{2,\alpha^\#}(\gamma(+0))=0$. *Maximum stress criterion* is to find α^{**} such that

$$\sigma_{\alpha^{**}} = \max_\theta \sigma_\theta \quad \text{and} \quad \sigma_{r\alpha^{**}} = 0, \tag{7}$$

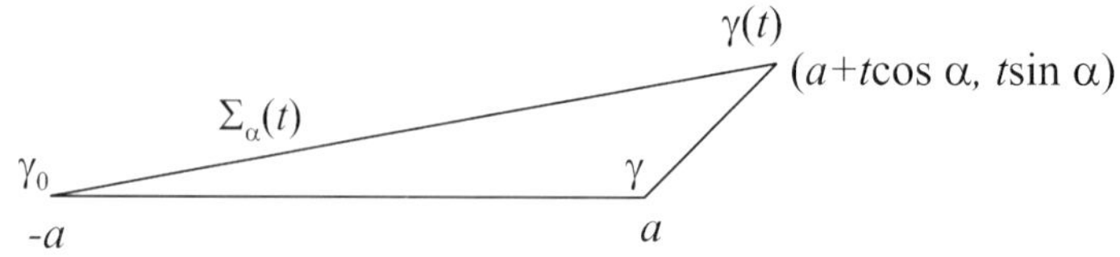

Fig. 2. Kinked crack $\Sigma^{\alpha}(t)$ and the straight crack $\Sigma_{\alpha}(t)$.

where

$$\sigma_{\theta}(\boldsymbol{v}) = \sigma_{11}(\boldsymbol{v})\sin^2\theta + \sigma_{22}(\boldsymbol{v})\cos^2\theta - \sigma_{12}(\boldsymbol{v})\sin 2\theta,$$

$$\sigma_{r\theta}(\boldsymbol{v}) = (\sigma_{22}(\boldsymbol{v}) - \sigma_{11}(\boldsymbol{v}))\sin\theta\cos\theta + \sigma_{12}(\boldsymbol{v})\cos 2\theta.$$

Here $\sigma_{\theta}=\sigma_{\theta}(\boldsymbol{u})$, and $\sigma_{r\theta}=\sigma_{r\theta}(\boldsymbol{u})$ with $\boldsymbol{u}=\boldsymbol{u}^{\alpha}(0)|_{\alpha=0}$. The displacement $\boldsymbol{u}$ at the initial crack ($t=0$) is expressed by (3) near γ with $t=0$, $r_0=r=|x-\gamma|$, $\theta_0=\theta$. Let us denote by K_l, $l=1,2$ the stress intensity factors at $t=0$. Here $\theta=\pm\pi$ expresses the upper/lower side of x_1-axis on $\Sigma\cap\omega=\{x;\ |x-\gamma|<\delta\}$ for a suitable number $\delta>0$. There are a lot of works to obtain the relationship between K_l and $K_{l,\alpha}$, $l=1,2$ by conformal mapping technique (see [2,6,1] for details).

Let us introduce the symbols

$$\tilde{K}_{1,\alpha}{:=}\lim_{r\to 0}(2\pi r)^{-1/2}\sigma_{\theta}(\boldsymbol{u})|_{\theta=\alpha},\quad \tilde{K}_{2,\alpha}{:=}\lim_{r\to 0}(2\pi r)^{-1/2}\sigma_{r\theta}(\boldsymbol{u})|_{\theta=\alpha}, \tag{8}$$

which is expressed as follows:

$$\tilde{K}_{l,\alpha} = \tilde{F}_{l1}(\alpha)K_1 + \tilde{F}_{l2}(\alpha)K_2,\quad l=1,2,$$

$$\tilde{F}_{11}(\theta) := \tfrac{3}{4}\cos(\theta/2) + \tfrac{1}{4}\cos(3\theta/2),\quad \tilde{F}_{12}(\theta) := -\tfrac{3}{4}\sin(\theta/2) - \tfrac{3}{4}\sin(3\theta/2),$$

$$\tilde{F}_{21}(\theta) := \tfrac{1}{4}\sin(\theta/2) + \tfrac{1}{4}\sin(3\theta/2),\quad \tilde{F}_{22}(\theta) := \tfrac{1}{4}\cos(\theta/2) + \tfrac{3}{4}\cos(3\theta/2). \tag{9}$$

From the definition of $\tilde{K}_{l,\alpha}$, the maximum stress criterion is equivalent to finding α^{**} which takes a maximum of $\tilde{K}_{1,\alpha}$ and $\tilde{K}_{2,\alpha^{**}}=0$. Moreover $(\mathrm{d}/\mathrm{d}\alpha)(\tilde{K}_{1,\alpha}{}^2+\tilde{K}_{2,\alpha}{}^2)=2\tilde{K}_{1,\alpha}(\mathrm{d}/\mathrm{d}\alpha)\tilde{K}_{1,\alpha}+2\tilde{K}_{2,\alpha}(\mathrm{d}/\mathrm{d}\alpha)$ $\tilde{K}_{2,\alpha}=0$ if $\alpha=\alpha^{**}$. This means that the three criteria are equivalent from (6) if $\tilde{K}_{l,\alpha}=K_{l,\alpha}$. So $K_{l,\alpha}-\tilde{K}_{l,\alpha}$ expresses the difference of the criteria. A check of the difference $K_{l,\alpha}-\tilde{K}_{l,\alpha}$ will be done by introducing the straight crack $\Sigma_{\alpha}(t)$ as shown in Fig. 2 with the same crack tips γ_0, $\gamma(t)$,

$$\Sigma_{\alpha}(t){:=}\{(x,y);\ y=(x+a)\zeta,\ -a\leqslant x\leqslant a+t\cos\alpha\},\quad \zeta=\frac{t\sin\alpha}{2a+t\cos\alpha}$$

and the displacement vector $\boldsymbol{u}_{\alpha}(t)$ is obtained by the variational method with undeformed state $\Omega_{\Sigma_{\alpha}(t)}=\Omega\backslash\Sigma_{\alpha}(t)$. By using the local polar coordinate system in (3), we now define $\tilde{K}_{l,\alpha}(\gamma(t))$ by

$$\tilde{K}_{1,\alpha}(\gamma(t)){:=}\lim_{r_t\to 0}\left.\frac{\sigma_{\theta_t}(\boldsymbol{u}_{\alpha}(t))}{\sqrt{2\pi r_t}}\right|_{\theta_t=0},\quad \tilde{K}_{2,\alpha}(\gamma(t)){:=}\lim_{r_t\to 0}\left.\frac{\sigma_{r_t\theta_t}(\boldsymbol{u}_{\alpha}(t))}{\sqrt{2\pi r_t}}\right|_{\theta_t=0}.$$

From the definitions, we have $\tilde{K}_{l,\alpha}(\gamma(t))\to\tilde{K}_{l,\alpha}$, $l=1,2$ as $t\to 0$. The difference $K_{l,\alpha}(\gamma(t))-\tilde{K}_{l,\alpha}(\gamma(t))$ is given by the effect of the stress concentration at γ. Since the stress concentration at γ decreases as $\alpha\to 0$, $\tilde{K}_{l,\alpha}$ and $K_{l,\alpha}$ will be close to each other near $\alpha=0$. The difference

$\tilde{K}_{l,\alpha}(\gamma(t)) - K_{l,\alpha}(\gamma(t))$ is estimated by $\boldsymbol{u}_{\alpha,t} - \boldsymbol{u}_t^{\alpha}$ in $\Omega \backslash V(\alpha,t)$ where $V(\alpha,t)$ stands for the triangle connecting the points γ_0, γ, $\gamma(t)$.

In the situation given in [1], $K_{l,\alpha} - \tilde{K}_{l,\alpha} = \mathrm{O}(\alpha^2)$, $l = 1,2$. In such a case, we can estimate the difference of the criteria with respect to K_2/K_1 exactly (see [5] for details). In the general case, the mathematical exact estimation is open.

References

[1] M. Amestoy, J.B. Leblond, Crack paths in plane situation—II, Internat. J. Solids Struct. 29 (1992) 465–501.

[2] H.Wu. Chien, Maximum-energy-release-rate criterion applied to a tension-compression specimen with crack, Elasticity 8 (1978) 183–205.

[3] K. Ohtsuka, J-integral and two-dimensional fracture mechanics, RIMS Kokyuroku, Kyoto University, Vol. 386, 1980, pp. 231–248.

[4] K. Ohtsuka, Generalized J-integral and three dimensional fracture mechanics I, Hiroshima Math. J. 11 (1981) 21–52.

[5] K. Ohtsuka, Theoretical and numerical analysis of fracture phenomena under dynamic loading, RIMS Kokyuroku, Kyoto University, Vol. 1145, 2000, pp. 179–189.

[6] Y. Sumi, A note on the first-order perturbation solution of a straight crack with slightly branched and curved extension under a general geometric and loading condition, Engrg. Fracture Mech. 24 (1986) 479–481.

ELSEVIER

Journal of Computational and Applied Mathematics 149 (2002) 341–349

JOURNAL OF COMPUTATIONAL AND APPLIED MATHEMATICS

www.elsevier.com/locate/cam

Fast linear equation solvers in high performance electromagnetic field analysis

Masaaki Shimasaki[a,*], Takeshi Iwashita[b], Takeshi Mifune[a]

[a]*Department of Electrical Engineering, Kyoto University, Yoshida Honmachi, Sakyo-ku, Kyoto 606-8501, Japan*
[b]*Data Processing Center, Kyoto University, Kyoto 606-8501, Japan*

Received 17 December 2001; received in revised form 4 February 2002

Abstract

Solving linear equations plays a crucial role in high performance electromagnetic field analysis. We describe forms and characteristics of a system of linear equations arising in electromagnetic field analysis with finite element method (FEM). Properties of ICCG and its parallelization are discussed in context of electromagnetic field analyses. Although current applicability of multigrid approach is rather limited in electromagnetic field analysis in comparison with ICCG, the multigrid method is important because it is quite fast when applied to very large-scale problems. We discuss the algebraic multigrid method in finite element electromagnetic field analysis.

Keywords: Finite element method; Electromagnetic field analysis; Parallelization of ICCG; Algebraic multigrid method

1. Introduction

Precise analysis of complex phenomena and detailed design optimization of electric and electronic devices lead to a large-scale finite element electromagnetic field analysis. Since most of total computation time is consumed in solution of linear equations, fast solvers of large-scale sparse linear equations play a crucial role in the analysis. Some points should be considered in solution of linear equations arising from finite element electromagnetic field analysis, especially indefiniteness caused by gauge conditions and complex coefficient matrices in frequency domain analyses. ICCG [16] is a standard and robust solver, but its parallelization is not necessarily easy and we discuss parallelization of ICCG. At present applicability of multigrid method is limited compared with ICCG but

* Corresponding author.
E-mail addresses: simasaki@kuee.kyoto-u.ac.jp (M. Shimasaki), take@kudpc.kyoto-u.ac.jp (T. Iwashita), mifune@fem.kuee.kyoto-u.ac.jp (T. Mifune).

PII: S0377-0427(02)00540-X

the multigrid method is very fast when applied to very large-scale problems. Extensive research has been carried out on geometric multigrid approach but geometric information is required in coarse grid generation and this restricts applicability of multigrid methods. Different from geometric multigrid methods, algebraic multigrid methods do not require geometric grid information in coarse grid generation and the method can be used as a black box solver for linear equations. We discuss algebraic multigrid algorithms in finite element electromagnetic field analysis.

This paper is organized as follows. Characteristics of linear equations arising from finite element electromagnetic field analysis are discussed in Section 2. In Section 3 we discuss parallelization of ICCG by ordering strategy of variables. Section 4 is devoted to discussion of algebraic multigrid algorithm and its parallelization in electromagnetic field analysis. Finally, conclusions are described in Section 5.

2. Characteristics of linear equations arising from finite element electromagnetic field analysis

2.1. $A-\phi$ method and related linear equations

The eddy current analysis is one of fundamental and very important problems in electromagnetic field analysis in electrical engineering. As shown in Fig. 1, the domain Ω of interest is divided into two subdomains Ω_1 where unknown eddy currents exist and Ω_2 where there is no eddy currents but there may be given source currents. Γ_{12} denotes the interface boundary between Ω_1 and Ω_2. Let us denote the unknown magnetic vector potential in the domain Ω_i $(i=1,2)$ by $\boldsymbol{A}_i$. We further denote the scalar potential in the domain Ω_1 by ϕ. The A–ϕ formulation of the problem leads to the following equations:

$$\operatorname{curl}\left(\frac{1}{\mu_1}\operatorname{curl}\boldsymbol{A}_1\right)+\sigma\left(\frac{\partial \boldsymbol{A}_1}{\partial t}+\operatorname{grad}\phi\right)=0 \quad \text{in } \Omega_1, \tag{1}$$

$$\operatorname{div}\left(\sigma\frac{\partial \boldsymbol{A}_1}{\partial t}+\sigma\operatorname{grad}\phi\right)=0 \quad \text{in } \Omega_1, \tag{2}$$

$$\operatorname{curl}\left(\frac{1}{\mu_2}\operatorname{curl}\boldsymbol{A}_2\right)=\boldsymbol{J}_S \quad \text{in } \Omega_2 \tag{3}$$

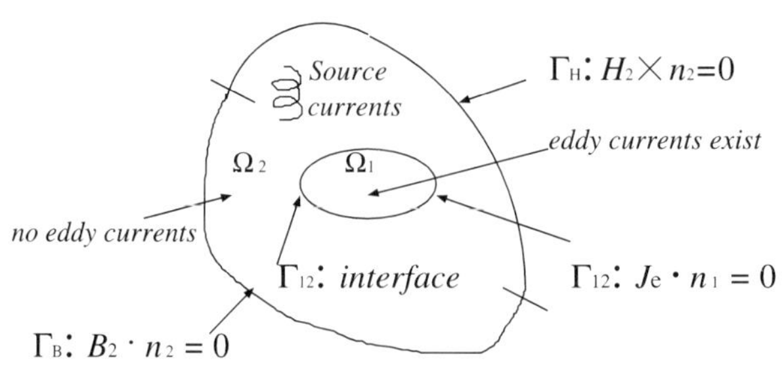

Fig. 1. Domains, boundary sections and interface in typical eddy current problems.

with boundary conditions

$$\boldsymbol{n}_2 \operatorname{curl} \boldsymbol{A}_2 = 0 \quad \text{on } \Gamma_B, \tag{4}$$

$$\boldsymbol{n}_2 \frac{1}{\mu_2} \operatorname{curl} \boldsymbol{A}_2 = 0 \quad \text{on } \Gamma_H \tag{5}$$

and with interface conditions:

$$\boldsymbol{n}_1 \operatorname{curl} \boldsymbol{A}_1 + \boldsymbol{n}_2 \operatorname{curl} \boldsymbol{A}_2 = 0 \quad \text{on } \Gamma_{12}, \tag{6}$$

$$\boldsymbol{n}_1 \frac{1}{\mu_1} \operatorname{curl} \boldsymbol{A}_1 + \boldsymbol{n}_2 \frac{1}{\mu_2} \operatorname{curl} \boldsymbol{A}_2 = 0 \quad \text{on } \Gamma_{12}, \tag{7}$$

$$\boldsymbol{n}_1 \boldsymbol{J}_\mathrm{e} = 0 \quad \text{on } \Gamma_{12}, \tag{8}$$

where $\boldsymbol{J}_\mathrm{e}$ denotes the eddy current and is given by

$$\boldsymbol{J}_\mathrm{e} = -\sigma \left(\frac{\partial \boldsymbol{A}_1}{\partial t} + \operatorname{grad} \phi \right) \tag{9}$$

and $\boldsymbol{J}_\mathrm{S}$ in the domain Ω_2, and given constant μ_i in the domain Ω_i denote the source current and permeability, respectively. The given constant σ denotes conductivity in the domain Ω_1. The vector $\boldsymbol{n}_i$ denotes the outward normal to the boundary Γ_{12} of the domain Ω_i $(i = 1,2)$.

The magnetic flux density $\boldsymbol{B}$ is of primary concern in the analysis and can be computed in the individual domain by

$$\boldsymbol{B}_i = \operatorname{curl} \boldsymbol{A}_i. \tag{10}$$

It should be noted that the magnetic field intensity $\boldsymbol{H}_i$ is given by

$$\boldsymbol{H}_i = \frac{1}{\mu_i} \boldsymbol{B}_i. \tag{11}$$

For a class of problems as those in numerical examples, the scalar potential ϕ can be neglected and the formulation becomes A-formulation.

There is indefiniteness in $\boldsymbol{A}$ and ϕ in Eq. (1) known as a 'gauge problem' [10], and it is necessary to specify a certain gauge condition to guarantee uniqueness of potentials. An appropriate implicit time difference scheme of Eq. (1) and finite element formulation yields a system of linear equations. Unless a gauge condition is specified, the resultant coefficient matrix becomes singular. The Coulomb gauge $\operatorname{div} \boldsymbol{A} = 0$ is one of traditional gauge conditions and in finite element analysis, the Coulomb gauge can be used as a constraint condition in weak formulation of the partial differential equation. Another gauge condition based on the notion of 'co-tree' of element meshes has been proposed to eliminate variables for the uniqueness of the solution [1,13]. Recently, however, it becomes clear that a singular system without gauge condition can be solved more rapidly by iterative methods than a nonsingular system with some gauge condition [7,14] and yet the magnetic field B can be computed from solution of the indefinite system. One of the most popular iterative methods is ICCG. Since the matrix is singular, ICCG may encounter breakdown in incomplete Choleskii factorization. The technique to avoid breakdown of ICCG [7,14] is as follows: The diagonal matrix D of the coefficient matrix is modified as $(1+\alpha)D$ in incomplete factorization, where α is a small parameter and the usual ICCG is applied. It should be noted that diagonal modification is only for preconditioning procedure and the equation to be solved remains singular. The basic idea for diagonal modification originates from MICCG [9] or from shifted ICCG [15].

2.2. Frequency domain analysis and related linear equations

For the steady-state analysis of alternating fields with the angular frequency ω, Eq. (1) is replaced by

$$\operatorname{curl}\left(\frac{1}{\mu_1}\operatorname{curl}\boldsymbol{A}_1\right)+\sigma(\mathrm{i}\omega\boldsymbol{A}_1+\operatorname{grad}\phi)=0, \tag{12}$$

where $\mathrm{i}^2=-1$, and the magnetic potential $\boldsymbol{A}_1$ and the scalar potential ϕ are complex variables. The coefficient matrix of a system of linear equations under investigation is complex symmetric but not Hermitian. The system of linear equations can be analyzed by the conjugate orthogonal conjugate gradient method, a CG like method [23]. The algorithm can be formally derived from the complex version of the ICCG algorithm and therefore occasionally called ICCG.

3. Parallelization of ICCG

3.1. Multi-color ordering method

ICCG is a very effective iterative solver for symmetric positive definite system of linear equations arising in the FE analysis in electrical engineering. Parallelization of the method is, however, difficult because of inherent sequential nature of forward and backward substitutions of the incomplete Choleski factorization. Thus, the problem is to find an efficient parallel preconditioner. There has been a vast amount of research for parallelization of ICCG including domain decomposition approach. Further references can be obtained in [4,18].

Besides domain decomposition approach, reordering of unknown variables is one of the most effective strategies for parallel processing [22,6] but reordering leads to an increase in the number of iteration steps [5], with respect to the standard lexicographic ordering. In order to cope with the trade-off between parallelization and convergence, S. Doi and A. Lichnewsky proposed the use of many colors, about 30 or more, in multi-color ordering [2,3]. Their “large numbered multi-color ordering” technique achieves a high degree of parallelism with a small increase of CG iterations. Since the method has been developed for the finite difference method with structured mesh, it cannot be directly applied to the finite element method with unstructured mesh. We are developing the algebraic multi-color ordering method [11] for unstructured finite element electromagnetic field computations. The crucial point of the method is selection of the colors of unknown variables. The color is automatically set so as to satisfy the following condition for parallel processing:

Condition (*MC: Multi-Color*). The unknowns with the same color do not have direct relationship with each other.

While several coloring methods for unstructured problems were previously presented [12,20], a new coloring technique has been proposed [11]. This is because the strategy for attaining better convergence in large-numbered multi-color ordering method, that is a use of many colors, cannot be effectively introduced into the previous methods, which are based on the independent set of the unknowns. Moreover, nodewise coloring which is discussed in [12,20] for CFD analyses with multiple degrees of freedom per node cannot be used for a linear system of equations arising in electromagnetic finite element analyses using both vector shape functions (edge elements) and scalar shape functions (nodal elements). On the other hand, since our coloring method is completely algebraic and its implementation is easy, it can be applied to many applications. In the method, the convergence can be improved by using many colors. The algorithm for coloring unknowns is described in the following way.

We denote the strictly lower triangular part of the coefficient matrix of linear equations by L. The column numbers of nonzero entries in the ith row of the matrix L are denoted by $\ln zc(i,j)$ $(j = 1,\ldots,\ln z(i))$, where $\ln z(i)$ is the number of nonzero entries in the ith row. We use the array $color(i)$ to denote the color number of the ith unknown variable. The number of colors $ncolor$ is set first. The array $color$ is initialized as 0. The color of the ith unknown is determined as follows:

Let $icolor$ be the current color (i.e. the color of the $(i-1)$th unknown is $icolor - 1$).

- *Step* 1: Check the colors of the $\ln zc(i,j)$th unknowns $(j=1,\ldots,\ln z(i))$. If $icolor$ is not assigned to these unknowns, set $icolor$ as the color of the ith unknown and move to the next $(i+1)$th unknown. Otherwise, move to Step 2.
- *Step* 2: Update $icolor$ to $\mathrm{mod}(icolor, ncolor) + 1$, and return to Step 1.

The number of colors $ncolor$ should be greater than or equal to the maximum value of $\ln z(i) + 1$ $(i = 1,\ldots,n)$. As long as $ncolor$ satisfies the condition, Step 1 above guarantees condition MC, which is required for parallel processing. The cost for setting the color is negligible compared with costs for the ICCG iterations. After coloring of the unknowns, the ICCG kernels including forward and backward substitutions are carried out in parallel.

3.2. Numerical example of multi-color ordering method

The method is applied to the IEEJ standard benchmark model of 3D eddy current analyses [19] and the model region is shown in Fig. 2.

The domain of interest is discretized by first-order brick-type edge elements. The electromagnetic field equations are formulated with A-formulation, and solved by the Galerkin method and the backward time difference method. The total number of unknowns is 1,011,920. The parallel performance is evaluated in one time step. The convergence criterion of ICCG is given by $\|r\|_2/\|b\|_2 < 10^{-7}$, where r is the residual vector of the *original linear system*. The modification factor $1+\alpha$ of diagonal elements in preconditioning is 1.07. Fig. 3 shows computed eddy current and experimental data. Testing various numbers of colors ranging from 33 to 100, the best performance is attained in the case of 60 for this problem. Table 1 shows the number of iterations, the computation time (elapsed time in seconds) of ICCG with multi-color ordering and those of Block ICCG, on Fujitsu VPP800 at Data Processing Center of Kyoto University (63 PE's, 8GB/PE). In this case the number of CG iterations is 390 with only 10% increase compared with the natural ordering method and it is much less than for Block ICCG. As a consequence, the computation time is greatly reduced compared with Block ICCG.

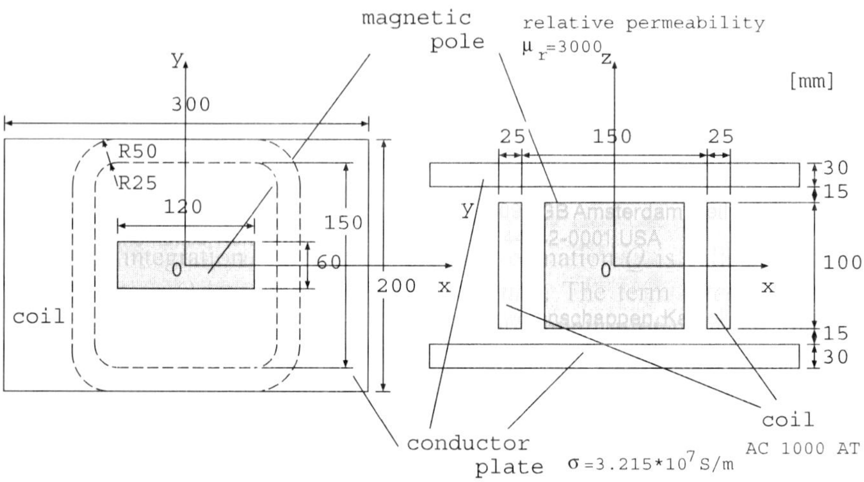

Fig. 2. The IEEJ standard benchmark model of 3D eddy current analysis.

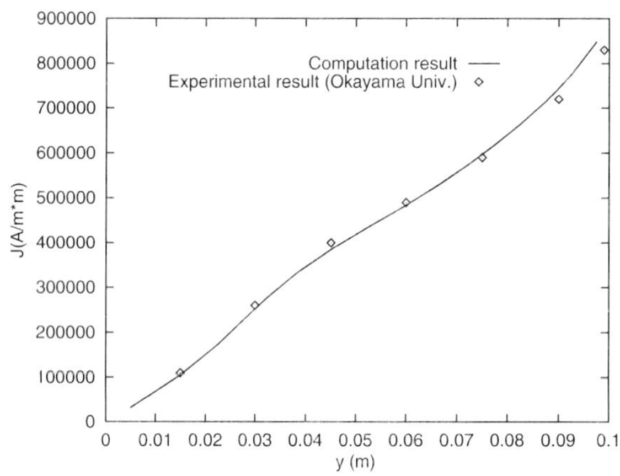

Fig. 3. Computed eddy current and experimental data.

4. Algebraic multigrid for finite element electromagnetic field analysis

We are developing the algebraic multigrid method for 2D eddy-current analysis [17]. The basic equation is given by (1) but in this two-dimensional case ϕ can be set to zero. The AMG algorithm consists of two phases, namely, the setup phase and the solution phase. The hierarchy of coarse grids and the interpolation operators are constructed in the setup phase. In the current implementation, we use the scalar algorithm described in [21].

In the solution phase, the standard MG iterations are performed. The Gauss–Seidel method is chosen as the smoother in this phase, and we use the (1,1) V-cycle in the AMG method [21], so that one smoothing is carried out before and after a coarse grid correction. In eddy current analysis,

Table 1
Comparison of computation time (elapsed time) and numbers of iterations

(a) ICCG on 1 P(natural ordering)

N_p	Elapsed time (s)	# of iterations
1	2119	366

(b) Multi-color ICCG and block ICCG

N_p	Multi-color ICCG		Block ICCG	
	Elapsed time(s)	# iterations	Elapsed time(s)	# iterations
2	1194.4	390	1394	530
4	606.3	390	1028	779
8	308.4	390	550.5	834
16	164.3	390	251.6	675
32	99.83	390	120.4	726

Table 2
Computation time (elapsed time) and number of iterations

N_p	AMG		ICCG($N_p = 1$), Block ICCG($N_p > 1$)	
	Elapsed time(s)	# of iterations	Elapsed time(s)	# of iterations
1	513	12.5	5219	340
2	273	12.6	3163	397
4	153	12.7	1748	407
8	97.2	12.9	998	423
16	67.5	13.2	630	440

a system of linear equations is solved only once in the first step. The coefficient matrix of the linear system is sparse and the Reverse Cuthill–McKee (RCM) algorithm [8] is applied to reduce the bandwidth of the matrix. In the setup phase, the selections of the coarse grid points are carried out sequentially. As an example, the electromagnetic field in an iron-core with a rectangular cross-section ($0.04\ \mathrm{m} \times 0.02\ \mathrm{m}$) is analyzed. The domain of the problem is discretized using an unstructured mesh with triangular nodal elements. The problem is 2D one without scalar potential and the coefficient matrix of the system of linear equations is positive definite. The number of degrees of freedom is 80,601. The performance of the AMG solver is compared with that of ICCG for $N_p = 1$, and those of conventional Block ICCG for $N_p > 1$ on a distributed memory parallel computer, HITACHI SR2201 at Department of Electrical Engineering of Kyoto University (31 PE's + PE for I/O, 0.5 GB memory for each PE).

Table 2 shows the computation time (elapsed time in seconds) and the number of iterations of AMG and those of ICCG solvers, where the number of AMG iterations indicates the number of V-cycles performed. For $N_p = 1$, sequential algorithms are used both for AMG and ICCG, and AMG is much faster than ICCG for this simple problem. Furthermore, the parallel AMG solver has also a good parallel performance.

5. Conclusion

In this paper, we discuss fast solvers for systems of linear equations arising in high performance electromagnetic field analyses. We investigate parallelization of ICCG as robust solvers with wide applicability and demonstrate the multi-color ordering strategy for parallelizable preconditioner for ICCG for finite element formulations of eddy current problems. As a very fast solver, we discuss the parallel algebraic multigrid method. In future it is necessary to investigate further applicability of the parallel algebraic multigrid method for more complex problems in finite element analysis.

References

[1] R. Albanese, G. Rubinacci, Solution of three dimensional Eddy current problems by integral and differential methods, IEEE Trans. Magn. 24 (1988) 98–101.

[2] S. Doi, A. Lichnewsky, A graph-theory approach for analyzing the effects of ordering on ILU preconditioning, INRIA Report, 1452, 1991.

[3] S. Doi, T. Washio, Ordering strategies and related techniques to overcome the trade-off between parallelism and convergence in incomplete factorization, Parallel Comput. 25 (1999) 1995–2014.

[4] J.J. Dongarra, I.S. Duff, D.C. Sorensen, H.A. van der Vorst, Numerical Linear Algebra for High Performance Computers, SIAM, Philadelphia, PA, 1998.

[5] I.S. Duff, G.A. Meurant, The effect of ordering on preconditioned conjugate gradients, BIT 29 (1989) 635–657.

[6] I.S. Duff, H.A. van der Vorst, Developments and trend in the parallel solution of linear systems, Parallel Comput. 25 (1999) 1931–1970.

[7] K. Fujiwara, T. Nakata, H. Ohashi, Improvement of convergence characteristic of ICCG method for the A–ϕ method using edge elements, IEEE Trans. Magn. 32 (1996) 804–807.

[8] A. George, J.W-H. Liu, Computer Solution of Large Sparse Positive Definite Systems, Prentice-Hall, Englewood Cliffs, NJ, 1981.

[9] I. Gustafsson, A class of first order factorization methods, BIT 18 (1978) 142–156.

[10] N. Ida, Engineering Electromagnetics, Springer, New York, 2000.

[11] T. Iwashita, M. Shimasaki, Algebraic multi-color ordering for parallelized ICCG solver in finite element analysis, IEEE Trans. Magn. 38 (2002) 429–432.

[12] M.T. Jones, P.E. Plassmann, The efficient parallel iterative solution of large sparse linear systems, in: A. George, J.R. Gilbert, J.W.H. Liu (Eds.), Graph Theory and Sparse Matrix Computations, IMA volumes in Mathematics and its Applications, Vol. 56, Springer-Verlag, New York, 1993, pp. 229–245.

[13] A. Kameari, Three dimensional current calculation using edge elements for magnetic vector potential, in: K. Miya, (Ed.), Applied Electromagnetics in Materials, Proceedings of the First International Symposium, Tokyo, 3–5 October 1988, Pergamon Press, New York, pp. 225–236.

[14] A. Kameari, K. Koganezawa, Convergence of ICCG method in FEM using edge elements without Gauge condition, IEEE Trans. Magn. 33 (1997) 1223–1226.

[15] T.A. Manteuffel, An incomplete factorization technique for positive definite systems, Math. Comput. 34 (1980) 473–497.

[16] J. Meijerink, H.A. van der Vorst, An iterative solution method for linear systems of which the coefficient matrix is a symmetric M-matrix, Math. Comput. 31 (1977) 148–162.

[17] T. Mifune, T. Iwashita, M. Shimasaki, A fast solver for FEM analyses using the parallelized algebraic multigrid method, IEEE Trans. Magn. 38 (2002) 369–372.

[18] M.M. Monga Made, H.A. van der Vorst, Parallel incomplete factorizations with pseudo-overlapped subdomains, Parallel Comput. 27 (2001) 989–1008.
[19] T. Nakata, N. Takahashi, T. Imai, K. Muramatsu, Comparison of various methods of analysis and finite elements in 3D magnetic field analysis, IEEE Trans. Magn. 27 (1991) 4073–4076.
[20] T. Osoda, K. Maruyama, T. Washio, S. Doi, S. Yamada, Vectorization and parallelization technique of block ILU preconditioning for unstructural problems, IPSJ Trans. HPS 41 (2000) 92–99 (in Japanese).
[21] J. Ruge, K. Stüben, Algebraic multigrid, in: St. McCormick (Ed.), Multigrid Method, SIAM, Philadelphia, 1986.
[22] H.A. van der Vorst, T.F. Chan, Parallel preconditioning for sparse linear equations, Z. Angew. Math. Mech. 76 (1996) 167–170.
[23] H.A. van der Vorst, J.B.M. Melissen, A Petrov–Galerkin type method for solving $Ax = b$, where A is symmetric complex, IEEE Trans. Magn. 26 (1990) 706–708.

JOURNAL OF COMPUTATIONAL AND APPLIED MATHEMATICS

ELSEVIER Journal of Computational and Applied Mathematics 149 (2002) 351–357

www.elsevier.com/locate/cam

A parallel algorithm for generating molecular integrals over MO basis sets

Kazuto Nakata[a], Tadashi Murase[a], Toshihiro Sakuma[b], Toshikazu Takada[c,*]

[a] *VALWAY Technology Center, NEC Soft, Ltd., 1-18-6 Shinkiba, Tokyo 136-8608 Japan*
[b] *Platform Software Division, Scientific Platform Department, NEC Informatec Systems, Ltd., Tsukuba, 305-8501, Japan*
[c] *Fundamental Research Laboratories, NEC Corporation, 34 Miyukigaoka, Tsukuba 305-8501, Japan*

Received 16 October 2001; received in revised form 6 December 2001

Abstract

In the post Hartree–Fock theories such as multi-configuration self consistent field and configuration interaction, two electron integral transformation to molecular orbital sets is the most time consuming process for large-scale calculations. Parallelization is key to minimize the computer time for it. Then, a parallel integral-driven algorithm is presented for the integral transformation.

1. Introduction

In the ab initio molecular orbital (MO) theories, the multi-configuration self consistent field (MC-SCF) and configuration interaction (CI) methods are well established as approaches to take electron correlations into account [10]. From the computational point of view, the most time consuming procedure in these theoretical frameworks is to transform two electron integrals from atomic orbital (AO) to MO basis sets. This transformation is inevitable, since their energy expressions are given in terms of these integrals. In this paper, a parallel integral-driven algorithm to transform two electron integrals from AOs to MOs on parallel computers such as PC clusters is presented. In the following sections, the conventional transformation algorithm is reviewed first and then the new method is introduced along with some benchmark results.

* Corresponding author.
E-mail address: takada@frl.cl.nec.co.jp (T. Takada).

PII: S0377-0427(02)00541-1

2. Conventional integral transformation scheme

Two electron integrals over AOs, χ, are defined as

$$(rs|tu) = \int_{-\infty}^{\infty}\int_{-\infty}^{\infty}\int_{-\infty}^{\infty}\int_{-\infty}^{\infty}\int_{-\infty}^{\infty}\int_{-\infty}^{\infty} \chi_r(r_1)\chi_s(r_1)\frac{1}{|r_1 - r_2|}\chi_t(r_2)\chi_u(r_2)\,\mathrm{d}r_1\,\mathrm{d}r_2. \tag{1}$$

By using the relations between AOs and MOs, that is,

$$\phi_a = \sum_r^N c_{ra}\chi_r, \tag{2}$$

the two electron integrals over MOs are simply generated to be

$$(ab|cd) = \sum_r^N\sum_s^N\sum_t^N\sum_u^N c_{ra}c_{sb}c_{tc}c_{ud}(rs|tu), \tag{3}$$

where N is the number of the atomic orbitals. With a primitive method having eight do loops for the indices of both AOs and MOs, the number of the numerical operations is N^8. But, a very efficient algorithm has been invented for the transformation, consisting of the four steps [10]. This idea has been implemented in major ab initio MO codes like Gaussian98 [3], GAMESS [2] and so on. The first step, for example, is written in Fig. 1 and similar procedures are repeated three more times for transforming from t to c and so on. In this algorithm, the number of the numerical operations is $4N^5$ in total. Conventionally, two electron integrals over AOs are stored on disks and sorted in proper sequences for transformations. But, as molecular sizes to be calculated are growing, the set of integrals to be stored is quickly beyond the storage capacity. On the other hand, in the Hartree–Fock scheme, the direct method [1] became common, in which these two electron integrals are evaluated repeatedly in SCF cycles and each of them is accumulated instantly into the Fock matrix by multiplying density matrix elements. However, since the energy expressions in MCSCF and CI are given in terms of the molecular integrals over MOs, the transformation is requisite.

Simple application of the conventional algorithm on parallel computers causes the following problems, that is,

(1) Since the two electron integrals over AOs are independent of one another, evaluations of these integrals are easily parallelized. But, all the integrals are needed for obtaining one single integral over MOs, causing serious bottleneck by heavy wall-to-wall transmissions among processors.

```
do for r, s and t
 do u
  do d
   (rs|td) = (rs|td) + c_ud (rs|tu)
  end do
 end do
end do
```

Fig. 1. First step of the conventional algorithm for transforming two electron integrals from AOs to MOs.

(2) To avoid such bottleneck, all the integrals over AOs must be calculated on each processor, which loses the benefit of the parallel computations of two electron integrals.

This conflict has been preventing usage of parallel computers for the post Hartree–Fock calculations. However, as is well recognized in HPC communities, parallelization is now key for large-scale simulations in any research field, no matter what kind of computers, vector or scalar parallel, are used. From this background, as seen below, a new transformation algorithm is invented here to overcome the difficulty, which makes it possible to carry out large-scale calculations even with the post Hartree–Fock theories on parallel computers.

3. A parallel algorithm for integral transformations

Fig. 2 schematically illustrates the new transformation algorithm. The indices of AO's, that is, r and s are assigned to a processor, which are denoted as R and S in the figure. That is, a pair of R and S, is distributed to a processor and all the combinations of t and u are considered there. After evaluating the set of the integrals, $(RS|\text{allall})$, the transformation from u to d is carried out in the

```
< first step >
do t
  do u
   do d
    (RS|td) = (RS|td) + c_ud (RS|tu)

   end do
  end do
 end do

< second step >
 do d
   do t
    do c
   (RS|cd) = (RS|cd) + c_tc (RS|td)

    end do
   end do
  end do

< third step >

do d
 do c
  do B
    (RB|cd) = (RB|cd) + c_SB (RS|cd)

   end do
  end do
 end do

< fourth step >
 do d
  do c
   do B
   do A
   (AB|cd) = (AB|cd) + c_RA (RB|td)

    end do
   end do
  end do
  end do
```

Fig. 2. Parallel algorithm for transforming two electron integrals from AOs to MOs.

Table 1
Numerical operation in conventional and present algorithms

	Conventional		Present[a]	
	Operations	Memory	Operations	Memory
First	N^4n	N^4+N^3n	N^2n	N^2+Nn
Second	N^3n^2	$N^3n+N^2n^2$	Nn^2	$Nn+N^2$
Third	N^2n^3	$N^2n^2+Nn^3$	n^3	n^2+n^3
Fourth	Nn^4	Nn^3+n^4	n^4	n^3+n^4
Gathering	—	—	N^2n^4	—

[a]For one processor.

first step. Then, the second step is performed. In the third step, the orbital, S, is transformed to all the MOs denoted by B. Note that only one component of S is involved in the $(RB|cd)$ integrals. Similarly, the transformation from R to A is carried out as the fourth step, where another do loop for B is needed. When all the transformations have been completed, the set of the integrals, $(AB|\text{allall})$ exists on every processor, which has only the contributions from the assigned orbitals of R and S. Therefore, these integrals must be gathered on one processor to obtain the correct two electron integrals $(ab|cd)$ over the molecular orbitals.

Another integral-driven transformation scheme has been presented in [9], in which N^6 operations are required, since pair-wise transformation for the overlap charges is carried out. For the direct CI matrix generation, a similar parallelization algorithm using only the index of r has been proposed by [6].

An MCSCF framework called complete active space SCF (CASSCF) has been proposed in [8], which gives a simple energy expression in the MCSCF methodology and consequently a smaller set of molecular orbitals is needed for the transformed integrals. Let n be the number of the CAS orbitals, which is at most nearly 10 in the recent calculations and then $n \ll N$ holds for large basis sets. In Table 1, the number of numerical operations and working memory sizes are summarized along with the conventional technique. The total operations for the conventional are $nN^4+n^2N^3+n^3N^2+n^4N$, while the present requires $N^2(nN^2+n^2N+n^3+n^4)+n^4N^2$ operations. As a result, $n^4N(N-1)+n^4N^2$ operations are extra. However, the cost for these operations is certainly negligible in comparing with the first step, if N is more than 200. The number of numerical operations for the first step is 10×200^4, while the extra operations are $10^4 \times (2 \times 200^2 - 200)$, which is only 5% of the former. Since permutations between r and s, and between (rs) and (tu) are no longer taken in the present algorithm, evaluation of the integrals over AOs is four times more than the conventional one.

It should be noticed that no transmissions among processors are required until all the transformations to obtain $(AB|\text{allall})$ integrals are completed and the number of two electron integrals to be calculated is still kept to be N^4 through the whole processors. Actually, it is a half of N^4. Therefore, it is concluded that the present algorithm solved the contradiction mentioned above.

This algorithm has been implemented in the CASSCF package of AMOSS-H13, which has been being developed by the NEC quantum chemistry group and is especially designed for large molecules like bio-related molecules.

Table 2
Elapsed time, speed up and parallelization ratio[a]

Basis set	Processors					
	1	2	4	8	16	32
MINI-4 (122 AO's)						
Elapsed time (s)	1890	950	495	252	139	76
Speed up	1	1.99	3.82	7.49	13.59	24.81
Ratio (%)	—	99.49	98.44	99.04	98.82	99.06
MIDI-4 (224 AO's)						
Elapsed time (s)	4417	2189	1130	595	296	161
Speed up	1	2.02	3.91	7.42	14.93	27.45
Ratio (%)	—	100.88	99.22	98.89	99.52	99.47

[a]Estimated from Amdahl's law.

4. Benchmarks

The speed-up measured on a PC cluster is shown in Table 2. The molecule used is artemisin ($C_{15}H_{22}O_5$). The basis sets are MINI-4 (122 AOs) and MIDI-4 (224 AOs) proposed in [7]. The computer system for the benchmark consists of 33 Pentium III (933MHz) processors with local memory of 1GB each and the OS and communication library are RedHat Linux 6.2 and LAM/MPI 6.5.4. One processor works as the master and the rest are used to be slaves. That is, the master processor spawns the tasks for integral evaluations and transformations to the slaves. In the CASSCF theory, the MOs are divided into three subspaces, that is, inner (I) for doubly occupied space, active (A) for CAS and secondly (S) for virtual. Consequently, for the augmented Hessian scheme based on the Newton–Raphson numerical procedure [4,11] to achieve rapid convergences, the following integral types appear: (AA|AA), (IA|AA), (SA|AA), (II|AA), (IS|AA), (SS|AA), (IA|IA), (IA|SA), (SA|SA), (II|SA), (IS|IA), (IS|SA), (SS|IA), (SS|II), (SI|SI). The elapsed times in Table 2 are for from the first to third steps shown in Fig. 1, while the integral evaluation itself is included. The fourth step is located deep inside a program unit with other functions for the performance efficiency and the separated measurement is rather difficult on the present code. From Table 2, it is clearly seen that the present algorithm works quite well on the parallel computer system. The parallelization ratios are estimated from Amdahl's law [5] using the measured speed up.

To show applicability of the present method, the transformations for the following molecules are carried out and the results are shown in Table 3. Up to now, the numbers of the AOs for the post-Hartree–Fock calculations are limited around several hundreds and therefore the data are showing that the present algorithm is promising.

5. Conclusions

Molecular simulations are expected to be key for the research fields like bioinformatics and nano-technology, since the most fundamental procedures are all chemical phenomena, which are

Table 3
Memory sizes and elapse time on PC cluster (32 processors)

Molecules	Numbers of AOs, I, A, S	Local memory (MB)[a]	Elapse time (min)
$C_{15}H_{22}O_5$	122 (MINI-4), 72, 8, 42	2	1.3
	224 (MIDI-4), 72, 8, 144	5	2.7
$C_{31}H_{43}NO_3$	218 (MINI-4), 126, 8, 84	6	2.6
	401 (MIDI-4), 126, 8, 267	18	12.3
$C_{30}H_{52}O_{26}$	332 (MINI-4), 216, 8, 108	14	17.5
	608 (MIDI-4), 216, 8, 384	51	96.2
$C_{86}H_{26}$	456 (MINI-4), 267, 8, 181	29	52.8
	826 (MIDI-4), 267, 8, 551	105	312.3
$C_{72}H_{76}N_8O_{12}Mg_2$	548 (MINI-4), 338, 8, 202	45	119.2
	994 (MIDI-4), 338, 8, 648	173	834.3

[a]For one processor and the estimations in Table 1 is for extreme cases that every different pair of r and s is assigned to different processors.

to be theoretically predicted by quantum mechanics. The ab initio MO theories are well established as numerical frameworks to solve the Schrodinger equation. The Hartree–Fock theory has been applied to large biological molecules, by taking advantage of its theoretical simplicity. But, as is seen in this article, applications of the post-Hartree–Fock theories have been impractical for the heavy integral transformations. To describe chemical reactions, there are some cases in which mixing of a few configuration state functions are essential to give proper dissociation limits or correct activation energies. In these calculations that may happen in biological molecules, the CASSCF is definitely useful.

A new parallel algorithm is presented to transform two electron integrals from AOs to MOs for the post-Hartree–Fock calculations. There are no transmissions needed among the processors until the half-transformed integrals are all generated. Gathering these integrals is completed at the end once and the data size to be transmitted is only in the order of n^4. Furthermore, the number of two electron integrals to be calculated is kept to be still an order of N^4 through the entire cluster system. From these advantages of the present algorithm, much larger calculations than present become practical soon on PC clusters.

The results of the benchmarks demonstrate high scalabilities of up to 32 processors. Since the parallelization is accomplished by the indices of the overlap charges denoted by r and s, the number of processor to be used reaches up to the range of millions for calculations even discussed here and benchmarks with more processors are necessary for testing applicability of the algorithm to such computers.

Acknowledgements

This project has been supported by “Research and Development for Applying Advanced Computational Science and Technology” of Japan Science and Technology Corporation.

References

[1] J. Almlof, K. Fraegri Jr., K. Korsell, J. Comput. Chem. 3 (1982) 385.
[2] GAMESS: http://www.msg.ameslab.gov/GAMESS/GAMESS.html.
[3] Gaussian 98: http://www.gaussian.com/.
[4] B.H. Lengsfield, J. Chem. Phys. 73 (1980) 382.
[5] T.G. Mattson, in: T.G. Mattson (Ed.), Parallel Computing in Computational Chemistry, ACS Symposium Series 592, American Chemical Society, Washington, DC, 1995, pp. 1–15.
[6] Y. Mochizuki, N. Nishi, Y. Hirahara, T. Takada, Theor. Chem. Acta 93 (1996) 211.
[7] J. Andzelm, M. Kłobukowski, E. Radzio-Andzelm, Y. Sakai, H. Tatewaki, Physical Sciences Data 16, in: S. Huzinaga (Ed.), Gaussian Basis Sets for Molecular Calculations, Elsevier, Amsterdam, 1984.
[8] B.O. Roos, P.R. Tayler, P.E.M. Seigbahn, Chem. Phys. 48 (1980) 157.
[9] P.R. Taylor, Internat. J. Quantum Chem. 31 (1987) 521.
[10] A.C. Wahl, G. Das, I, Shavitt, in: H.F. Schaefer III (Ed.), Methods of Electronic Structure Theory, Plenum Press, New York, 1977, pp. 51–78, 189–275.
[11] D.R. Yarkony, Chem. Phys. Lett. 77 (1981) 634.

ELSEVIER

Journal of Computational and Applied Mathematics 149 (2002) 359–371

JOURNAL OF COMPUTATIONAL AND APPLIED MATHEMATICS

www.elsevier.com/locate/cam

Finite element simulation of single crystal growth process using GSMAC method

H. Kohno, T. Tanahashi*

Department of Mechanical Engineering, Keio University, 3-14-1, Hiyoshi, Kohoku-ku, Yokohama 223-8522, Japan

Received 7 October 2001; received in revised form 15 March 2002

Abstract

The characteristics of the melt convection and the melt deformation process in two crystal growth techniques, Czochralski (CZ) and floating zone (FZ) methods, are investigated with three-dimensional numerical simulation using generalized simplified marker and cell—finite element method (GSMAC-FEM). Large eddy simulation (LES) with the Smagorinsky model is applied to calculate the melt convection with high Rayleigh numbers in the CZ method, and level set method (LSM) is adopted for the computation of liquid bridge deformation. First, the CZ (LSM) melt convections are investigated with the boundary conditions in terms of the free surface and the rotational rate changed. As a result, it is found that the heat radiation and the Marangoni effect on the free surface encourage the instability of flows, and the Coriolis' force can largely suppress the natural convection. Second, a series of simulations in terms of liquid bridge deformation are carried out with the coil position, coil shape, and the magnitude of surface tension varied. After that, it turns out that those are important factors in the melt shape. Moreover, it is proved numerically that the deformation of the liquid bridge is kept small by the skin effect and large surface tension in the real FZ crystal growth.

Keywords: CZ method; FZ method; Crystal growth; Free surface; GSMAC-FEM; LES; LSM

1. Introduction

Digital machines, such as personal computer, hand phone, and television, become familiar to us by the rapid progress of the information technology. The vital parts of the digital goods, like CPU and memory, are made of semiconductors, and they are put on the wafer obtained by slicing a silicon single crystal. So the quality of the wafer is very important because it determines their performance. Today, there are two main methods to produce the silicon single crystal. One is the Czochralski

* Corresponding author.
E-mail addresses: kohno@tana.mech.keio.ac.jp (H. Kohno), taka@mech.keio.ac.jp (T. Tanahashi).

PII: S0377-0427(02)00543-5

(CZ) method and the other is the floating zone (FZ) method. These methods are separately used according to the characteristics of the products.

The CZ method is widely adopted as a crystal growth technique whose product is superior in heat and mechanical strength. It is important to control the melt flow for the products of good quality, especially to control the condition of the crystal/melt interface. The effect of heat radiation becomes more noticeable in the large radius of a crucible because of the expansion of the free surface area. Besides, the Coriolis' force generated by the crucible rotation works on the flow in the melt strongly [1]. So it is considered that the melt convection becomes more unstable and complex than that obtained for a smaller crucible. Thus, in this study, characteristics of the melt convections with different boundary conditions in terms of the free surface and the change of the rotational rate are investigated with three-dimensional numerical simulations.

The FZ method is a crystal growth technique whose product realizes high purity and a high resistance rate. In this method, the growth process is carried out without contact with a crucible by supplying the Joule's heat by induction locally to a silicon polycrystal rod. So the molten crystal is sustained by its surface tension. For this instability, the melt is often spilled out from the heating zone in the crystal growing process. However, this problem can be almost settled if the growth process is conducted under microgravity. There are two possible factors which work in the melt there, that is, Marangoni convection [2,3] and the melt deformation. Although both effects should be investigated thoroughly, the latter is only considered in this study. For this reason, three-dimensional numerical simulations are carried out to analyze large deformation problems of an electrically conductive liquid bridge. The hybrid finite element method-boundary element method (FEM-BEM) [4,5] is utilized for the purpose of calculation with the open boundary problems such as an infinite electromagnetic field economically. The computation of a free surface with large deformation process is stably conducted by adopting level set method (LSM) [5,6].

2. Numerical approach for the CZ method

2.1. Mathematical model and governing equations

The measurements of the crucible radius and the depth at the center are 3.19×10^{-1} and 4.04×10^{-1} m, respectively. In the calculation, the hexahedron elements are used, and the melt is divided into 34 in height, 37 in the radial direction, and 56 in the azimuthal direction. Temperatures of the sidewall, the crystal/melt interface, and the ambient gas are fixed at 1700, 1685, and 1273 K, respectively. Temperature on the bottom wall decreases 5 K linearly from the side to the center. The top melt surface is assumed to keep the shape flat.

At the initial stage, the melt is at rest with the melting point temperature. Then, two kinds of calculations are carried out with different purposes. First, the radiant and Marangoni effects are investigated as Case 1. Here, both crystal and crucible are kept static. Second, the effect of the Coriolis' force is investigated with different crucible rotational rates (0, 5, 10, and 20 rpm; Taylor numbers are 0, 9.96×10^{10}, 3.99×10^{11}, and 1.59×10^{12}) as Case 2 after the calculation is performed with the crystal and the crucible static until 200 s. Here, the condition on the free surface is adiabatic, and the Marangoni effect is not taken into consideration. The computation is conducted in the rotational coordinate system, and the rotational direction of the crystal and the crucible are set the same for simplicity.

Table 1
Physical properties of silicon melt

$\rho = 2.52 \times 10^3$ kg m^{-3}
$\nu = 3.00 \times 10^{-7}$ m^2 s^{-1}
$\beta = 1.40 \times 10^{-4}$ K^{-1}
$\sigma_T = -7.00 \times 10^{-5}$ N m^{-1} K^{-1}
$\alpha = 2.64 \times 10^{-5}$ m^2 s^{-1}
$\varepsilon = 3.00 \times 10^{-1}$
$C_p = 1.00 \times 10^3$ J kg^{-1} K^{-1}
$T_m = 1685$ K

The governing equations in the velocity and temperature fields are described as follows:

(1) *In the melt*:

$$\nabla \cdot \mathbf{v} = 0, \tag{1}$$

$$\frac{\partial \mathbf{v}}{\partial t} + \mathbf{v} \cdot \nabla \mathbf{v} = -\frac{1}{\rho} \nabla P + \nu \nabla \cdot (\mathbf{v}\nabla + \nabla \mathbf{v}) - 2(\boldsymbol{\omega} \times \mathbf{v}) + \beta \mathbf{g}(T - T_m), \tag{2}$$

$$\frac{\partial T}{\partial t} + (\mathbf{v} \cdot \nabla) T = \alpha \nabla^2 T. \tag{3}$$

(2) *On the free surface*:

$$[\mathbf{t}^{(n)}]_2' = \sigma_T \nabla_{\|} T, \tag{4}$$

$$q_{out} = \varepsilon\sigma(T^4 - T_a^4), \tag{5}$$

where $\mathbf{v}$, ρ, P, ν, $\boldsymbol{\omega}$, β, $\mathbf{g}$, T, T_m, α, $\mathbf{t}^{(n)}$, T_a, σ_T, $\nabla_{\|}$, and ε are the velocity, density, sum of the pressure and centrifugal force, kinematic viscosity, angular velocity, thermal expansion coefficient, gravitational acceleration, temperature, melting point, thermal diffusivity, stress vector, ambient gas temperature, thermal coefficient of surface tension, surface nabla operator, and surface emissivity, respectively. Those equations are discretized by the FEM with the generalized simplified marker and cell (GSMAC) algorithm [3]. Physical properties of silicon melt are listed in Table 1.

2.2. Large eddy simulation

In this study, large eddy simulation (LES) with the Smagorinsky model, which accounts for the effect of buoyancy, is applied to solve the high Rayleigh number problem ($Ra = 8.44 \times 10^7$). Then Eqs. (2) and (3) are rewritten with the grid scale as follows:

$$\frac{\partial \bar{\mathbf{v}}}{\partial t} + \bar{\mathbf{v}} \cdot \nabla \bar{\mathbf{v}} = -\frac{1}{\rho} \nabla \bar{P} + (\nu + \nu_{SGS}) \nabla \cdot (\bar{\mathbf{v}}\nabla + \nabla \bar{\mathbf{v}}) - 2(\boldsymbol{\omega} \times \bar{\mathbf{v}}) + \beta \mathbf{g}(\bar{T} - T_i), \tag{6}$$

$$\frac{\partial \bar{T}}{\partial t} + (\bar{\mathbf{v}} \cdot \nabla)\bar{T} = (\alpha + \alpha_{\mathrm{SGS}})\nabla^2 \bar{T}, \tag{7}$$

$$Pr_{\mathrm{SGS}} = \frac{v_{\mathrm{SGS}}}{\alpha_{\mathrm{SGS}}}, \tag{8}$$

$$v_{\mathrm{SGS}} = (C_{\mathrm{S}}\Delta)^2 |\bar{S}|_B, \tag{9}$$

$$|\bar{S}|_B = \left(|\bar{S}|^2 - \frac{g\beta}{Pr_{\mathrm{SGS}}} \frac{\partial \bar{T}}{\partial z} \right), \tag{10}$$

where v_{SGS}, α_{SGS}, Pr_{SGS} are the Smagorinsky kinematic viscosity, thermal diffusivity, Prandtl number, and C_{S}, Δ, S are the Smagorinsky constant, grid size, and rate-of-deformation tensor, respectively. The values of C_{S} and Pr_{SGS} are set at 0.1 and 1.0 in the whole calculation.

3. Numerical approach for the FZ method

3.1. Mathematical model and governing equations

A cylindrical model applies to the computational area of LSM and a coil in which alternating current flows is wound on it. The position and the width of a coil can be changed freely. The air–silicon area is divided into 24 in height, 30 in the radial direction, and 72 in the azimuthal direction, and the coil is done into 1–2 in height and 60 in the azimuthal direction. The radius and the height of a liquid bridge are 8.97×10^{-3} and 2.5×10^{-3} m, respectively, and it is set under microgravity (set $0g$ in this calculation). The effect the Joule's heat gives to the flow field is not considered. In this study, three kinds of calculations are carried out. First, the liquid bridge deformations are investigated as Case 1 with the coil position and shape varied with the low frequency alternating current (2 kHz). In the next place, the deformations depending on the magnitude of surface tension are compared as Case 2. Lastly, the numerical analysis corresponding to the real FZ growth condition is conducted with the high frequency alternating current (2 MHz) as Case 3. The current effective value is set at 1.25×10^2 A in three cases. The governing equations are composed of the equation of continuity and the unsteady Navier–Stokes equation in the velocity field:

$$\nabla \cdot \mathbf{v} = 0, \tag{11}$$

$$\frac{\partial \mathbf{v}}{\partial t} + (\mathbf{v} \cdot \nabla)\mathbf{v} = -\frac{1}{\rho}\nabla p + v\nabla^2 \mathbf{v} + \frac{1}{\rho}\mathbf{J} \times \mathbf{B} - \frac{1}{\rho}\sigma_{\mathrm{s}}(\nabla \cdot \mathbf{n})\mathbf{n}\delta_{\varepsilon}, \tag{12}$$

where p, $\mathbf{J}$, $\mathbf{B}$, σ_{s}, $\mathbf{n}$, and δ_{ε} are the pressure, electric current density, magnetic flux density, surface tension, unit vector at the interface pointing to outside, and approximate Delta function. These equations are also discretized by FEM and solved by the GSMAC algorithm. Physical properties of silicon melt and air are listed in Table 2.

3.2. Hybrid FEM-BEM

BEM is superior in the calculation of an infinite area to FEM. On the other hand, FEM is effective to the calculation of an object with a complicated shape when an appropriate boundary condition is

Table 2
Physical properties of silicon melt and air

Silicon melt
$\rho_1 = 2.33 \times 10^3$ kg m^{-3}
$\nu_1 = 4.30 \times 10^{-7}$ m^2 s^{-1}
$\sigma_{m1} = 1.06 \times 10^6$ S m^{-1}
$\sigma_s = 7.35 \times 10^{-3}$–$6.00 \times 10^{-1}$ N m^{-1}
$\mu_r = 1.00$
Air
$\rho_2 = 2.08 \times 10^{-1}$ kg m^{-3}
$\nu_2 = 3.84 \times 10^{-4}$ m^2 s^{-1}
$\sigma_{m2} = 0.00$ S m^{-1}

given. In this study, the hybrid FEM-BEM formed by the combination of the merits of two methods is applied to the computation of the magnetic field physical quantities by the $A - \phi$ method [1]. FEM is adopted in the area applying LSM, and BEM is done in the infinite area including air and a coil.

4. Results

4.1. CZ method

Fig. 1 shows the comparison of the isotherms on the horizontal plane between the adiabatic condition and the condition in which the radiation and the Marangoni effect are taken into consideration in Case 1. It is found that the temperature distribution in Fig. 1(b) is complex and some lumps with high temperature are formed near the free surface, while the isotherms are not observed except under the crystal/melt interface in Fig. 1(a). This phenomenon is derived from the instability of the melt flow caused by the radiation and Marangoni effect [7]. If a small fluctuation occurs near the free surface, it is finally promoted and stabilized with the help of both density and surface tension gradients. That is why the difference between two cases is so clear, and this is also reflected in the incoming heat flux defined as $\int \mathrm{d}T/\mathrm{d}n\,\mathrm{d}S$ through the crystal/melt interface in Fig. 2. When the heat radiation on the free surface is considered, the magnitude of heat flux to the crystal is reduced because the melt is cooled on the way from the crucible to the crystal. However, it is found from the average values that the Marangoni effect does not influence the melt flow largely. This tendency can be simply evaluated with two dimensionless numbers, that is the Rayleigh number ($Ra = g\beta L^3 \Delta T/\alpha\nu$) and the Marangoni number ($Ma = |\sigma_T| L \Delta T/\rho\alpha\nu$). If the representative length L becomes twice, the Rayleigh number increases by 8 times while the Marangoni number does by twice. So the natural convection becomes dominant as the size of a crucible becomes large, and the Marangoni effect is relatively weakened by this convection.

Fig. 3 shows the comparison of the isothermal surfaces between four cases in Case 2. As the increase of the crucible rotational rate, the Coriolis' force grows stronger. As a result, the natural convection is largely suppressed and the melt condition finally changes from the three-dimensional

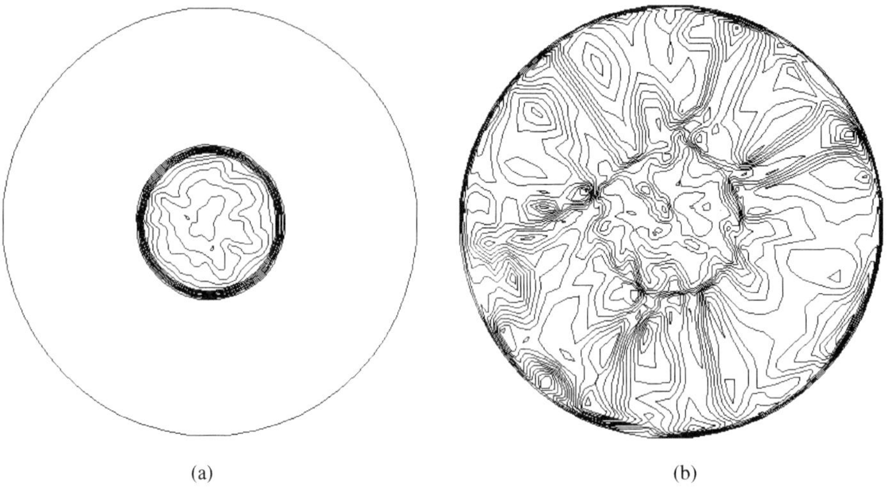

Fig. 1. Isotherms on the horizontal planes ($z = 4.00 \times 10^{-1}$ m) with $\Delta T = 1$ K at 400 s in Case 1. (a) Adiabatic condition and (b) radiation+Marangoni effect.

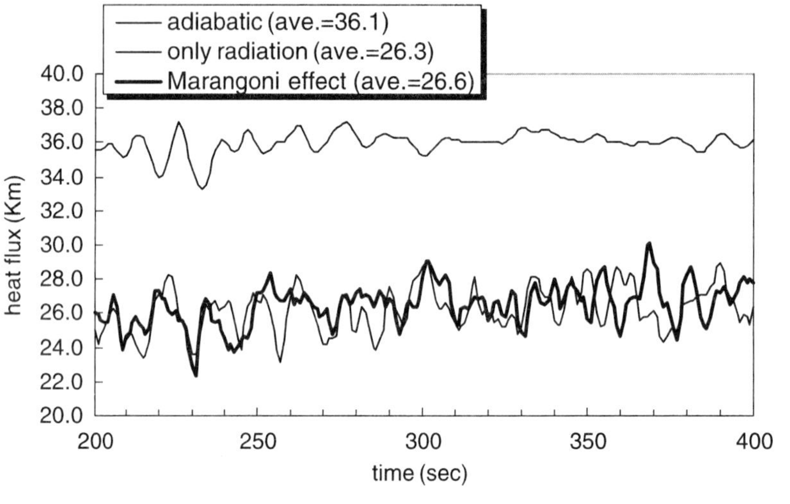

Fig. 2. Comparison of the incoming heat flux through the crystal/melt interface in Case 1.

asymmetrical turbulence to the two-dimensional axisymmetrical flow as shown in Fig. 3(d). Fig. 4 shows the comparison of the histories of temperature fluctuation at a point just under the crystal/melt interface. These graphs reflect the characteristics of the flow field well. When the rotational rate is set small, the flow structure gets complicated as shown in Fig. 3(b) because of the competition between the buoyant force and the Coriolis' force. So the amplitude of the temperature fluctuation becomes large as shown in Fig. 4(a). However, as the Coriolis' force gets dominant, it is observed that the fluctuation is sufficiently controlled in Fig. 4(b).

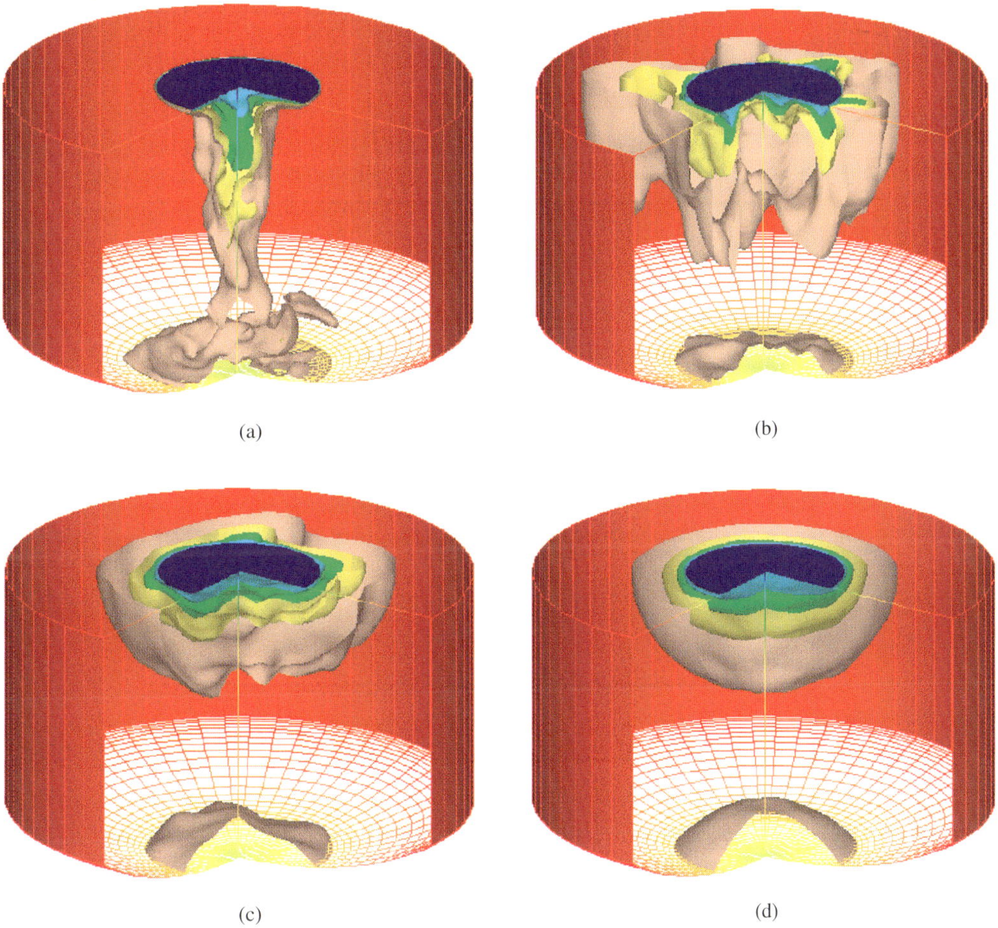

Fig. 3. Comparison of the isothermal surfaces in Case 2. (a) Case 2.1 (0 rpm); (b) Case 2.2 (5 rpm); (c) Case 2.3 (10 rpm); (d) Case 2.4 (20 rpm).

4.2. FZ method

Fig. 5 shows the comparison of the deformation processes between four cases with the surface tension fixed at 7.35×10^{-3} N m^{-1}. The coil positions of Cases 1.1, 1.2, and 1.3 are set at 2.00×10^{-2}, 1.50×10^{-2}, and 1.00×10^{-2} m from the bottom, respectively, and their coil widths in height are 5.00×10^{-3} m. On the other hand, the coil width in height is twice in Case 1.4 with the current effective value kept the same, and it is set at 1.50×10^{-2} m. When the alternating current is imposed on the silicon melt, the induced current occurs and it flows in the direction of easing the variation of the magnetic flux density. This is called eddy current, and it generates the inward Lorentz force. So in any case, the liquid part located near the circular coil yields to pressure generated by the Lorentz force and dents inside. From the Figs. 5 and 6, it is observed that the liquid shape and the deformation quantity depend on the relative coil position and the coil width. In this study, the deformation quantity in Case 1.2 is the largest. While the liquid parts at the coil position are curved

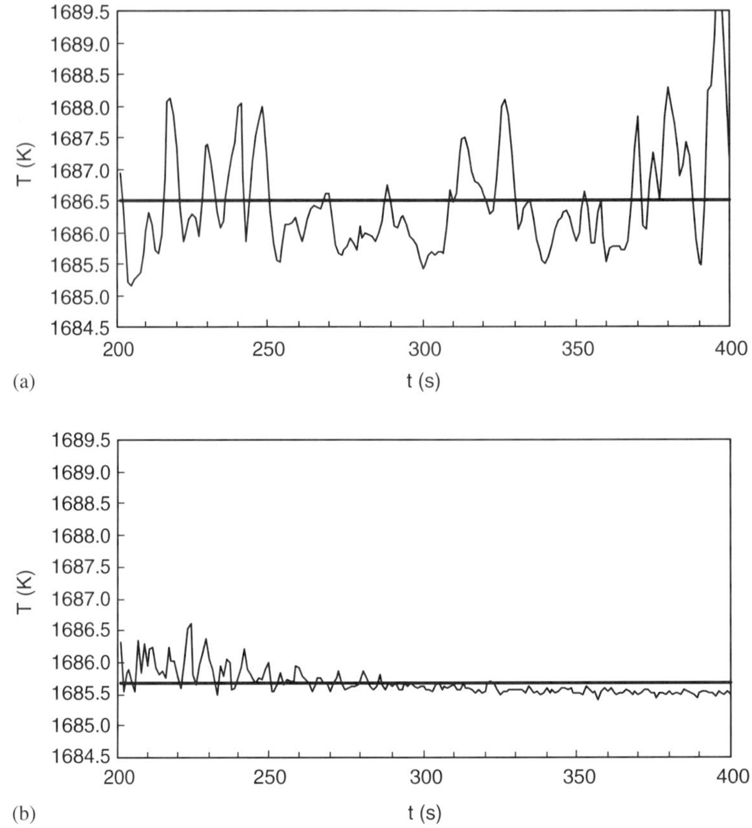

Fig. 4. Histories of temperature fluctuation and its average value just under the crystal ($x = 0.00$ m, $y = 0.00$ m, $z = 4.00 \times 10^{-1}$ m) in Case 2. (a) 5 rpm and (b) 20 rpm.

smoothly in Cases 1.1, 1.2, and 1.4, this liquid part is swelled a little in Case 1.3 because of the equilibrium of the Lorentz force worked on the horizontal plane.

Fig. 7 shows the results in which the magnitude of surface tension is changed intentionally. Each surface tension in Fig. 7(a)–(c) is increased to 2, 5, 10 times as large as the one in Fig. 5(b) (Case 2.1), and Fig. 7(d) shows two-dimensional distribution of the level set function and velocity on the vertical plane which passes the central axis in Case 2.4. The deformation is restricted as the increase of the magnitude of surface tension. This is also confirmed by two-dimensional surface lines in Fig. 8 quantitatively. The quantity of deformation becomes reduced as the magnitude of surface tension increases. This phenomenon indicates that the Lorentz force which turns inside a liquid bridge is balanced with the outward force in proportion to the magnitude of curvature and surface tension as shown in Eq. (12). It is also found that the deformation quantities of Cases 2.3 and 2.4 are not so different. So the difference of the deformation velocity is gradually reduced as the magnitude of surface tension increases. In Fig. 7(d), it is observed that the nature of the distance function is well kept by the appropriate reinitialization process [5,6]. The velocity distribution near

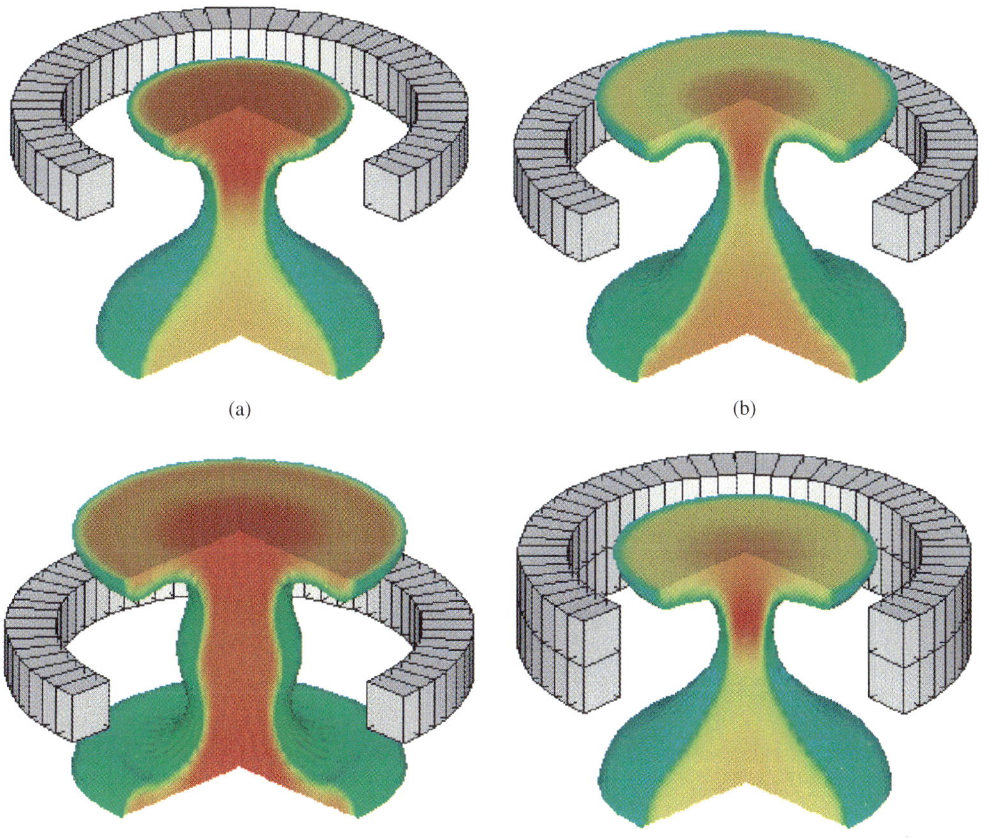

(a) (b)

(c) (d)

Fig. 5. Comparison of the deformation processes and pressure distributions between four cases in Case 1. (a) Case 1.1, P_{max} : 39.8 Pa; (b) Case 1.2, P_{max} : 62.3 Pa; (c) Case 1.3, P_{max} : 50.3 Pa; (d) Case 1.4, P_{max} : 61.7 Pa.

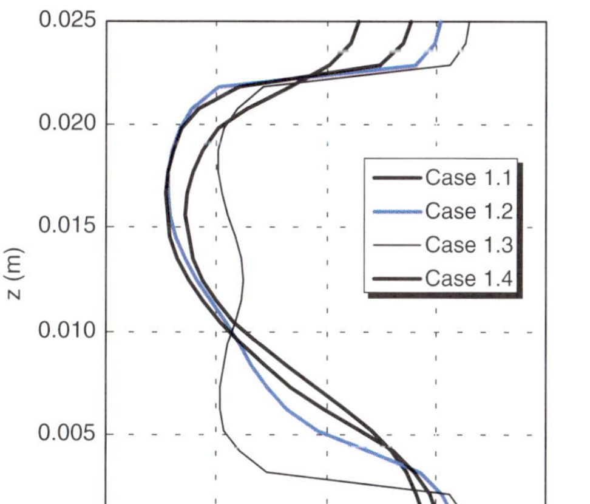

Fig. 6. Comparison of the transient liquid bridge shapes at 0.20 s between four cases in Case 1.

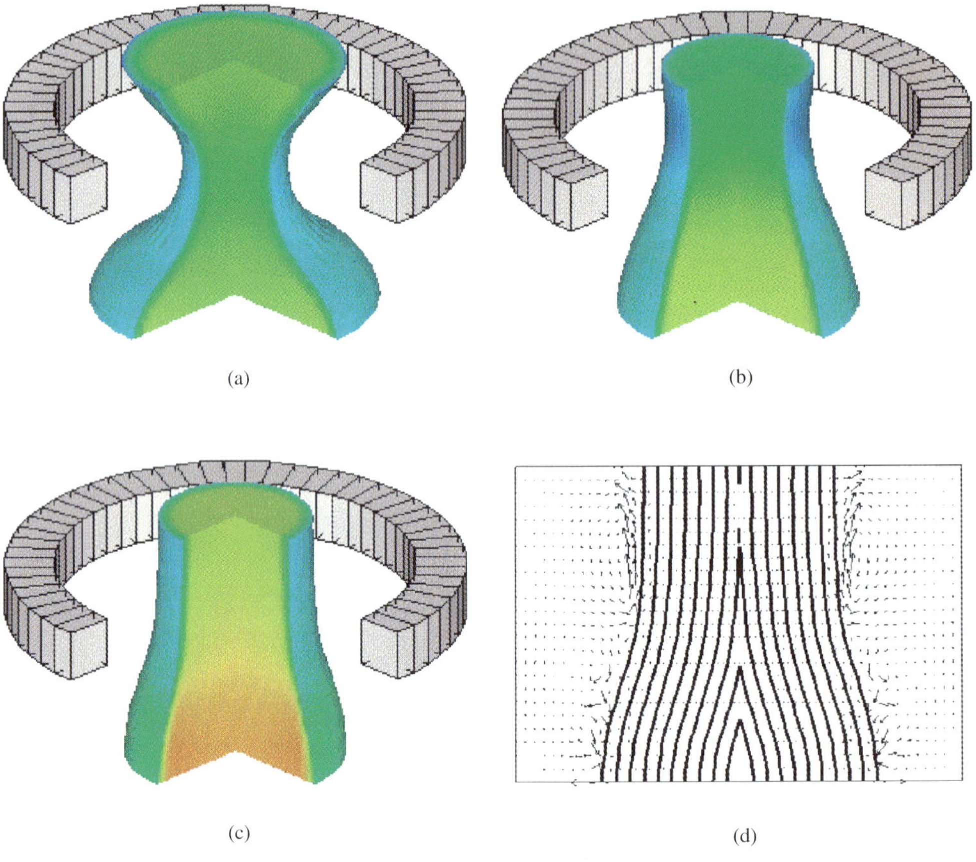

Fig. 7. Comparison of the deformation processes and pressure distributions between three cases (a)–(c), and the velocity distribution (d) in Case 2. (a) Case 2.2, $P_{\max}$: 40.8 Pa; (b) Case 2.3, $P_{\max}$: 42.1 Pa; (c) Case 2.4, $P_{\max}$: 57.2 Pa; (d) $V_{\max} = 8.32 \times 10^{-2}$ m s^{-1}.

the free surface is almost parallel to the surface line. That means the deformation process has almost finished and the shape of the liquid bridge is determined.

In the real FZ process, the frequency of the alternating current is much higher and the surface tension is larger than that in Case 2.4. According to this fact, the numerical simulation corresponding to the real FZ growth condition is finally carried out with high frequency (2 MHz) and the real surface tension (6.00×10^{-1} N m^{-1}). Fig. 9 shows three-dimensional view of Joule's heat and the distributions of magnetic flux density and velocity on the vertical plane including the central axis. In Fig. 9(a), it is observed that the magnetic flux density does not penetrate into the silicon liquid bridge although it is distributed throughout the air area. To be precise, this physical quantity is concentrated on the thin layer including the surface. The cause of this phenomenon is the skin effect which is appeared remarkably in the alternating magnetic field with high frequency. The skin depth, which shows the skin effect quantitatively, is described in the following equation:

$$\delta = \sqrt{\frac{2}{\sigma_m \mu_0 f}}, \tag{13}$$

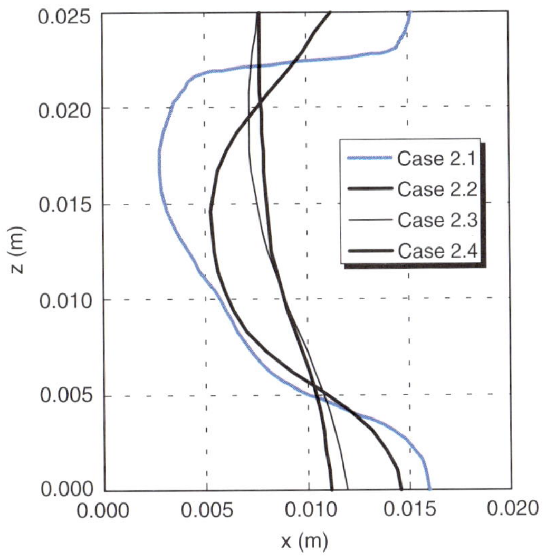

Fig. 8. Comparison of the transient liquid bridge shapes at 0.20 s between four cases in Case 2.

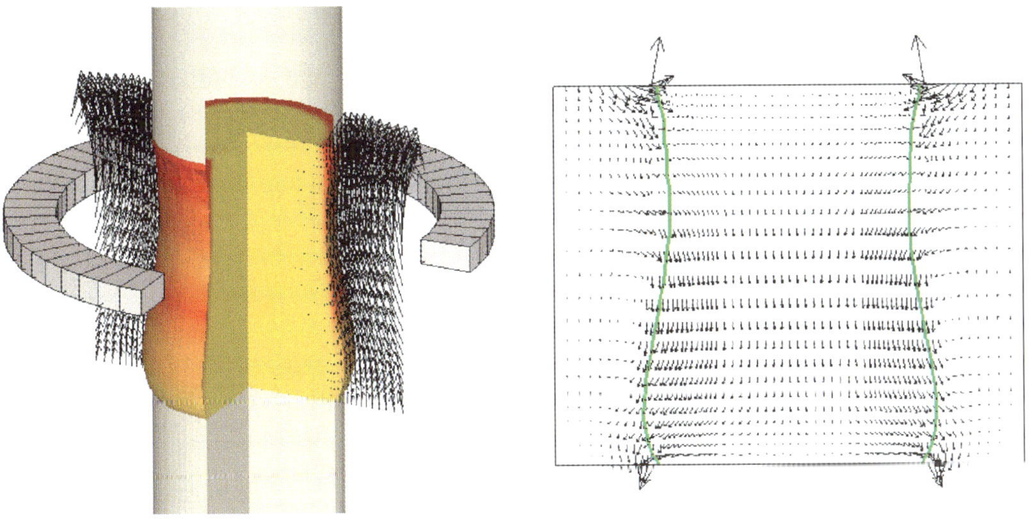

(a) (b)

Fig. 9. 3D view of Joule's heat and magnetic flux density (a) and 2D distribution of velocity (b) in Case 3. (a) $B_{max} = 6.46 \times 10^{-3}$ T; (b) $V_{max} = 3.34$ m s^{-1}.

where μ_0 and f are the magnetic permeability in a vacuum and frequency. In this study, δ is calculated at 8.66×10^{-4} m, about one-tenth of the radius. So, the Joule's heat is generated only in the thin layer by this effect, and the Lorentz force as a body force becomes weak. In addition, the force of repulsion to the deformation is strong because of the large magnitude of surface tension. Thus, the deformation quantity is restricted to be small as shown in Fig. 9(a). This is also confirmed by the velocity distribution in Fig. 9(b). The direction of the velocity vector in the vicinity of the free surface is almost parallel to the surface, which indicates that the liquid shape is determined.

5. Conclusions

Three-dimensional numerical simulations are carried out to investigate the characteristics of the silicon melt convection and the deformation process in two main crystal growth techniques, CZ and FZ methods. In the computation of the CZ melt convection, LES with the Smagorinsky model is applied to solve the high Rayleigh and Reynolds number problems. In the FZ melt convection under microgravity, LSM is adopted to calculate the large deformation problem stably, and hybrid FEM-BEM is utilized for the calculation of the electromagnetic field economically. The results obtained in this study are summarized as follows:

5.1. CZ method

1. The radiation and the Marangoni effect on the free surface encourage the instability of the melt flow by promoting the initial fluctuations. For that reason, some lumps with high temperature are observed near the free surface.
2. In comparison with the effect of heat radiation, the Marangoni effect does not play an important role to the melt convection in a large-scale CZ crucible.
3. The natural convection is suppressed by the Coriolis' force as the increase of the crucible rotational rate. As a result, the melt condition changes to the two-dimensional axisymmetrical flow.

5.2. FZ method

1. It is confirmed numerically that the deformation of a liquid bridge caused by the Lorentz force depends on the relative coil position, coil shape, and the magnitude of surface tension under microgravity.
2. In the real FZ process, the magnetic field physical quantities are concentrated in a tiny range in depth from the free surface because of the skin effect. So, the deformation of a liquid bridge is kept small with the help of large surface tension.

References

[1] Y. Kishida, K. Okazawa, Geostrophic turbulence in CZ silicon crucible, J. Cryst. Growth 198/199 (1999) 135–140.
[2] Z. Zeng, H. Mizuseki, K. Higashino, Y. Kawazoe, Direct numerical simulation of oscillatory Marangoni convection in cylindrical liquid bridges, J. Cryst. Growth 204 (1999) 395–404.
[3] H. Kohno, T. Tanahashi, Three-dimensional numerical simulation of unsteady Marangoni convection in the CZ method using GSMAC-FEM, CMES 2 (2) (2001) 155–170.

[4] C.H. Jung, T. Minowa, T. Tanahashi, Numerical analysis on molten metal in cold crucible using GSMAC-ALE method, Trans. Japan Soc. Mech. Eng. 66 (650) (2000) 2521–2528 (in Japanese).
[5] H. Kohno, T. Tanahashi, Single crystal growing problem in the FZ method under microgravity (three-dimensional numerical simulation with the hybrid FEM-BEM applying LSM on the free surface), Trans. Japan Soc. Mech. Eng. 67 (658) (2001) 1408–1415 (in Japanese).
[6] M. Sussman, P. Smereka, S. Osher, A level set approach for computing solutions to incompressible two-phase flow, J. Comput. Phys. 114 (1994) 146–159.
[7] K.W. Yi, K. Kakimoto, M. Eguchi, M. Watanabe, T. Shyo, T. Hibiya, Spoke patterns on molten silicon in Czochralski system, J. Cryst. Growth 144 (1994) 20–28.